T0192173

Lecture Notes in Artificial Intelligence 13589

Subseries of Lecture Notes in Computer Science

More information about this subseries at https://link.springer.com/bookseries/1244

Leszek Rutkowski · Rafał Scherer ·
Marcin Korytkowski · Witold Pedrycz ·
Ryszard Tadeusiewicz ·
Jacek M. Zurada (Eds.)

Artificial Intelligence and Soft Computing

21st International Conference, ICAISC 2022
Zakopane, Poland, June 19–23, 2022
Proceedings, Part II

 Springer

Editors
Leszek Rutkowski ⓘ
Systems Research Institute of the Polish
Academy of Sciences
Warsaw, Poland

The Institute of Computer Science
AGH University of Science and Technology
Kraków, Poland

Marcin Korytkowski ⓘ
Częstochowa University of Technology
Częstochowa, Poland

Ryszard Tadeusiewicz ⓘ
AGH University of Science and Technology
Kraków, Poland

Rafał Scherer ⓘ
Częstochowa University of Technology
Częstochowa, Poland

Witold Pedrycz ⓘ
University of Alberta
Edmonton, AB, Canada

Jacek M. Zurada ⓘ
University of Louisville
Louisville, KY, USA

ISSN 0302-9743 ISSN 1611-3349 (electronic)
Lecture Notes in Artificial Intelligence
ISBN 978-3-031-23479-8 ISBN 978-3-031-23480-4 (eBook)
https://doi.org/10.1007/978-3-031-23480-4

LNCS Sublibrary: SL7 – Artificial Intelligence

This Springer imprint is published by the registered company Springer Nature Switzerland AG
The registered company address is: Gewerbestrasse 11, 6330 Cham, Switzerland

Preface

This volume constitutes the proceedings of the 21st International Conference on Artificial Intelligence and Soft Computing ICAISC 2022, held in Zakopane, Poland, on June 19–23, 2022. The conference was organized by the Polish Neural Network Society in cooperation with the Department of Intelligent Computer Systems at the Częstochowa University of Technology, the University of Social Sciences in Łódź, and the IEEE Computational Intelligence Society, Poland Chapter. The conference was held under the auspices of the Committee on Informatics of the Polish Academy of Sciences.

Previous conferences took place in Kule (1994), Szczyrk (1996), Kule (1997) and Zakopane (1999, 2000, 2002, 2004, 2006, 2008, 2010, 2012, 2013, 2014, 2015, 2016, 2017, 2018, 2019, 2020 and 2021) and attracted a large number of papers and internationally recognized speakers: Lotfi A. Zadeh, Hojjat Adeli, Rafal Angryk, Igor Aizenberg, Cesare Alippi, Shun-ichi Amari, Daniel Amit, Plamen Angelov, Sanghamitra Bandyopadhyay, Albert Bifet, Piero P. Bonissone, Jim Bezdek, Zdzisław Bubnicki, Jan Chorowski, Andrzej Cichocki, Swagatam Das, Ewa Dudek-Dyduch, Włodzisław Duch, Adel S. Elmaghraby, Pablo A. Estévez, João Gama, Erol Gelenbe, Jerzy Grzymala-Busse, Martin Hagan, Yoichi Hayashi, Akira Hirose, Kaoru Hirota, Adrian Horzyk, Tingwen Huang, Eyke Hüllermeier, Hisao Ishibuchi, Er Meng Joo, Janusz Kacprzyk, Nikola Kasabov, Jim Keller, Laszlo T. Koczy, Tomasz Kopacz, Jacek Koronacki, Zdzislaw Kowalczuk, Adam Krzyzak, Rudolf Kruse, James Tin-Yau Kwok, Soo-Young Lee, Derong Liu, Robert Marks, Ujjwal Maulik, Zbigniew Michalewicz, Evangelia Micheli-Tzanakou, Kaisa Miettinen, Krystian Mikołajczyk, Henning Müller, Christian Napoli, Ngoc Thanh Nguyen, Andrzej Obuchowicz, Erkki Oja, Nikhil R. Pal, Witold Pedrycz, Marios M. Polycarpou, José C. Príncipe, Jagath C. Rajapakse, Šarunas Raudys, Enrique Ruspini, Roman Senkerik, Jörg Siekmann, Andrzej Skowron, Roman Słowiński, Igor Spiridonov, Boris Stilman, Ponnuthurai Nagaratnam Suganthan, Ryszard Tadeusiewicz, Ah-Hwee Tan, Dacheng Tao, Shiro Usui, Thomas Villmann, Fei-Yue Wang, Jun Wang, Bogdan M. Wilamowski, Ronald Y. Yager, Xin Yao, Syozo Yasui, Gary Yen, Ivan Zelinka and Jacek Zurada.

The aim of this conference is to build a bridge between traditional artificial intelligence techniques and so-called soft computing techniques. It was pointed out by Lotfi A. Zadeh that "soft computing (SC) is a coalition of methodologies which are oriented toward the conception and design of information/intelligent systems. The principal members of the coalition are: fuzzy logic (FL), neurocomputing (NC), evolutionary computing (EC), probabilistic computing (PC), chaotic computing (CC), and machine learning (ML). The constituent methodologies of SC are, for the most part, complementary and synergistic rather than competitive".

These proceedings present both traditional artificial intelligence methods and soft computing techniques. Our goal is to bring together scientists representing both areas of research. This volume is divided into five parts:

- Neural Networks and Their Applications
- Fuzzy Systems and Their Applications
- Evolutionary Algorithms and Their Applications
- Pattern Classification
- Artificial Intelligence in Modeling and Simulation

I would like to thank our participants, invited speakers and reviewers of the papers for their scientific and personal contribution to the conference. The advice and constant support of the Honorary Chair of the conference Prof. Ryszard Tadeusiewicz is acknowledged with many thanks. Finally, I thank my co-workers Łukasz Bartczuk, Piotr Dziwiński, Marcin Gabryel, Rafał Grycuk, Marcin Korytkowski and Rafał Scherer, for their enormous efforts to make the conference a very successful event. Moreover, I would like to acknowledge the work of Marcin Korytkowski who was responsible for the Internet submission system.

June 2022   Leszek Rutkowski

Organization

ICAISC 2022 was organized by the Polish Neural Network Society in cooperation with the University of Social Sciences in Łódź and the Institute of Computational Intelligence at Częstochowa University of Technology.

ICAISC Chairpersons

General Chair

Leszek Rutkowski, Poland

Co-chair

Rafał Scherer, Poland

Technical Chair

Marcin Korytkowski, Poland

Financial Chair

Marcin Gabryel, Poland

Area Chairs

Fuzzy Systems

Witold Pedrycz, Canada

Evolutionary Algorithms

Zbigniew Michalewicz, Australia

Neural Networks

Jinde Cao, China

Computer Vision

Dacheng Taom, Australia

Machine Learning
Nikhil R. Pal, India

Artificial Intelligence with Applications
Janusz Kacprzyk, Poland

International Liaison
Jacek Żurada, USA

ICAISC Program Committee

Hojjat Adeli, USA
Cesare Alippi, Italy
Shun-ichi Amari, Japan
Rafal A. Angryk, USA
Robert Babuska, Netherlands
James C. Bezdek, Australia
Piero P. Bonissone, USA
Bernadette Bouchon-Meunier, France
Jinde Cao, China
Juan Luis Castro, Spain
Yen-Wei Chen, Japan
Andrzej Cichocki, Japan
Krzysztof Cios, USA
Ian Cloete, Germany
Oscar Cordón, Spain
Bernard De Baets, Belgium
Włodzisław Duch, Poland
Meng Joo Er, Singapore
Pablo Estevez, Chile
David B. Fogel, USA
Tom Gedeon, Australia
Erol Gelenbe, UK
Jerzy W. Grzymala-Busse, USA
Hani Hagras, UK
Saman Halgamuge, Australia
Yoichi Hayashi, Japan
Tim Hendtlass, Australia
Francisco Herrera, Spain
Kaoru Hirota, Japan
Tingwen Huang, USA
Hisao Ishibuchi, Japan
Mo Jamshidi, USA

Robert John, UK
Janusz Kacprzyk, Poland
Nikola Kasabov, New Zealand
Okyay Kaynak, Turkey
Vojislav Kecman, USA
James M. Keller, USA
Etienne Kerre, Belgium
Frank Klawonn, Germany
Robert Kozma, USA
László Kóczy, Hungary
Józef Korbicz, Poland
Rudolf Kruse, German
Adam Krzyzak, Canada
Věra Kůrková, Czech Republic
Soo-Young Lee, Korea
Simon M. Lucas, UK
Luis Magdalena, Spain
Jerry M. Mendel, USA
Radko Mesiar, Slovakia
Zbigniew Michalewicz, Australia
Javier Montero, Spain
Eduard Montseny, Spain
Kazumi Nakamatsu, Japan
Detlef D. Nauck, Germany
Ngoc Thanh Nguyen, Poland
Erkki Oja, Finland
Nikhil R. Pal, India
Witold Pedrycz, Canada
Leonid Perlovsky, USA
Marios M. Polycarpou, Cyprus
Danil Prokhorov, USA
Vincenzo Piuri, Italy

Sarunas Raudys, Lithuania	Hideyuki Takagi, Japan
Olga Rebrova, Russia	Dacheng Tao, Australia
Vladimir Red'ko, Russia	Vicenç Torra, Spain
Raúl Rojas, Germany	Burhan Turksen, Canada
Imre J. Rudas, Hungary	Shiro Usui, Japan
Norihide Sano, Japan	DeLiang Wang, USA
Rudy Setiono, Singapore	Jun Wang, Hong Kong
Jennie Si, USA	Lipo Wang, Singapore
Peter Sincak, Slovakia	Paul Werbos, USA
Andrzej Skowron, Poland	Bernard Widrow, USA
Roman Słowiński, Poland	Kay C. Wiese, Canada
Pilar Sobrevilla, Spain	Bogdan M. Wilamowski, USA
Janusz Starzyk, USA	Donald C. Wunsch, USA
Jerzy Stefanowski, Poland	Ronald R. Yager, USA
Vitomir Štruc, Slovenia	Xin-She Yang, UK
Ron Sun, USA	Gary Yen, USA
Johan Suykens, Belgium	Sławomir Zadrożny, Poland
Ryszard Tadeusiewicz, Poland	Jacek Zurada, USA

ICAISC Organizing Committee

Rafał Scherer
Łukasz Bartczuk
Piotr Dziwiński
Marcin Gabryel (Finance Chair)
Rafał Grycuk
Marcin Korytkowski (Databases and Internet Submissions)

Contents – Part II

Various Problems of Artificial Intelligence

Bioinformatics, Biometrics and Medical Applications

Contents – Part I

Fuzzy Systems and Their Applications

Evolutionary Algorithms and Their Applications

Pattern Classification

Artificial Intelligence in Modeling and Simulation

Computer Vision, Image and Speech Analysis

Unsupervised Pose Estimation by Means of an Innovative Vision Transformer

Nicolo' Brandizzi[1], Andrea Fanti[1], Roberto Gallotta[1], Samuele Russo[2], Luca Iocchi[1], Daniele Nardi[1], and Christian Napoli[1(✉)]

[1] Department of Computer, Automation and Management Engineering, Sapienza University of Rome, via Ariosto 25, 00185 Roma, Italy
{brandizzi,iocchi,nardi,cnapoli}@diag.uniroma1.it
[2] Department of Psychology, Sapienza University of Rome, via dei Marsi 78, 00185 Roma, Italy
samuele.russo@uniroma1.it

Abstract. Attention-only Transformers [34] have been applied to solve Natural Language Processing (NLP) tasks and Computer Vision (CV) tasks. One particular Transformer architecture developed for CV is the Vision Transformer (ViT) [15]. ViT models have been used to solve numerous tasks in the CV area. One interesting task is the pose estimation of a human subject. We present our modified ViT model, *Un-TraPEs* (UNsupervised TRAnsformer for Pose Estimation), that can reconstruct a subject's pose from its monocular image and estimated depth. We compare the results obtained with such a model against a ResNet [17] trained from scratch and a ViT finetuned to the task and show promising results.

Keywords: Computer vision · Image understanding · Pose estimation · Visual transformers · Artificial intelligence and applications

1 Introduction

The field of Pose Estimation (PE) studies how to use sensory data information, both visual and non, to identify human motion and, specifically, joint configurations. While a PE system can be built around sensory data from wearable devices [5,10], this field sees many publications and applications in conjunction with computer vision. Indeed, datasets made of static images or video sequences are commonly available, and the ones used in this tasks are slowly increasing in size [1,6,20,29]. Consequently, many applications in a broad spectrum of domains greatly rely on PE systems. Some examples are intelligent home monitoring, surveillance, human-computer interaction [4,31,36] in which robots greatly benefit from the additional information about human behavior. More advanced applications are tied to the field of robotics, where researchers deploy a PE pipeline to teach a simulated robot to mimic complex movements [26,32].

One of the main challenges when dealing with pose estimation is the intra-class and interclass similarities. Indeed, while a pose such as "standing" can be

L. Rutkowski et al. (Eds.): ICAISC 2022, LNAI 13589, pp. 3–20, 2023.
https://doi.org/10.1007/978-3-031-23480-4_1

expressed as a rigorous set of joint movements, other human actions are represented by diverse body movements by different users. On the other hand, some activities may share significant similarities and can be challenging to differentiate. While the latter are issues related to the PE task, more problems are inherited from computer vision. Indeed, PE datasets often include images taken "in the wild" where a diverse set of conditions such as background, lightness, scaling, and point of view represent significant limitations as well. To overcome these limitations, we work with Vision Transformers [15], which are highly robust to severe occlusions, perturbations, and domain shifts. Moreover, ViT can also grasp general features in the image context, allowing them to perform accurate segmentation without pixel supervision [25]. Although transformers allow for better generalization, they also require more data than canonical convolutional neural networks. For this reason, in this work, we deploy a training technique known as self-supervising learning (SSL), which allows us to efficiently make use of the available dataset and still obtain comparable results.

Indeed, recent works study the affinity between SSL and transformers in the field of speech processing [38], molecular propriety prediction [11], and computer vision [3,21]. This technique allows the encoder-decoder architecture to train on a vast amount of information without relying on data augmentation techniques like the CNN counterpart. On the other hand, SSL and vision transformers work mainly in image classification [16] and pixel prediction [9] while neglecting other computer vision-related tasks. This work applies SSL and ViT in a novel pipeline to tackle the pose estimation problem and show promising results compared with traditional CNN approaches and a fine-tuned version of [15].

The paper is structured as follows. Section 2 provides a background on the methods and discusses the state-of-the-art for attention mechanism, transformers for computer vision and pose estimation. Section 3 reviews one of the most influential PE datasets, Multimodal Gesture Recognition: Montalbano V2 (ECCV' 14) [8], and describe the *Un-TraPEs* model together with the training pipeline over the last decade. Section 4 presents experiments, baseline comparison, and their relative results. Finally, in Sect. 5, we discuss conclusions and future directions.

2   Related Works

Transformers were first introduced in the Natural Language Processing domain in [34], as a more performant and efficient alternative for sequence-to-sequence tasks. In contrast with previous approaches, which employed a mixture of attention mechanisms and recurrent architectures, as well as other artificial intelligence based techniques and neural network based classifiers [7,22,35,37], Transformers rely solely on attention; this allows the model to draw global dependencies without regard to their distance in the input or output sequence.

Attention is a mechanism to map a query vector and a set of key-value vector pairs to an output vector. More specifically, the original Transformer used the

Scaled Dot–Product Attention function, computed simultaneously over multiple queries, yielding

$$\text{Attention}(Q, K, V) = \text{softmax}\left(\frac{QK^T}{\sqrt{d_k}}V\right)$$

where:

- $Q \in \mathbb{M}^{n \times d_k}$ contains the set of n queries, each of dimension d_k;
- $K \in \mathbb{M}^{m \times d_k}$ contains the n keys, each of dimension d_k;
- $V \in \mathbb{M}^{m \times d_v}$ contains the n values, each of dimension d_v;
- the scaling factor $\sqrt{d_k}$ is used to avoid vanishing gradients in the softmax.

The transformer uses this function as the basis for the so–called *Multi–Head Attention* layers, which compute multiple learned Scaled Dot–Product Attention operations, concatenating and projecting the results; this results in the following layer function:

$$\text{MultiHead}(Q, K, V) = \text{Concat}(\text{head}_1, \text{head}_2, \dots)W^0$$

$$\text{head}_i = \text{Attention}(QW_i^Q, KW_i^K, VW_i^V)$$

Where the parameters W_i^Q, W_i^K, W_i^V, and W^O are learned.

These multi-head attention layers are deployed differently throughout the model:

- self–attention in the encoder, where queries, keys, and values come from the previous layer of the encoder; each position here can thus attend to all positions in the previous layer;
- self–attention in the decoder, where queries, keys, and values come from the previous layer of the decoder; however, differently from the encoder self–attention, masking is applied here to preserve the auto-regressive property, meaning that each position here can attend only to all previous positions up to and including that position;
- encoder-decoder attention in the decoder, where queries come from the previous decoder layer, while keys and values come from the output of the encoder; this allows every position in the decoder to attend overall positions in the input sequence

The whole diagram of the original architecture (directly from [34]) is reported in Fig. 1.

Transformers for Computer Vision. While models entirely based on Transformers for NLP tasks soon became the de–facto standard, they were used sparingly for computer vision tasks, often simply replacing a small part of CNN-based architectures. The first work that applies a Transformer–only model to computer vision straightforwardly and efficiently is the Vision Transformer (ViT) [15], which showed that the reliance on convolution-based models is not needed

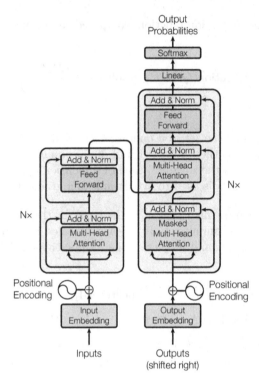

Fig. 1. Architecture of the original Transformer. Image taken from [34].

to achieve good performance. The key idea of this work is the method used to apply the original Transformer encoder from [34] to images instead of text sequences.

More specifically, to feed images of dimensions $W \times H \times C$ to the Transformer encoder, they are first split into N square 2D patches of identical dimensions $P \times P \times C$, with P being the patch size. These patches are projected to N tokens, each of size D, by a learnable linear layer. These tokens are referred to as the *patch embeddings*, and their dimension D is equal to the latent vector size of the Transformer encoder. Additionally, position embeddings are added to the patch embeddings before being fed to the encoder to retain positional information in the original sequence of images patches.

Models inspired by ViT seem to perform best when trained on extensive labeled datasets or using teacher networks (usually CNN-based) [23,33]. A recent notable application of ViT that instead uses an unsupervised learning approach is [3], which introduces SiT, a self–supervised vision transformer model. Their system trains an autoencoder transformer model in a self–supervised manner on a large (but unlabeled) dataset on a pretext task and then uses the frozen encoder as a feature extractor for a classification head, trained on a much smaller labeled dataset. This eliminates the need for manual labeling of the larger dataset used in the unsupervised learning phase, still retaining good performance and even outperforming other self–supervised learning methods.

Unsupervised Learning. In supervised learning, the loss function used to train the model requires each sample's ground truth (labels) to be known. Instead, an *unsupervised* learning pipeline does not require any additional information. A notable case of unsupervised learning is *self–supervised* learning, in which labels for a task are generated automatically for each sample. Self–supervised tasks are often used to train a feature extractor to be used later for a supervised task. More specifically, a *pretext* self-supervised scheme is typically used to train an autoencoder model to learn a meaningful representation of a dataset. This allows using the resulting encoder as a fixed module for a (usually small) model, later trained in a supervised fashion on a smaller labeled dataset. This super-vised task is referred to as the *downstream* task in this context. Typical pretext tasks include image reconstruction, transformation recognition, and contrastive learning in computer vision [19].

Pose Estimation. In computer vision, the general task of estimating the spatial configuration of an entity is called *pose estimation.* The specific representation of this pose depends on the type of targeted bodies and the space considered.

In general, the task is to estimate the pose of either:

- single bodies, such as everyday objects or tools; in this case, a single position–orientation pair is estimated [2, 24];
- multi-body entities, usually humans, robots, or animals; these are generally modeled as linked joints so that the objective is to estimate the position and orientation of each joint (usually relative to the parent joint) [12, 40].

The domain of the actual positions and orientations usually is one or more of:

- the 2D space; the most common choice here is to use planar Cartesian coor-dinates for positions and a single angle for the orientations;
- the 3D space; in this case, Cartesian coordinates or 3D meshes [18, 28] are used for the positions, while various representations, such as Euler angles, quaternions, etc., can be used for the orientations;
- the image space; usually, the pose is given as an ordered list of keypoint coordinates in the image space.

Additional variations depend on auxiliary information given with the image: depth values and/or mask images may be available together with the 3 RGB channels. Typically, manual annotation provides this information or from a sep-arate model.

3   Methodology

This section overviews the *Un-TraPEs* model, the dataset, and the training procedure.

3.1   Un-TraPEs Model

The *Un-TraPEs* model is based on the modified ViT model [15], SiT [3], a Vision Transformer for Self-Supervised Learning (SSL). The original SiT model's task is image reconstruction of altered pictures[1]. The decoder's output is fed to a convolution layer that projects the latent vector into the image space. On the other hand, the ViT architecture used a classifier to discern the altered pictures.

Our model is instead composed of three components:

1. The Encoder: this component is the same encoder described in the SiT paper.
2. The ImageDecoder: this component is active only during the self-supervised learning (SSL) training. It projects the encoder's output back into the image space, effectively reconstructing the image, similar to SiT's image reconstruction approach.
3. The PoseEstimator: this component is active only during the fully-supervised learning (FSL) training and projects the encoder's output into the human pose space, effectively predicting the human pose.

We can define the ImageDecoder and PoseEstimator as different heads of the model attached to the encoder. Then, our model can be graphically represented as in Fig. 2.

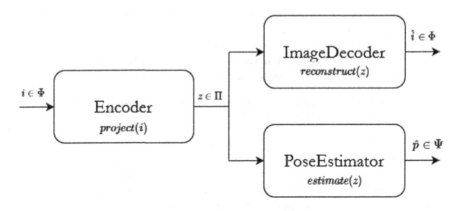

Fig. 2. Diagram model of *Un-TraPEs*

We then defined a wrapper model to *Un-TraPEs* that we used during training. This wrapper allowed us to switch between the two possible heads, controlling the output of the forward pass according to the training mode (SSL or FSL). We added additional features to the wrapper model for sugar-coding, such as internal training and testing loop, a prediction visualizer, and saving/loading model states from checkpoints or finalized model files.

---

[1] Images are augmented via random drops, random replace, color distortion, blurring, and grey-scaling.

3.2   Dataset Analysis

In our experiment, the dataset we chose to use is the Multimodal Gesture Recognition: Montalbano V2 (ECCV '14) [8]. The dataset, already partitioned into train and test subsets, contains over 14000 gestures from a vocabulary of 20 Italian sign gesture categories. An overview is given in Table 1.

Table 1. Overview of the Montalbano dataset [8] with its constituents.

Data type	Format	Description
Video	.mp4	1. RGB data 2. Depth data 3. Subject segmentation 4. Skeleton information[a]
Table	.csv	1. General information for sample 2. Ground truth labels

[a]The skeleton is made of 20 joints: HipCenter, Spine, ShoulderCenter, Head, ShoulderLeft, ElbowLeft, WristLeft, HandLeft, ShoulderRight, ElbowRight, WristRight, HandRight, HipLeft, KneeLeft, AnkleLeft, FootLeft, HipRight, KneeRight, AnkleRight, and FootRight. Nine values encode each join: W_x, W_y, W_z are the world coordinates, R_x, R_y, R_z are the rotation values, and P_x, P_y, P_z are the pixel coordinates.

We preprocess the provided dataset to fit it to our task. We extracted every ten frames for each sample, as most of the frames provided no additional information and were thus considered superfluous. Next, we saved the RGB data, the Depth data, and the Skeleton data separately. Since our goal is Pose Estimation, our ground truth is the user's Skeletons, so we do not need to extract any other information from the original dataset. We apply this extraction to both the training and the testing datasets, resulting in 63415 and 32850 samples for the training and testing sets.

When accessing a sample from the newly constructed dataset, we apply the user masking to both the RGB image and the Depth image and concatenate the results. Formally, given a mask m, a RGB image i_{rgb} and a depth image i_d, we obtain the new image sample i_{rgbd} as:

$$i_{rgbd} = \left(\bigcup i_c \cdot m, \text{for } c \in i_{rgb} \right) \cup (i_d \cdot m)$$

where $\cdot$ is the element-wise (here, pixel-wise) product.

Data Augmentation. Moreover, we apply two kinds of data augmentation to each RGBD image. The first is Gaussian Noise, where random Gaussian Noise is summed onto the RGB channels of the image: given an image i with pixels value $\in [0, 1]$, we augment it as

$$i_{\text{aug}} = \min(1, \max(0, i_{rgb} + \delta)) \cup i_d$$

where :

$$\delta = \mathcal{N}(0, 1)$$

has the same dimensions as i_{rgb}.

The second augmentation is the Random Erasure [39] technique that makes the model robust to occlusions by removing parts of the image randomly, differently from [13, 27]. Random Erasure sets the pixel values in a randomly defined rectangular area of the image to 0^2. Formally, this is accomplished by first defining an area of the image to erase:

$$x_1, x_2 = \text{rand}(0, W)$$

$$y_1, y_2 = \text{rand}(0, H)$$

$$area = \text{rectangle}((x_1, y_1), (x_2, y_1), (x_2, y_2), (x_1, y_2))$$

where H, W are the height and width of the image, rand is a function that returns a random integer value between the specified range and rectangle defines the rectangular area described by the four 2D vertices. The augmented image is obtained by setting the pixel values of the original image in the generated area to 0:

$$i_{aug,x,y} = 0 \quad \text{if} \quad (x, y) \in area, \quad i_{x,y} \quad \text{otherwise}$$

$$\forall(x, y) \in ([0, W], [0, H])$$

We note that we originally planned to apply color jittering to the RGB channels of the augmented image as an additional data augmentation technique. In earlier experiments, however, we found that it was not yielding sufficiently different performances on the same model that could justify the increased computational time needed to apply it.

3.3 Training Process

We now overview the training process for both the *Un-TraPEs* and the other baseline models.

We first define the following spaces:

- The space of the sample data is Φ, with dimension $|\Phi| = C \times H \times W$, where C is the number of channels in an image.
- The space of the latent representation of the sample is Π with dimension $|\Pi| = \text{num\_patches} \times \text{latent\_dim}$

---

[2] In the original paper, this value could be specified arbitrarily or even randomly per pixel. In our application, however, we want to simulate an occluded subject, so the user mask would reflect this as implemented.

- The dimension of the skeleton's joints space is Ψ with dimension $|\Psi| = $ num_joints $\times$ joints_dim

In our experiment, these spaces have the following dimensions:

- $|\Phi| = 4 \times 240 \times 320 = 307200$
- $|\Pi| = 300 \times 768 = 230400^3$
- $|\Psi| = 20 \times 9 = 180$

We also define the *Un-TraPEs* parameters vector as

$$\Theta_{untrapes} = \theta_{enc} \cup \theta_{dec} \cup \theta_{est}$$

and the other models parameters vector as Θ_{base}.

For the *Un-TraPEs* model, we first train it to encode the altered input samples vector X_{alt} in the latent space. We obtain X_{alt} from augmenting the original samples vector X as previously mentioned:

$$X_{alt} = \text{aug}(X)$$

In order to ensure an accurate projection in the latent space of the sample and its subsequent decoding, we optimize the model to minimize the reconstruction loss

$$\mathcal{L}_{rec}(Y, \hat{Y}) = \frac{1}{n} \sum_i^n |y_i - \hat{y}_i| \quad \forall (x_i, \hat{y}_i) \in (Y, \hat{Y})$$

where

$$|Y| = n, \quad y \in \mathbb{R}^\Phi, \quad y \subset X$$
$$|\hat{Y}| = n, \quad \hat{y} \in \mathbb{R}^\Phi, \quad \hat{y} \subset Y$$

$\mathcal{L}_{rec}$ is a standard L1 loss applied between the original unaltered sample Y and the reconstructed sample $\hat{Y} = \Theta_{untrapes}(X_{aug})$. We only optimize the vision encoder and decoder parameters during this part of the training. As such, this training loop optimizes the image-to-latent prediction and latent-to-image reconstruction. The encoder handles the first part, and the second is by the decoder. Formally, the two steps can be expressed as:

$$project(x) : \mathbb{R}^\Phi \rightarrow \mathbb{R}^\Pi$$

and

$$reconstruct(z) : \mathbb{R}^\Pi \rightarrow \mathbb{R}^\Phi$$

To train the model to correctly estimate the pose of the human subject we instead minimize the estimation loss

$$\mathcal{L}_{est}(Y, P) = \frac{1}{n} \sum_i^n (y_i - p_i)^2 \quad \forall (y, p) \in (Y, P)$$

---

[3] The number of patches is determined by the size of each patch and the dimension of the image; as we have set the size of a patch to 16×16 pixels, we end up with 300 patches.

where

$$|Y| = n, \quad y \in \mathbb{R}^{\Phi}, \quad y \subset Y$$
$$|P| = n, \quad p \in \mathbb{R}^{\Psi}, \quad p \subset P$$

$\mathcal{L}_{est}$ is a simple MSE loss applied between the original unaltered sample Y and the predicted skeleton pose $P = \text{untrapes}_{\Theta}(X_{aug})$. In this training loop we only optimize the parameters of the estimator network. Thus, we can write this loop formally as the step

$$estimate(z) : \mathbb{R}^{\Pi} \rightarrow \mathbb{R}^{\Psi}$$

For clarity, we report in Fig. 3 which parts of the *Un-TraPEs* model are active during which part of the training process and which are training ($\boldsymbol{T}$) and which are frozen ($\boldsymbol{F}$).

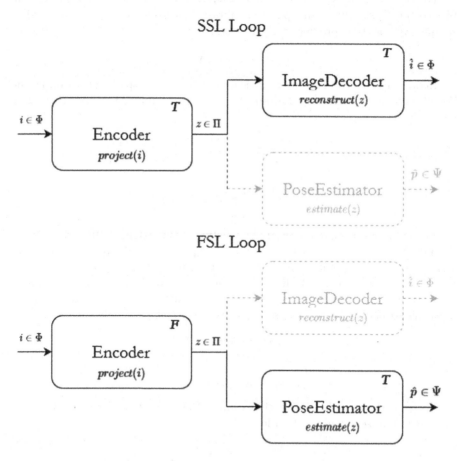

Fig. 3. Overview of *Un-TraPEs* modules during training. Semi-transparent modules are not active, **T** indicates that the module is Trainable and **F** suggests that the module is Frozen.

Training Parameters. The optimizer for *Un-TraPEs* was AdamW with a fixed learning rate of $5e - 4$ and weight decay set to 0.05.

We trained *Un-TraPEs*'s encoder and decoder for 30 epochs. We then switched to fully-supervised learning and trained the pose estimator for another 30 epochs (while keeping the encoder's parameters frozen). Prediction samples obtained with this configuration can be found under **Experimental results**.

4   Results

We report the loss values obtained after training *Un-TraPEs* for 30 epochs both for the unsupervised training and the supervised training in Table 2.

As shown from Fig. 4, the model does indeed learn to do reconstruction and perform in-painting, filling the missing parts of the altered image with realistic patches. This is no surprise since it is the task of SiT, but it still is impressive how quickly this in-painting ability showed up during training. Moreover, the training and test loss presented show how the architecture can learn in just 15 epochs, reaching a lower bound of 0.0003 and 0.0038, respectively.

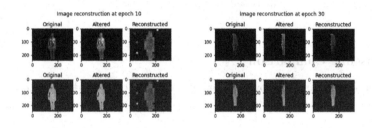

(a) Reconstruction at epoch 10   (b) Reconstruction at epoch 30

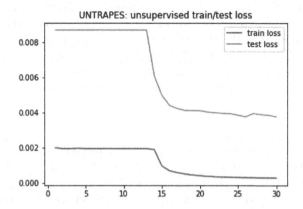

(c) Train and test losses for *Un-TraPEs* during unsupervised learning

Fig. 4. Various reconstructions using *Un-TraPEs* at different training epochs.

On the other hand, for the pose estimation sub-task, we noticed how the model converges on prediction for the pose estimation sub-task, which shows similar poses after epoch 11. We assume that a local minimum was encountered, and the pixel values of the joints kept being predicted around this locus. Indeed, by analyzing the dataset further, we noticed how this minimum falls close to the mean distribution of the joints. This hypothesis is also validated by the loss curves reported in Fig. 5c. The test loss is very high, though decreasing, whereas the training loss is much lower but very noisy. This behavior indicates overfitting, which can be handled with more data augmentation or introducing robust techniques such as Dropout [30]. Moreover, we experimented with different seed initialization at epoch ten and observed either diverging behaviors where the model predicts unnatural poses or convergence to another similar local minimum.

Table 2. *Un-TraPEs* loss metrics for training type.

Training type	Train loss	Test loss
Unsupervised	0.0003	0.0038
Supervised	0.0302	0.0439

4.1   Baseline Comparison

To better understand the overfitting problem, we compared the latter results with other baseline architectures.

Conv-PEs. Our first comparison is performed against a pure CNN-based architecture. The latter is a pre-trained ResNet-18 to which we include a convolution layer that reduces the number of channels from 4 (RGBD) to 3 (RGB) and appended a fully connected layer to predict the human pose. The CNN and FC layers can learn while the ResNet-18 module is kept frozen during training. From Fig. 6c, we can see that the test loss increases with some noisy spikes while the training loss decreases steadily. Moreover, the predictions Fig. 6 indicate a similar behavior to the one presented in the previous section.

ViT-PEs. Our next step was to finetune a pre-existing ViT model. We created a wrapper that would compute the ViT's latent predictions and then forward them through the PoseEstimator module. The employed ViT is pretrained on the ImageNet dataset [14] for 401 epochs. Due to the incompatibility between input types, we also resize the image to be $(224, 224)^4$ and, since i_{aug} is a four-channel image (RGBD), we multiply the D channel over the RGB channels

---

[4] This has no repercussion on the estimation of the pose as the skeleton is normalized before any transformation and denormalized only when showing the prediction on image.

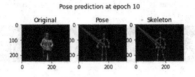

(a) Reconstruction at epoch 10

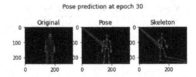

(b) Reconstruction at epoch 30

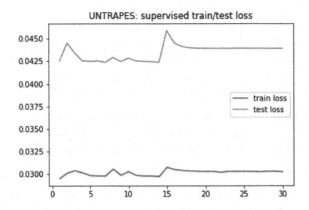

(c) Train and test losses for *Un-TraPEs* during supervised learning

Fig. 5. Various reconstructions using *Un-TraPEs* at different training epochs.

to incorporate the depth information. This step does not alter the pixel values bound ($[0, 1]$) as both the D channel and the RGB channels are bounded in the same $[0, 1]$ interval.

We still report an overfitting trend with this architecture, as shown in Fig. 7c. Indeed,7 we see how the same local minimum is predicted with the pose estimation module.

Overfitting. Although we did not overcome the overfitting problem, we show how it affects both *Un-TraPEs* and the baseline architecture for the supervised task. We speculate that this behavior is due to the nature of the task, being

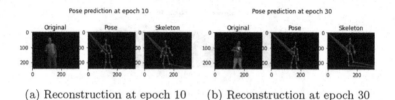

(a) Reconstruction at epoch 10   (b) Reconstruction at epoch 30

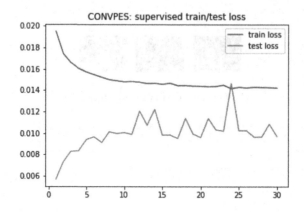

(c) Train and test losses for Conv-PEs during supervised learning

Fig. 6. Various reconstructions using Conv-PEs at different training epochs.

a complex problem, and on the dataset itself, which does not contain enough variety to be treated with these models. For convenience, we report the various losses per model in Table 3.

Table 3. Metrics comparison across tested models for pose estimation only.

Model	Train loss ($\downarrow$)	Test loss ($\downarrow$)
Un-TraPEs	0.0302	0.0439
Conv-PEs	**0.0142**	**0.0097**
ViT-PEs	0.0297	0.0424

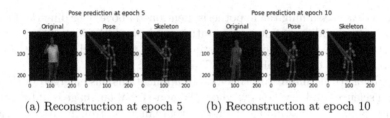

(a) Reconstruction at epoch 5   (b) Reconstruction at epoch 10

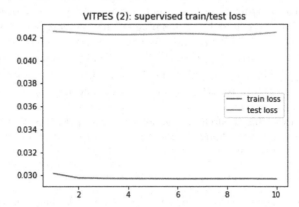

(c) Train and test losses for ViT-PEs during supervised learning (second method)

Fig. 7. Various reconstructions using ViT-PEs at different training epochs.

5   Conclusions

While we found that a CNN-based architecture achieves decent performance, we believe that it could be outperformed by a Transformer-based model, though our results seem to suggest otherwise. One reason both *Un-TraPEs* and the ViT-based model perform worse than the Conv-PEs (though similarly) could be the PoseEstimator module itself. It is possible that such module either

1. is too simple and therefore can't generalize well enough or
2. is intrinsically unable to generalize (though this is the least likely reason).

It is also possible that the optimization landscape of estimating the pose from the latent space is more complex to traverse than the optimization landscape of estimating the pose from the image space; however counter-intuitive.

Finally, we would like to note that, although the final numerical results indicate otherwise, it seems that the predictions using the pretrained ViT model more closely resemble the original pose. In contrast, the *Un-TraPEs* and Conv-PEs estimations share the same problem. We are unsure why this would be the case since the loss computation is the same across models.

Future Works. Assuming the hypothesis mentioned above regarding the PoseEstimator holds validity, an interesting future work would be integrating the user segmentation mask and the depth estimation in the model itself, either as a separate network or internally to the Transformer.

Finally a strong interest in this kind of applications has emerged in the field of social sciences and psychology, where the non-verbal language is of fundamental importance. Understanding the non verbal language trough pose estimation could allows us to interpret human's intentions and the presence of incongruity between verbal language and body language. This can enable us to improve social interactions for social robots and human-interactive interfaces, as well as to support all the possible applications where a human operator (e.g. a psychologist during a therapy session) could be supported by a system that can automatically register and report such kind of information.

Acknowledgments. This research was supported by the HERMES (WIRED) project within Sapienza University of Rome Big Research Projects Grant framework 2020.

References

1. Andriluka, M., Pishchulin, L., Gehler, P., Schiele, B.: 2D human pose estimation: new benchmark and state of the art analysis. In: IEEE Conference on Computer Vision and Pattern Recognition (CVPR) (2014)
2. Anguelov, D., Srinivasan, P., Koller, D., Thrun, S., Rodgers, J., Davis, J.: Scape: shape completion and animation of people. In: ACM SIGGRAPH 2005, pp. 408–416 (2005)
3. Atito, S., Awais, M., Kittler, J.: Sit: self-supervised vision transformer (2021)
4. Avanzato, R., Beritelli, F., Russo, M., Russo, S., Vaccaro, M.: Yolov3-based mask and face recognition algorithm for individual protection applications, vol. 2768, pp. 41–45 (2020)
5. Baldi, T.L., Farina, F., Garulli, A., Giannitrapani, A., Prattichizzo, D.: Upper body pose estimation using wearable inertial sensors and multiplicative kalman filter. IEEE Sens. J. **20**(1), 492–500 (2019)
6. Brandizzi, N., Bianco, V., Castro, G., Russo, S., Wajda, A.: Automatic RGB inference based on facial emotion recognition, vol. 3092, pp. 66–74 (2021)
7. Capizzi, G., Lo Sciuto, G., Napoli, C., Tramontana, E., Wozniak, M.: A novel neural networks-based texture image processing algorithm for orange defects classification. Int. J. Comput. Sci. Appl. **13**(2), 45–60 (2016)
8. Chalearn: Montalbano v2 dataset, eCCV 2014 (2014)
9. Chen, M., et al.: Generative pretraining from pixels. In: International Conference on Machine Learning, pp. 1691–1703. PMLR (2020)
10. Chen, W., et al.: A survey on hand pose estimation with wearable sensors and computer-vision-based methods. Sensors **20**(4), 1074 (2020)
11. Chithrananda, S., Grand, G., Ramsundar, B.: Chemberta: large-scale self-supervised pretraining for molecular property prediction. arXiv preprint arXiv:2010.09885 (2020)
12. Choutas, V., Pavlakos, G., Bolkart, T., Tzionas, D., Black, M.J.: Monocular expressive body regression through body-driven attention. In: Vedaldi, A., Bischof, H., Brox, T., Frahm, J.-M. (eds.) ECCV 2020. LNCS, vol. 12355, pp. 20–40. Springer, Cham (2020). https://doi.org/10.1007/978-3-030-58607-2_2

13. Das, S., Kishore, P.S.R., Bhattacharya, U.: An end-to-end framework for unsupervised pose estimation of occluded pedestrians (2020)
14. Deng, J., Dong, W., Socher, R., Li, L.J., Li, K., Fei-Fei, L.: Imagenet: a large-scale hierarchical image database. In: 2009 IEEE Conference on Computer Vision and Pattern Recognition, pp. 248–255. IEEE (2009)
15. Dosovitskiy, A., et al.: An image is worth 16x16 words: transformers for image recognition at scale (2021)
16. Dosovitskiy, A., et al.: An image is worth 16x16 words: transformers for image recognition at scale. arXiv preprint arXiv:2010.11929 (2020)
17. He, K., Zhang, X., Ren, S., Sun, J.: Deep residual learning for image recognition (2015)
18. Honari, S., Constantin, V., Rhodin, H., Salzmann, M., Fua, P.: Unsupervised learning on monocular videos for 3D human pose estimation (2021)
19. Jaiswal, A., Babu, A.R., Zadeh, M.Z., Banerjee, D., Makedon, F.: A survey on contrastive self-supervised learning (2021)
20. Lin, T.-Y., et al.: Microsoft COCO: common objects in context. In: Fleet, D., Pajdla, T., Schiele, B., Tuytelaars, T. (eds.) ECCV 2014. LNCS, vol. 8693, pp. 740–755. Springer, Cham (2014). https://doi.org/10.1007/978-3-319-10602-1_48
21. Liu, A.T., Li, S.W., Lee, H.Y.: Tera: self-supervised learning of transformer encoder representation for speech. arXiv preprint arXiv:2007.06028 (2020)
22. Liu, J., Wang, G., Hu, P., Duan, L.Y., Kot, A.C.: Global context-aware attention LSTM networks for 3D action recognition. In: Proceedings of the IEEE Conference on Computer Vision and Pattern Recognition, pp. 1647–1656 (2017)
23. Liu, Z., et al.: Swin transformer: hierarchical vision transformer using shifted windows. arXiv preprint arXiv:2103.14030 (2021)
24. Loper, M., Mahmood, N., Romero, J., Pons-Moll, G., Black, M.J.: SMPL: a skinned multi-person linear model. ACM Trans. Graph. (TOG) **34**(6), 1–16 (2015)
25. Naseer, M., Ranasinghe, K., Khan, S., Hayat, M., Khan, F.S., Yang, M.H.: Intriguing properties of vision transformers. arXiv preprint arXiv:2105.10497 (2021)
26. Peng, X.B., Abbeel, P., Levine, S., van de Panne, M.: Deepmimic: example-guided deep reinforcement learning of physics-based character skills. ACM Trans. Graph. (TOG) **37**(4), 1–14 (2018)
27. Perla, S., Das, S., Mukherjee, P., Bhattacharya, U.: Cluenet: a deep framework for occluded pedestrian pose estimation. In: 30th British Machine Vision Conference, pp. 1–15 (2019)
28. Rhodin, H., Salzmann, M., Fua, P.: Unsupervised geometry-aware representation for 3D human pose estimation (2018)
29. Sigal, L., Black, M.J.: Humaneva: synchronized video and motion capture dataset for evaluation of articulated human motion. Brown Univertsity TR 120(2) (2006)
30. Srivastava, N., Hinton, G., Krizhevsky, A., Sutskever, I., Salakhutdinov, R.: Dropout: a simple way to prevent neural networks from overfitting. J. Mach. Learn. Res. **15**(1), 1929–1958 (2014)
31. Starczewski, J.T., Pabiasz, S., Vladymyrska, N., Marvuglia, A., Napoli, C., Woźniak, M.: Self organizing maps for 3D face understanding. In: Rutkowski, L., Korytkowski, M., Scherer, R., Tadeusiewicz, R., Zadeh, L.A., Zurada, J.M. (eds.) ICAISC 2016. LNCS (LNAI), vol. 9693, pp. 210–217. Springer, Cham (2016). https://doi.org/10.1007/978-3-319-39384-1_19
32. Starke, S., Zhao, Y., Zinno, F., Komura, T.: Neural animation layering for synthesizing martial arts movements. ACM Trans. Graph. (TOG) **40**(4), 1–16 (2021)

33. Touvron, H., Cord, M., Douze, M., Massa, F., Sablayrolles, A., Jégou, H.: Training data-efficient image transformers & distillation through attention. In: International Conference on Machine Learning, pp. 10347–10357. PMLR (2021)
34. Vaswani, A., et al.: Attention is all you need (2017)
35. Wang, Y., Huang, M., Zhu, X., Zhao, L.: Attention-based LSTM for aspect-level sentiment classification. In: Proceedings of the 2016 Conference on Empirical Methods in Natural Language Processing, pp. 606–615 (2016)
36. Wozniak, M., Polap, D., Kosmider, L., Napoli, C., Tramontana, E.: A novel approach toward X-ray images classifier, pp. 1635–1641 (2015). https://doi.org/10.1109/SSCI.2015.230
37. Wozniak, M., Polap, D., Napoli, C., Tramontana, E.: Graphic object feature extraction system based on cuckoo search algorithm. Expert Syst. Appl. **66**, 20–31 (2016). https://doi.org/10.1016/j.eswa.2016.08.068
38. Xie, Z., et al.: Self-supervised learning with swin transformers. arXiv preprint arXiv:2105.04553 (2021)
39. Zhong, Z., Zheng, L., Kang, G., Li, S., Yang, Y.: Random erasing data augmentation (2017)
40. Zhou, Y., Habermann, M., Habibie, I., Tewari, A., Theobalt, C., Xu, F.: Monocular real-time full body capture with inter-part correlations. In: Proceedings of the IEEE/CVF Conference on Computer Vision and Pattern Recognition, pp. 4811–4822 (2021)

CamCarv - Expose the Source Camera at the Rear of Seam Insertion

Muhammad Irshad(ID), Ngai Fong Law(✉)(ID), and Ka Hong Loo(ID)

Department of Electronic and Information Engineering, The Hong Kong Polytechnic University, Hung Hom, Hong Kong
ngai.fong.law@polyu.edu.hk

Abstract. It is well known that photo response non-uniformity (PRNU) noise based source attribution helps to verify the camera used to take an image. Recent advances in content-aware image resizing method such as seam carving allow an image to be resized while the critical content is retained. In this paper, we propose identifying the source camera from seam inserted images using blocks as small as 20 × 20. In particular, the correlation is computed between the noise residue of the seam inserted image and the camera PRNU constructed using different numbers of im-ages. We found that different correlation patterns with the camera PRNUs are ob-served, depending on whether the image is taken by that camera or not. Addition-ally, based on this observation, features are extracted from the correlation patterns which are then weighted and combined to form a decision metric for source cam-era identification. We demonstrate by our experimental results that our approach is effective in identifying the source camera in seam insertion images.

Keywords: Source camera identification · Seam insertion · Correlation pattern

1 Introduction

Over the last few decades, digital images have become an essential part of our lives [1,2]. A high level of development in image editing technologies has resulted from the growing demands. As a result, digital images can be easily altered on a computer. The authenticity of digital images is therefore crucial in today's digital world [3]. Forensic imaging has gained popularity as a tool to combat counterfeiting and reveal the truth. To detect duplicated regions in an image, double JPEG compression detection can be used to detect whether an image has repeatedly been compressed by JPEG; and computer graphics identification can help differentiate computer generated graphics from photographic images [4,5].

In the past two decades, significant progress has been made on many forensic topics, such as source identification, which aims at identifying the device that took the image [6]. As part of digital multimedia forensics is attribution of images, a very important task. Investigators fighting such hideous crimes as

L. Rutkowski et al. (Eds.): ICAISC 2022, LNAI 13589, pp. 21–34, 2023.
https://doi.org/10.1007/978-3-031-23480-4_2

terrorism and pornography will have an advantage in identifying which device acquired a particular image [7,8]. For the most part, device identification relies on the photo response non-uniformity (PRNU) noise, a kind of fingerprint left by the camera when it takes a picture [9]. A PRNU pattern is caused by random imperfections that appear in all acquired images as a deterministic result of subtle camera sensor imperfections [10,11]. By analyzing a large set of images taken by the device itself, With sophisticated processing steps, it is possible to estimate the PRNU for any given device. In ideal circumstances, image attribution is relatively easy and extremely reliable be-cause of the robustness of the camera PRNU pattern. However, performance drops when the operating conditions are unfavorable. For example, the performance may be severely impaired if the image size is small or if only a small image patch is available. The performance would also be affected if only a limited number of images are available for estimating the device PRNU [12]. Besides, if the image is scaled or seam-carved, the performance will be deteriorated greatly [13].

The seam carving algorithm is an image resizing algorithm other than cropping and resampling, which has more flexibility than image retargeting [14,15]. Many retargeting techniques have been developed [16,17], such as mesh-deformation-based methods [18] and combination methods [19], but seam carving remains an effective tool due to its simplicity and scalability, Video frame sequences [20], stereo images, image compression [21], and high-concentration of important image features [22] are a few examples of seam carving applications. An eight-pixel path with the least amount of energy from top to bottom (or from left to right), referred to as a seam, is deleted or inserted one by one. In other words, pixels with higher energy remain untouched, so important structure is not affected in image resizing. Seam carving was originally carried out using gradient magnitude as the energy function in order to avoid seams crossing edges [14]. By accumulating energy, the seam carving involves determining which seams should be deleted or duplicated.

This paper considers the problem of source camera identification from a single seam-insertion image. The seam-insertion image would make the image lose the positional correspondence with the PRNU pattern. As a result, the PRNU could not identify the source camera from a seam-insertion image. Moreover, we study the correlation of the noise residue in a seam-insertion image and PRNU. Our study finds that the correlation behaves differently in matched and unmatched cases. Matching refers to taking the test image from the corresponding camera. Based on the difference in the correlation pattern, five types of features are extracted from the pattern to identify the source camera. The technique selects appropriate blocks from the test image to identify the source camera. Earlier research has shown how successful source identification depends on the presence of an uncarved 50×50 block from a seam-carved image [23]. Our method still correctly identifies the source camera even though there is only one 20×20 uncarved block.

The rest of this paper is organized as follows. Section 2 provides an overview of seam insertion resizing method. Section 3 presents a new approach to defeat seam-insertion based counter measures from a single seam-insertion image. Experimental results are presented in Sect. 4. Finally, Sect. 5 concludes the paper and discusses future work.

2   Review on Seam Insertion

The seam carving algorithm is an intelligent method of resizing images that preserves the quality of important objects without distorting the images [20]. Using gradient information and dynamic programming, seam carving can be effective at resizing images or videos by removing or duplicating unimportant seams A seam is a series of eight connected pixels from left to right or top to bottom. Dynamic programming is used to calculate the cumulative energy of each seam and eliminates the one with the lowest cumulative energy. Mathematically, the absolute gradient magnitude $\mathbf{e(I)}$ of an $\mathbf{n_x \times m_y}$ and $\mathbf{I}$ is defined as,

$$e(I) = \left| \frac{\partial}{\partial x} I \right| + \left| \frac{\partial}{\partial y} I \right| \tag{1}$$

where $\mathbf{I}$ represent an image, $\frac{\partial \mathbf{I}}{\partial \mathbf{x}}$ and $\frac{\partial \mathbf{I}}{\partial \mathbf{y}}$ denote the partial derivatives of the image with respect to x and y directions, and $|.|$ means absolute value. A large value of $\mathbf{e(I)}$ represents regions such as highly texture areas that should not be scaled. On the other hand, a small value of refers to smooth regions that can be scaled without any obvious visual distortions. Let x be a mapping indicating the relationship between the row index and the y coordinate of the seam, i.e. $x : [1, \ldots, n] \rightarrow [1, \ldots, m]$. The vertical seam is defined as,

$$s^x = \{s_i^x\}_{i=1}^n = \{(i, x(i))\}_{i=1}^n \text{ such that } |x(i) - x(i-1)| \leq 1, \forall i \tag{2}$$

Similarly, let y be a mapping indicating the relationship between the column index and the x coordinate of the seam, i.e., $y : [1, \ldots, n] \rightarrow [1, \ldots, m]$. The horizontal seam is defined as,

$$s^y = \{s_j^y\}_{j=1}^m = \{(y(j), j)\}_{j=1}^m \text{ such that } |y(j) - y(j-1)| \leq 1, \forall j \tag{3}$$

The optimal seam can be found using dynamic programming. That is, the image is traversed from the second row to the last row by searching the cumulative minimum absolute gradient value among all possible connected paths. Let be the cumulative minimum absolute gradient value for the pixel (i, j) in an image I. It can be expressed as,

$$M = \cdots + e(i-1, j) + \min \begin{pmatrix} e(i, j-1), \\ e(i, j), e(i, j+1) \end{pmatrix} + \cdots \tag{4}$$

The path called seam corresponds to the set of points achieving the minimum value of M. Similarly, the original images can be enlarged by inserting seams from bottom-up or from right to left. The seam insertion pattern will repeat until the desired image size is achieved. To insert a seam, a seam has to be removed first and then two pixels are inserted at the seam position. Figure 1 shows the process of seam insertion. The pixel $\mathbf{I_{i,j}}$ (in blue color) is first removed. Then two pixels $\mathbf{I'_{i,j}}$ and $\mathbf{I'_{i,j+1}}$ (in red color) are inserted. Note that the distance between $\mathbf{I'_{i,j}}$ and $\mathbf{I'_{i,j-1}}$ are farther away when compared to the distance between $\mathbf{I_{i,j}}$ and $\mathbf{I'_{i,j}}$, similarly the distance between $\mathbf{I_{i,j}}$ and $\mathbf{I'_{i,j+1}}$ is greater than that between $\mathbf{I_{i,j}}$ and $\mathbf{I'_{i,j+1}}$ which can be more explained by following Eqs. 5 and 6.

$$I_{i,j} = round(I_{i,j-1} + I_{i,j}/2) \tag{5}$$

$$I'_{i,j+1} = round(I_{i,j} + I_{i,j+1})/2) \tag{6}$$

In Eqs. (5) and (6), the corresponding pixel $\mathbf{I_{i,j}}$ in every selected seam is replaced by two pixels $\mathbf{I_{i,j}}$ and $\mathbf{I'_{i,j+1}}$, which are computed by averaging $\mathbf{I_{i,j}}$ with their left and right neighbors. This will increase image size and due to its altered dimensions and content, image is considered a tampered image. Moreover, the noise residue of a seam insertion image would lose position correspondence with the camera sensor noise pattern PRNU as shown in Fig. 1. Thus the seam insertion method can be used as a way to attack the PRNU-based source identification method. As a result, the problem of detecting seam insertion is critical for image forensics.

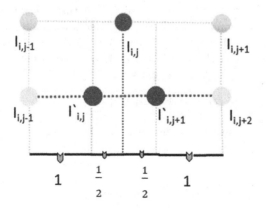

Fig. 1. Seam insertion. The first row shows the original pixel positions in an image. The second row shows the removal of the original pixel $\mathbf{I_{i,j}}$ and the insertion of two pixels $\mathbf{I'_{i,j}}$ and $\mathbf{I'_{i,j+1}}$ (Color figure online)

3   Proposed Approach

Suppose seams are inserted for image resizing. In that case, the pixel-to-pixel correspondence between the noise residue and the camera fingerprint (PRNU) is destroyed. Consequently, the noise pattern from a seam inserted image would not be aligned with the PRNU pattern from the corresponding source camera. This complicates the task of attribution for seam inserted images using PRNU.

Table 1. List of camera devices, including their brand, model, image resolution, number of flat images (#Flat), number of natural images (#Nat), file format and the total number of images

S. No	Camera model	ID	Labeled	Image resolution	#Flat	#Nat	Image format	Total no. of images
1	LG_D290	D04	C_1	3264 × 2448	70	50	JPEG	120
2	Canon_Ixus_55_0	D06	C_2	2592 × 1944	50	70	JPEG	120
3	Sony_XperiaZ1	D12	C_3	5248 × 3936	50	70	JPEG	120
4	Asus_Zenfone2Laser	D23	C_4	3264 × 2763	50	70	JPEG	120
4	Asus_Zenfone2Laser	D23	C_4	3264 × 2763	50	70	JPEG	120
5	Samsung_NV15_0	D64	C_5	3648 × 2763	70	50	JPEG	120

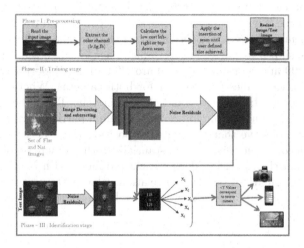

Fig. 2. Operational flowchart of the proposed model

This study seeks to identify the source camera of seam inserted images by examining changes in correlation patterns in when the number of images N used to obtain the camera PRNU increases. The proposed dataset consists of models of DSLR cameras, compact cameras, and mobile phones as shown in Table 1. Suppose seams are inserted for image resizing. In that case, the pixel-to-pixel correspondence between the noise residue and the camera fingerprint (PRNU) is destroyed. Consequently, the noise pattern from a seam inserted image would not be aligned with the PRNU pattern from the corresponding source camera. This complicates the task of attribution for seam inserted images using PRNU. This study seeks to identify the source camera of seam inserted images by examining changes in correlation patterns in Corr $(K_A \otimes f(I), I - f(I))$ when the number of images N used to obtain the camera PRNU increases. The proposed dataset consists of models of DSLR cameras, compact cameras, and mobile phones. A flowchart of the proposed algorithm is shown in Fig. 2. In Phase II, the camera PRNU is constructed from a number of images taken by that camera. Let $\mathbf{K_A(N)}$ denotes the PRNU of camera A estimated from N number of images taken by A. The number of images N used to obtain $\mathbf{K_A}$ should be large to have a reliable estimation of the camera PRNU. However, it may not be possible to have so many images. In the literature [14], N should be at least 30. As a tradeoff, we have kept the minimum number of images, i.e, **N = 12** so that the experiment can be judged from the lowest possible number of images. Assume that X is a testing image. We found that as N increases, the correlation coefficient between X and $\mathbf{K_A(N)}$ exhibits different patterns depending on whether camera A captures the test image X. For image X taken from camera A, i.e. case 1, the correlation values increase with N as shown in Fig. 3. For example, when N is 20, the correlation values for LG_D290 and Samsung_NV15_0 are 20 and 73 respectively. When N is increased to 40, the values become 24 and 90. When N is increased further to 120, they become 50 and 100. On the other hand, i.e. for case 2 with different cameras, the correlation coefficient between X and other camera PRNU $\mathbf{K_A(N)}, \mathbf{C} \neq \mathbf{A}$ exhibits random fluctuations when N increases as shown in Fig. 4. For example, when N is 20, the correlation values for LG_D290 and Samsung_NV15_0 are .25 and .26 respectively. When N is increased to 40, the values become $-.1$ for both cameras. When N is increased further to 120, they become $-.1$ and .1 respectively. Thus, the trend can be used to determine if the testing image comes from a particular camera device or not. Besides, we observe that the matched position (i.e., the position with the highest correlation value) changes substantially with an increase of N in case 2. However, the matched position does not change much with N in case 1. These two pieces of information, i.e., the correlation value patterns and the matched location pattern, can be used to provide a reliable source attribution of seam inserted images.

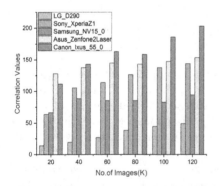

Fig. 3. The plot shows the correlation values from the same device.

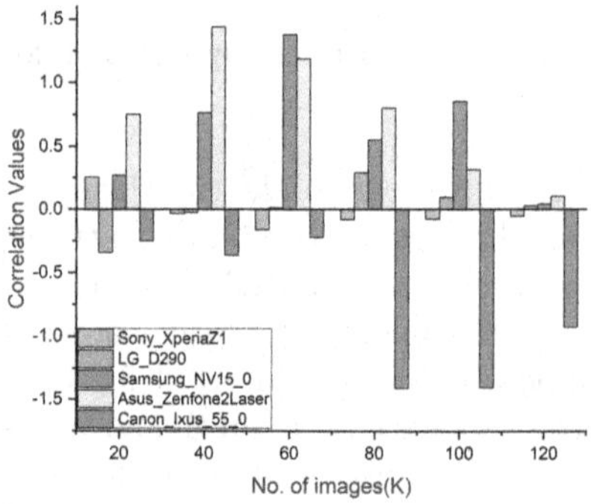

Fig. 4. The plot shows the correlation values from different devices.

4 Proposed Approach

4.1 Proposed Framework for Forensic Investigation

In this paper, a close-set scenario is considered, i.e., the seam inserted images must come from one of the cameras in the dataset. Figures 5 shows an example image captured by Samsung_NV15_0. The original size of the images is 3648 × 2763. After deleting and inserting 200 seams horizontally and 200 vertically, the size of the images becomes 3848 × 2936, respectively. The critical point that needs to be understood here is that these images are carved without visual distortion. However, content creators can achieve the purpose of hiding the source of the camera images. The experiment environment is Windows 10 system, 32G DDR3, Intel®UHD graphic card 630 with 1G memory, MATLAB R2021a, Spyder (Python 3.8), and Ubuntu 64-bit version.

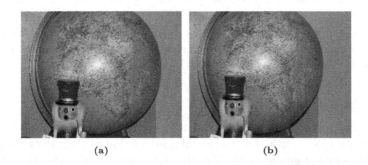

(a)   (b)

Fig. 5. (a) Original Image and (b) the seam-inserted image with 200 seams inserted

4.2   Correlation Patterns Study

Let $\mathbf{K_c(rm)}$ denotes the PRNU of camera C estimated from rm number of images taken by C. The correlation coefficients against r for a test image X from SamsungNV15-0 and different PRNUs is shown in Fig. 6. It can be seen that the obtained correlation coefficients of the test image (i.e seam inserated) with PRNU $\{\mathbf{K_c(m), K_c(2m)}, ...\mathbf{K_c(rm)}\}$ from the camera (Samsung_NV15_0). Although the correlation coefficients show an increasing trend, because of the manupolation we do not see the same trend as in Fig. 3. On the other hand, for different camera PRNUs, the correlation coefficients show random patterns when r increases. It is important to note that distortion of the image leads to significant differences in its correlation value. As you can see in Fig. 3, where the test image was not altered and the correlation index ranges between 1 and 100. In contrast, the range of the correlation index for altered image in Fig. 6 is between 1 and 50.

The proposed method of source camera identification in seam inserted images is composed of two parts: (1) uncarved block selection and (2) decision metric calculation. In the following sections, details are discussed.

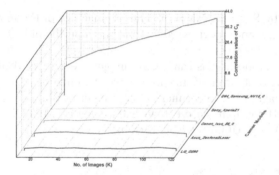

Fig. 6. The plot of the obtained correlation coefficients with PRNUs of different cameras against contrasted from different numbers of images rm used to construct the PRNUs where the testing image X is taken from camera

4.3   Block Selection

The proposed method would first locate a 128 × 128 block which likely contains uncarved sub-blocks such that the standard deviation of the values inside this block is the smallest among all the other blocks of the same size. Such block determines the smoothness of pixels' pattern (i.e., locating a block whose pixels have not been skewed by seam insertion). Typically, this may be located in regions of high energy gradient where seams could not pass frequently. Figure 7 illustrates an example of the block "P", which fulfills the requirement. The noise residue of block P is denoted as $\mathbf{R_p}$. It is obtained through denoising as $mathbf\ R_p = I_p - \hat{I}_P^{(0)}$ where $\hat{\mathbf{I}}_P^{(0)} = f(I_P)$. It will be used to identify the source camera.

Fig. 7. The block P with a size of 128 × 128 such that the standard deviation of the values in this block is the smallest among all the blocks.

4.4   Decision Metric Calculation

The set $\{\mathbf{C_l} \in \mathbf{C}, l = 1, 2, ...\mathbf{L}\}$ contains all cameras under testing, where is the total number of cameras. Assume that we have a total of M images taken by camera C. These M images are divided into r groups randomly, with m number of images in a group, i.e., $M = r*m$. A set of camera fingerprints constructed from different number of images can be obtained in the offline camera fingerprint construction step as $\{K_{cl}(m), K_{cl}(2m), ..., K_{cl}(rm)\}$. As discussed above, an uncarved block (P) is extracted from the seam inserted image. A search will then be performed within the searching window in the camera fingerprint to find the potentially corresponding position between P and the camera fingerprint. We define maximum correlation coefficient as $Corr_{j,cl}$ and the matched position $(x^*, y^*)_{jcl}$ as the location that maximizes $Corr_{j,c_i}$, where j is the location of the block within a test image of camera C, i.e.,

$$Corr_{j,c1} = \max corr(A_{j,c1}(x, y) \otimes f(I_P), R_P) \tag{7}$$

$$(\overset{*}{x}, \overset{*}{y}) = \arg\max corr(A_{j,c1}(x, y) \otimes f(I_P), R_P) \tag{8}$$

where argmax returns the value of block index with maximum gradient magnitude. $A_{j,c_i}(x, y)$ is a sub-block inside $K_{c_i}(jm)$ centered at (x, y). The size of $A_{j,c_i}(x, y)$ is the same as that of R_P. Having obtained a set of correlation coefficients $\{Corr_{j,c_i}, j = 1, 2, ..., r\}$ and matched positions $\{(x^*, y^*)_{jcl}, j = 1, 2, ..., r\}$ for $\{C_l \in C, l = 1, 2, ..., L\}$, features are extracted from them to devise robust decision metric. The decision metric determines which camera device in $\{C_l \in C, l = 1, 2, ..., L\}$ was used to capture testing photo that has been altered by seam insertion. For each camera device, five types of features are extracted. They are:

(1) The sum of correlation coefficients from all camera fingerprints of the camera C_l i.e.,

$$x_{1,c_l} = \sum_{j=1}^{r} corr_{j,c_l} \tag{9}$$

(2) The change of correlation coefficients,

$$x_{2,c_l} = corr_{r,c_l} - corr_{1,c_l} \tag{10}$$

(3) The absolute difference of the change of the location (x^*, y^*) with respect to the number of images used to construct the camera PRNU, i.e.,

$$x_{3,c_l} = \sum_{j=1}^{r-1} \left| (x^*, y^*)_{j+1,c_l} - (x^*, y^*)_{j,c_l} \right| \tag{11}$$

(4) Let M be a 20×20 sub-block of P. Where β_j feature is obtained as $[x_{4,c_l} = \sum_{j=1}^{r} \beta_j]$,

$$\beta_j = \sum_{x=1}^{20} \sum_{y=1}^{20} \frac{(corr_{j,c_l} - M(x, y))^2}{(corr_{j,c_l})} \tag{12}$$

(5)

$$x_{5,c_l} = \beta_r - \beta_1 \tag{13}$$

For each camera $\mathbf{C_1}$, a set of features $\{\mathbf{X_{i,c_1}}, \mathbf{i} = 1, 2, ..., 5\}$ is obtained. This set of features is normalized between 0 and 1 as follows,

$$y_{i,c_l} = \frac{x_{i,c_l} - \min\limits_{l' \in \{1,\cdots,L\}} x_{i,c_{l'}}}{\max\limits_{l' \in \{1,\cdots,L\}} x_{i,c_{l'}} - \min\limits_{l' \in \{1,\cdots,L\}} x_{i,c_{l'}}} \tag{14}$$

Furthermore, entropy is used to set the weights for desired results. To obtain entropy, probabilities have to be used, i.e.,

$$P_{i,c_l} = \frac{y_{i,c_l}}{\sum\limits_{l' \in \{1,\cdots,L\}} y_{i,c_{l'}}} \tag{15}$$

Then the scaled entropy of the ith feature, which is a function of entropy weight method (EWM) can be obtained as,

$$e_i = -\frac{1}{\ln 5} \sum_{l' \in \{1,\cdots,L\}} P_{i,c_{l'}} \ln(P_{i,c_{l'}}) \tag{16}$$

A Large difference among the feature values means a more accurate result. The weight is set as,

$$W_i = \frac{1 - e_i}{\sum\limits_{i=1}^{5} 1 - e_i} \tag{17}$$

4.5   Feature Integration

Five features are extracted from the correlation patterns and the matched locations for source identification in our proposed method, as discussed in Sect. 4.4. Features x1 and x2 characterize the correlation pattern. In particular, x1 considers the total correlation values from all different PRNUs constructed by k number of images while x2 considers the fluctuations of the correlation values for different k. The feature x3 is about the change of the matched location. The feature x4 considers the total change of the correlation value with respect to a 20×20 sub-block for different k while the feature x5 characterizes the fluctuation of this change for different k. After applying normalization, the weighting $\mathbf{W_1}, \mathbf{W_2}, \mathbf{W_3}, \mathbf{W_4}, \mathbf{and} \mathbf{W_5}$ are obtained. After the calculation of $\mathbf{W_i}$ i.e. i = 1...5 for each feature, an overall decision metric y is obtained as follows.

$$y = W_1 x_1 + W_2 x_2 - W_3 x_3 + W_4 x_4 + W_5 x_5 \tag{18}$$

4.6   Decision Matrices Calculation

Decision matrices are calculated for the testing seam inserted image X from different cameras. Tables 2, 3, 4, 5 and 6 show the feature values for all cameras regarding seam insertion accordingly. The largest 'y' value indicates that the test image is taken from that particular source camera.

Table 2. Feature values obtained from the PRNUs of different cameras when the test image X is taken from D04_ LG_D290 camera.

D04	X1	X2	X3	X4	X5	Y
D04	220.3524	23.5397	0	369.9572	−0.0001	116.0571
D06	−0.5613	−0.2244	201	−2317.6165	−7.12E−05	−801.3286
D12	−6.2043	−0.2388	0	−2457.2593	5.672E−06	−540.8667
D23	−0.4796	−0.0088	0	−2701.4912	−0.0002593	−931.7697
D64	−0.7101	−0.0251	119	−3717.7948	−6.86E−06	−987.1610

Table 3. Feature values obtained from the PRNUs of different cameras when the test image X is taken from D06_Canon_Ixus_55_0 camera.

D06	X1	X2	X3	X4	X5	Y
D04	−25.9392	−1.7703	333	−554.3998	−6.79E−05	−191.3155
D06	18.8015	3.9480	0	338.3334	−9.85E−05	71.9129
D12	−5.5871	0.9521	0	−96.8547	−9.18E−05	−22.1749
D23	13.9066	−0.7008	630	−315.2618	−0.00024	−59.2742
D64	13.8698	−0.7116	1274	270.3477	−9.45E−05	−196.3888

Table 4. Feature values obtained from the PRNUs of different cameras when the test image X is taken from D12_Sony_XperiaZ1camera.

D12	X1	X2	X3	X4	X5	Y
D04	−11.0814	−3.5095	140	−624.4292	−6.79E−05	−167.1125
D06	−0.9692	−0.0749	2	−3510.9369	−5.33E−05	−771.5897
D12	170.0986	25.0946	0	10.8195	5.67E−06	40.4304
D23	−1.5014	−0.2281	0	−3267.7101	−0.00025	−717.8333
D64	−36.9991	−4.3620	13	−481.4884	−6.33E−05	−113.8142

Table 5. Feature values obtained from the PRNUs of different cameras when the test image X is taken from D23_Asus_Zenfone2Laser camera.

D12	X1	X2	X3	X4	X5	Y
D04	−32.6380	−5.7244	0	−185.0077	−0.00017	−48.1352
D06	−16.9144	−1.6751	99	−301.2149	−3.40E−05	−69.7766
D12	14.2105	3.2304	381	1275.9207	5.67E−06	180.6204
D23	1982.9065	101.0801	0	198.3625	0.002513	450.9818
D64	−9.9063	0.4425	65	−471.8606	−6.33E−05	−105.4670

4.7   Performance Analysis

The Tables 2, 3, 4, 5 and 6 clearly shows that the Y value can indicate the origin of the seam-inserted testing image. Some features are sensitive to changes in matched location resulting from seam insertion, and others are dependent on the texture or intensity of the test images. Normalization is performed by decimal scaling of the five features. A decimal scale moves the decimal point of the attribute's value, as shown in Table 7. Therefore, the D-scale can reliably identify the source camera of those seam-inserted images.

Table 6. Feature values obtained from the PRNUs of different cameras when the test image X is taken from D64_SamsungNV15_0 camera.

D12	X1	X2	X3	X4	X5	Y
D04	9.0054	1.1768	984	846.0746	$-6.79\text{E}-05$	-24.9802
D06	16.8836	-0.9916	1183	738.9027	$-5.33\text{E}-05$	-84.0695
D12	13.9345	0.5615	306	272.7137	$5.67\text{E}-06$	-3.5651
D23	7.3446	-0.4432	608	369.2084	-0.000251339	-45.5490
D64	95.6016	4.4758	0	346.3521	$1.38\text{E}-04$	86.0243

Table 7. Shows the decimal scaled values for seam insertion.

	D-Scale (C_1)	D-Scale (C_2)	D-Scale (C_3)	D-Scale (C_4)	D-Scale (C_5)
C_1	0.12	-0.80	-0.54	-0.93	-0.29
C_2	-0.80	0.07	-0.02	-0.06	-0.21
C_3	-0.17	-0.77	0.04	-0.72	-0.09
C_4	-0.05	-0.07	0.18	0.45	0.17
C_5	-0.02	-0.08	0.004	0.05	0.09

Table 8. Confusion matrix for test images taken from cameras.

Labels	(C_1)	(C_2)	(C_3)	(C_4)	(C_5)
C_1	83.79	1.76	4.84	9.61	0.00
C_2	0.00	96.93	1.58	1.49	0.00
C_3	0.19	0.00	95.48	0.00	4.33
C_4	0.03	0.05	18.61	81.31	0.00
C_5	1.09	0.00	6.75	0.00	92.16
Average(%)	89.93				

In Table 8, five random test images selected from the dataset are used for the computation of the confusion matrix for each camera. Test images were selected from a variety of backgrounds (light, dark, object orientations, etc.). Using this algorithm, five trials were completed to compute an average accuracy. The average identification accuracy is 89.93.

5   Conclusion

The aim of this paper is to develop a method for identifying the source camera from seam inserted images, using only a block of the seam inserted images and the camera PRNUs. Correlation is performed between the block of the seam inserted images and the camera PRNU. Our proposed approach is based on two observations on the correlation pattern: as the number of images used to construct the PRNU increases, the correlation values increase if the seam inserted image is taken by the corresponding camera. If the corresponding camera is not used, the correlation value will appear random and the matched position will change substantially. From these observations, we extracted five types of features to identify the source camera. By combining these five features, we construct a decision metric. Experimental results show that the decision metric is effective in identifying the source camera of seam inserted images.

Acknowledgement. This work was supported by the GRF Grant 15211720, (project code: Q79N), of the Hong Kong SAR Government. Muhammad Irshad Ibrahim would like to thank the postdoctoral fellowship support from the Hong Kong Polytechnic University (GYW4X).

References

1. Jacoby, M., Usländer, T.: Digital twin and internet of things-current standards landscape. Appl. Sci. **10**(18), 6519 (2020)
2. Haider, S.A., et al.: The inclusive analysis of ICT ethical issues on healthy society: a global digital divide approach. Procedia Comput. Sci. **183**, 801–806 (2021)
3. Garba, A., et al.: A digital rights management system based on a scalable blockchain. Peer-to-Peer Netw. Appl. **14**(5), 2665–2680 (2021)
4. Liu, Q.: An approach to detecting jpeg down-recompression & seam carving forgery under recompression anti-forensics. Pattern Recogn. **65**, 35–46 (2017)
5. Murthy, A., et al.: Internet of things, a vision of digital twins and case studies. In: IoT and Spacecraft Informatics, pp. 101–127. Elsevier (2022)
6. Bernacki, J.: A survey on digital camera identification methods. Forensic Sci. Int. Digit. Investig. **34**, 300983 (2020)
7. Armstrong, J., Mellor, D.: Internet child pornography offenders: an examination of attachment and intimacy deficits. Leg. Criminol. Psychol. **21**(1), 41–55 (2016)
8. Irshad, M., et al.: City vision: CCTV images based public surveillance model. In: 2021 International Conference on Electronic Information Technology and Smart Agriculture, pp. 416–420. IEEE (2021)
9. Zhao, Y., Zheng, N., Qiao, T., Xu, M.: Source camera identification via low dimensional PRNU features. Multimedia Tools Appl. **78**(7), 8247–8269 (2019)

10. Ye, C.H., Lee, D.: CMOS image sensor: characterizing its PRNU (photo-response non-uniformity). In: Optical Data Storage 2018: Industrial Optical Devices and Systems, vol. 10757, p. 107570A. International Society for Optics and Photonics (2018)
11. Aziz, S., Jiang, H., Peng, J., Ruan, J., Wang, H.: Optimization of base operation points of MTDC grid for improving transition smooth. In: 2017 IEEE Conference on Energy Internet and Energy System Integration (EI2), pp. 1–6. IEEE (2017)
12. Chan, L.-H., Law, N.-F., Siu, W.-C.: A confidence map and pixel-based weighted correlation for PRNU-based camera identification. Digit. Investig. **10**(3), 215–225 (2013)
13. Zhang, W.-N., Liu, Y.-X., Zhou, J., Yang, Y., Law, N.-F.: An improved sensor pattern noise estimation method based on edge guided weighted averaging. In: Chen, X., Yan, H., Yan, Q., Zhang, X. (eds.) ML4CS 2020. LNCS, vol. 12487, pp. 405–415. Springer, Cham (2020). https://doi.org/10.1007/978-3-030-62460-6_36
14. Rubinstein, M., Shamir, A., Avidan, S.: Improved seam carving for video retargeting. ACM Trans. Graph. (TOG) **27**(3), 1–9 (2008)
15. Aziz, S., et al.: Anomaly detection in the internet of vehicular networks using explainable neural networks (XNN). Mathematics **10**(8), 1267 (2022)
16. Liu, Y., Zou, Z., Yang, Y., Law, N.F.B., Bharath, A.A.: Efficient source camera identification with diversity-enhanced patch selection and deep residual prediction. Sensors **21**(14), 4701 (2021)
17. Shi, C., Law, N.F., Leung, F.H., Siu, W.C.: A local variance based approach to alleviate the scene content interference for source camera identification. Digit. Investig. **22**, 74–87 (2017)
18. Michael, R., Ariel, S., Shai, A.: Improved seam carving for video retargeting. ACM Trans. Graph (TOG) **27**(16), 1–9 (2008)
19. Rashid, A., Peng, Y., Muhammad, T., Muhammad, I.: Combination of total variation and robust bilateral filter in image denoising. In: Information Technology and Intelligent Transportation Systems, pp. 127–141. IOS Press (2019)
20. Basha, T.D., Moses, Y., Avidan, S.: Stereo seam carving a geometrically consistent approach. IEEE Trans. Pattern Anal. Mach. Intell. **35**(10), 2513–2525 (2013)
21. Rashid, A., et al.: Image denoising using wavelet transform. In: Information Technology and Intelligent Transportation Systems: Proceedings of the 3rd International Conference on Information Technology and Intelligent Transportation Systems, Xi'an, China, 15–16 September 2018, vol. 314, p. 142. IOS Press (2019)
22. Frankovich, M., Wong, A.: Enhanced seam carving via integration of energy gradient functionals. IEEE Signal Process. Lett. **18**(6), 375–378 (2011)
23. Taspinar, S., Mohanty, M., Memon, N.: PRNU based source attribution with a collection of seam-carved images. In: 2016 IEEE International Conference on Image Processing (ICIP), pp. 156–160. IEEE (2016)

Text Line Segmentation in Historical Newspapers

Ladislav Lenc[1,2(✉)], Jiří Martínek[1,2], and Pavel Král[1,2]

[1] Department of Computer Science and Engineering, Faculty of Applied Sciences,
University of West Bohemia, Plzeň, Czech Republic
{jimar,pkral}@kiv.zcu.cz
[2] NTIS - New Technologies for the Information Society, Faculty of Applied Sciences,
University of West Bohemia, Plzeň, Czech Republic
llenc@kiv.zcu.cz

Abstract. This paper deals with page segmentation into individual text lines used as an input of a line-based OCR system. This task is usually solved in one step which directly identifies text lines in whole documents. However, a direct approach may jeopardize the reading order of the lines and thus deteriorate the overall transcription result.

We propose a novel approach which decomposes this problem into two steps: text-block and text-line segmentation. The particular tasks are handled by algorithms based on fully convolutional neural networks.

The proposed method is evaluated on two standard corpora, Europeana and RDCL 2019, and on a novel dataset created from data available in Porta fontium portal. This dataset is freely available for research purposes.

Keywords: Document image segmentation · Layout analysis · Fully convolutional network · FCN

1 Introduction

Preservation of historical documents stored in various archives is very important. Many efforts have been invested into digitisation of such archival documents. Nevertheless, the digitisation is just the first step in the process of making the documents accessible and exploitable.

The goal of the subsequent processing is to convert the documents into a text form and allow efficient indexing, searching, or even more sophisticated tasks from the natural language processing (NLP) field, such as classification or summarisation.

The first step of document image processing, which determines the success of all the following tasks is the segmentation. Therefore, the main goal of this paper consists in proposing an efficient and accurate segmentation method that also preserves a reading order.

The final outcome of many page segmentation methods is a segmentation mask for text regions, eventually also for other document elements such as images

L. Rutkowski et al. (Eds.): ICAISC 2022, LNAI 13589, pp. 35–48, 2023.
https://doi.org/10.1007/978-3-031-23480-4_3

or tables, and do not consider cropping the regions and further processing of them. We want to go further and create a complete method which converts the input image into an ordered set of text regions and text-line images suitable for processing by an optical character recognition (OCR) system. Our work is dedicated to the processing of archival documents from the Czech-Bavarian border area stored in Porta fontium portal[1]. Namely, we process newspapers from the end of the nineteenth century printed in Fraktur script.

Current OCR algorithms usually rely on neural networks that recognise whole text lines [2,23,24]. We thus aim at segmentation of pages into individual text lines. The task can be carried out directly on whole pages. However, in this case, it is difficult to determine the reading order of the extracted lines which is crucial for further processing of page transcription obtained by an OCR engine. It is obvious especially in the case of complex page layouts with more columns. To be able to find the reading order, it is necessary to determine the page structure and first create an ordered set of text regions which can be further split into text lines. Therefore, the presented approach combines the results of three partial algorithms based on fully convolutional neural networks (FCNs):

1. Text and non-text segmentation on a pixel level to differentiate foreground and background;
2. Separator detection and subsequent cropping of text regions and reading order estimation;
3. Baseline detection within the cropped text regions and extraction of text-line images.

The main strength of the proposed approach is that the particular tasks complement each other and can reduce final error rate of the whole method. One example is the erroneous merging of text regions by the text segmentation method. Such merging can be solved by detecting separators that divide the regions, and thus the final error is reduced. We must also mention that if we apply the baseline detection algorithm directly on the whole page it may lead to merged lines over a separator. It typically occurs in low quality scans with text regions lying very close to each other. In this case, the overall reading order of the page is incorrect. The presented method solves the above mentioned issues which is the main contribution of our work.

The approach is evaluated on a newly created dataset collected from Porta fontium. This dataset is freely available for research purposes[2] and represents another contribution of this work. We further evaluate the system on Europeana and RDCL 2019 corpora.

2   Related Work

Most of nowadays methods treat the document segmentation as a pixel labelling problem and use deep learning for this task. An approach aiming at historical documents proposed by Chen et al. [3] is based on super-pixel calculation

---

[1] http://www.portafontium.cz/.

[2] http://ocr-corpus.kiv.zcu.cz/.

using simple linear iterative clustering [1]. The patches around super-pixels are classified by a convolutional neural network (CNN). This method outperformed a previously presented method [4] which solves document segmentation by convolutional auto-encoders.

Fully convolutional networks (FCNs) were developed for semantic image segmentation. The concept of FCNs is different from the above mentioned approaches. The training objective is to directly apply several convolutional layers followed by de-convolutions or up-sampling of the input image and obtain the final pixel-level segmentation as a result. A pioneering work was presented by Long et al. [15]. The authors took concepts of classification networks such as VGG net [25] or AlexNet [12] and incorporated it into the fully convolutional networks. The architecture has outperformed state-of-the-art results in the semantic segmentation task.

A well known example of an FCN is U-Net [21]. The method was first applied on bio-medical data segmentation. However, it can be also used for document image segmentation. Another FCN-based model in the biomedical domain was proposed by Novikov et al. [17]. It performs a multi-class segmentation of anatomical organs in chest radiographs (x-rays), namely for lungs, clavicles, and heart.

Sherrah [22] applies an FCN to carry out semantic segmentation in aerial images. He performed a fine-tuning of a pre-trained CNN to make better image features for high-resolution aerial images where it is crucial to find boundaries.

A modification of U-net architecture proposed by Wick and Puppe [26] was successfully applied to document image segmentation. The convolutional layers use kernel size of 5 and padding is used to keep image dimension. The network improved state-of-the-art results on several document segmentation datasets.

U-Net architecture enriched with spatial attention (A) and residual structures (R) was proposed by Grüning et al. in [11]. The network was named ARU-Net and it was designed mainly for the task of baseline detection. However, it is possible to use it for arbitrary pixel-labelling problem when appropriate training data are available. An adaptive version of U-Net was utilized for text-line segmentation in [16]. It is a modified and optimized version of U-Net. Smaller number of convolutional filters are used in this work and up-sampling operation in the decoder part is replaced by 2D de-convolution.

Another segmentation method based on an FCN was presented in [27]. First 5 convolutional layers are taken from the VGG network [25]. It is followed by three additional layers with kernel size 3×3. All de-convolution layers have kernels of size 2×2 and stride 2. The approach was tested on DIVA-HisDB and achieved pixel-level accuracy of 99%. The input images are not used directly. Instead, smaller crops with size 320×320 pixels are utilized.

A complex document segmentation and evaluation method was proposed by Li et al. [14]. The label pyramid network (LPN) utilizes an FCN as a core. The label map pyramid is transformed from region class label-map by distance transformation and multi-level thresholding. One single probability map is obtained by summing up the outputs of the LPN. The authors use intersection over union (IoU) and ZoneMap metric [9] for document region segmentation evaluation.

Zhong et al. [28] propose a large dataset for document layout analysis with several deep learning algorithms for baseline evaluation. The authors compare different pre-trained F-RCNN and M-RCNN models for fine tuning using the data from this dataset with interesting results. For evaluation they present mean average precision (MAP) and IoU of bounding boxes.

3   Document Image Segmentation

The presented method is composed of two main sub-tasks. The first one is to distinguish text from background and other non-text content such as images and illustrations and to extract the text blocks. We rely on FCN networks trained for text/background segmentation.

Text segmentation is complemented by separator segmentation within this sub-task. Generally, we differentiate black and white separators. We denote the black ones as explicit and the white ones as implicit separators. The white separators are actually the gaps between text regions. The goal of the separator segmentation step is to detect black separators within the processed page. Such separators determine the overall page structure and are used to define the reading order. It would be possible to omit the separator segmentation step and define the layout only according to white separators inferred from the text segmentation mask in the same matter as X-Y cuts and similar methods. However, our preliminary experiments have shown that the black separators can help to divide regions where text segmentation mask erroneously merges the regions. This happens mainly in cases where two text columns are very close to each other. We use another FCN for the separator segmentation task.

Combining the separator segmentation and the result of the text segmentation allows us to define the page layout and to divide the page into several text blocks. Additionally, we determine the reading order of the extracted regions which is very important for further processing of the page content.

The second sub-task is segmentation of text lines within the detected text blocks. For this task, we utilize an FCN model trained for baseline detection. The baseline positions together with connected component analysis are then directly used for line images extraction.

The architecture of the presented approach is shown in Fig. 1. We describe the above-mentioned tasks in more detail in the followings sections.

3.1   Text-Block Segmentation

Segmentation using an FCN can be seen as a mapping of an input image to several maps of probabilities with the same dimensions as the input image. Each

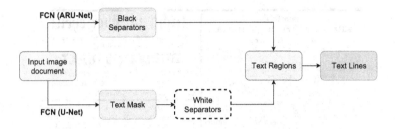

Fig. 1. Architecture of the presented approach

map indicates probabilities of pixel membership in a given class. In our case, only one map indicating probabilities that a pixel belongs to a text region is considered. The probability map contains values between 0 and 1 and we thus need to threshold it to obtain a segmentation mask. Threshold value is set to standard value of 0.5 in all cases.

Ground-truths (GT) for such a network are presented in a form of binary images where the text content is marked by 1 (white) and the background pixels by 0 (black). We do not distinguish between the background and the foreground pixels within a text block area. The ground-truth images are constructed according to the GT stored in the PAGE XML format [18].

The trained network is used to predict the probability map. The thresholded output then forms a segmentation mask. Using a connected components analysis we can separate masks for individual text regions. By multiplying the obtained region mask and the original image we obtain the segmented text region and use it for further processing. In our preliminary experiments, we have compared the performance of two state-of-the-art candidates from the FCN family, namely U-Net [21] and high performance FCN (HP-FCN) [26]. The experiments have shown that the performances of both networks are very similar. Due to the fact, that the HP-FCN has lower number of parameters and is faster, we decided to use this one in the system. It is basically an adaptation of U-Net designed for processing of historical documents. The main difference is that it does not use the skip connections.

3.2   Separator Segmentation

Separators split the page into smaller logical units (columns or smaller sections depending on the level of the separators). In all but a few cases, they can prevent merging of two or more text regions that can occur if we segment the page only according to a mask obtained in the text segmentation step.

A fragment of the visualisation of two considered regions: text regions (blue surrounding polygons) and separator regions (bold black lines) is depicted in Fig. 2.

Within this task, we first apply an FCN trained for separator segmentation on the input image. Even though we can use any of the above-mentioned FCNs for the separator segmentation task, our preliminary experiments have shown

Fig. 2. Examples of detected text regions (blue surrounding polygons) and separator regions (black bold lines) in a document (Color figure online)

that the most suitable candidate is ARU-Net [11] which extends U-Net with residual blocks and attention mechanism and is designed for detection of line objects.

The GT images for network training are constructed in the same way as the ones for text segmentation. The separator regions are marked according to the corresponding PAGE XML file. The result of the network prediction is a map that indicates the probability that a given pixel belongs to the separator class.

The second step is the post-processing of the map. We apply morphological opening in order to remove noise. It is followed by closing with a rectangular structural element which should remove small gaps in separator lines.

Then we detect the separator regions by the means of connected component analysis. Some separators may be still torn apart into several pieces. We therefore try to merge the small regions into larger ones using a simple clustering algorithm and get a final set of horizontal and vertical separators. In this task, we solve the explicit (black) separators. The white ones are inferred from the text segmentation map according to gaps between the regions.

The layout of processed pages is considered to be hierarchical. We thus apply a recursive region splitting algorithm which starts with one region containing the whole page. We first search for horizontal separators. If no horizontal ones are detected we search for the vertical ones. The separators are used for dividing the region into several parts and the same algorithm is then applied on each of the resulting parts. If we cannot divide the region further, the algorithm is finished (see Fig. 3).

The result is a set of text blocks sorted with respect to assumed reading order.

3.3 Text-Line Segmentation

The goal of this task is to extract images of individual text lines. For baseline prediction, we utilize ARU-Net which has proven to be very successful mainly on the task of baseline detection in handwritten text.

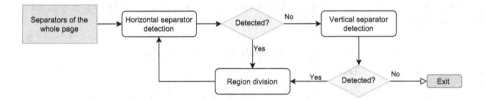

Fig. 3. Flowchart of the recursive region splitting algorithm

Baseline (a line where most characters rest upon and descenders extend below) representation of a text line is very simple and it has a number of advantages. The baseline representation is able to tackle skew and/or orientation as well as multi-column layout. Last but not least, the annotation of baselines is more manageable than annotating surrounding polygons which is another example of a text-line representation.

Because of the lack of annotated data, we utilize the original model which was trained on handwritten documents. Our preliminary experiments have shown that it is sufficient also for printed data.

The trained ARU-Net predicts a baseline map and its output needs to be post-processed in order to get exact baseline coordinates and to be able to extract the text-line image. Using the connected components statistics we derive the x-height (height of small letters without ascenders) and based on it and the baseline position we can extract the line image.

4   Experimental Setup

In this section, we first describe the data collections that were used for our experiments. We also report the evaluation criteria that we utilised for measuring the performance of the individual tasks and for the overall evaluation.

4.1   Corpora

Porta Fontium. This dataset was created from document images digitised within the Porta fontium project[3]. We have selected a newspaper called *"Ascher Zeitung"*. This newspaper dates back to the second half of the nineteenth century and it is printed in German with Fraktur font. We collected 25 pages in total. The pages were annotated using Aletheia [6] and the layout is saved in the PAGE XML format. The transcription was utilised using tools presented in [13].

All pages are annotated on the paragraph level by bounding polygons. The test set contains also text lines with corresponding baselines.

Table 1 shows statistical information of this dataset. Regions include the number of text regions and separator regions.

---

[3] http://www.portafontium.eu/.

Table 1. Porta fontium dataset statistical information (the number of documents, regions and text-lines)

Part	Documents	Regions	Text-lines
Train	10	338	1,156
Test	10	292	1,267
Validation	5	110	651

Europeana. The goal of the Europeana project [5] is allowing access to digitised European cultural heritage. It contains a huge amount of various historical documents. There are also scanned newspapers among others.

We have selected a number of newspaper pages with similar characteristics as the data we are processing. The resulting set contains 95 pages written in German with highly variable page layouts. The pages used in this paper have corresponding ground-truths stored in the PAGE format. We have split the pages into train, validation and test parts containing 68, 9 and 18 pages respectively.

RDCL 2019. RDCL 2019 was designed for the ICDAR 2019 Competition on recognition of documents with complex layouts. This dataset contains scanned pages from contemporary journals. The character of images differs significantly from the two datasets described above. The images are of high quality and easy to binarize because they contain nearly no noise. We use the example set which contains 15 images with corresponding ground truths for testing.

4.2   Evaluation Criteria

In this section, we summarise the metrics we use to measure the performance and how we evaluated all parts of the approach as well as the overall evaluation.

Text Segmentation. We use two pixel-based metrics, namely F1-score [19] and intersection over union (IoU) also known as Jaccard index [20] that are standard metrics in the segmentation task.

Both metrics are computed for each class (foreground and background) separately and the final result is an average value.

PRIMA Evaluator. PRIMA Evaluator framework [7] was utilised both for the overall evaluation and for the evaluation of the separator detection algorithm. It offers a nice visualisation of segmentation errors and has many possibilities how to set the configuration (including different weights for different types of error: merge, split, false detection, miss or partial miss) according to the desired scenario.

To measure the performance of the separator detection algorithm we kept the default error weights and we filtered the success rate particularly for separator regions.

The overall evaluation was performed with a standard *General Document Recognition* evaluation profile. The result thus takes into consideration errors in text regions and in separator regions as well. Reading order is also incorporated into the overall score.

Baseline Detection. We decided to completely follow the evaluation scheme presented by Grüning et al. in [10]. This scheme was utilized for ICDAR 2017 [8] and ICDAR 2019[4] competition on baseline detection (cBAD). The evaluation is carried out by a standalone JAR[5].

The *R-value* indicates how reliably the text is detected – ignoring layout issues while the *P-value* indicates how reliable the structure of the text lines (layout) is [10].

5 Experiments

The goal of the performed experiments is to evaluate the suitability of the above described networks for the tasks of text segmentation, separator segmentation and, above all, baseline detection. We also evaluate the whole presented method and compare the results with the state of the art.

5.1 Text-Block Segmentation

In this section, we first evaluate and compare text segmentation based on the FCN network using three training scenarios:

1. Training from scratch on Europeana (E);
2. Training from scratch on Porta fontium (P);
3. Training from scratch on Europeana and then fine-tuning on Porta fontium (E+P).

We show the impact of three different input sizes on the segmentation results. The aim of this experiment is to find the most suitable configuration of the network. Table 2 shows the results on the validation part of Porta fontium dataset. All results are measured five times and we report the average and standard deviation values. During training we utilise early stopping based on validation loss.

Based on the results, we can conclude that larger dimensions of the input are beneficial for text segmentation. The pre-training on Europeana and fine-tuning on Porta fontium brings only slight improvement. The resolution of images is set to 960×672. We always train the model on Europeana and fine-tune it on the target dataset train part if available. Table 3 shows results of our system obtained on the three utilised datasets.

---

[4] https://scriptnet.iit.demokritos.gr/competitions/11/.

[5] https://github.com/Transkribus/TranskribusBaseLineEvaluationScheme.

Table 2. Text segmentation results (best results in bold)

Size	Data	F1	IoU
320×224	E	93.6 ± 0.16	88.1 ± 0.28
320×224	P	95.2 ± 0.08	91.0 ± 0.14
320×224	E+P	95.7 ± 0.05	91.8 ± 0.10
640×448	E	94.7 ± 0.11	90.0 ± 0.19
640×448	P	97.1 ± 0.07	94.4 ± 0.12
640×448	E+P	97.3 ± 0.02	94.8 ± 0.04
960×672	E	95.1 ± 0.08	90.8 ± 0.16
960×672	P	97.4 ± 0.03	95.0 ± 0.05
960×672	E+P	$\mathbf{97.4 \pm 0.04}$	$\mathbf{95.1 \pm 0.08}$

Table 3. Text segmentation results obtained with HP-FCN network and image size 960×672

	Europeana	Porta	RDCL 2019
F1	92.6	97.3	89.4
IoU	86.8	94.9	82.3

Next we employ the separator segmentation. We measure the performance of our separator detection algorithm (see Sect. 3.2). We consider the black (explicit) separators in this task. Finally, we evaluate the whole text-block segmentation algorithm. We use the PRIMA Evaluator tool for separator detection evaluation as well as for the final text-block segmentation evaluation.

For each testing page, we detect the separator regions and compare them with the corresponding ground truth using PRIMA Evaluator. We report the average results over 10 testing pages (see left part of Table 4). We present arithmetic mean (AM) of the separator success rates.

The right part of Table 4 shows the final evaluation of the whole text-block segmentation algorithm (see Sect. 3). We report the success rates obtained by PRIMA Evaluator with *document structure* evaluation profile as mentioned in Sect. 4.2.

Table 4. Separator detection results of the proposed approach (left) and the performance of the overall system (right) on three different datasets; average values of arithmetic mean (AM) are used

Database	Separator detection	Overall evaluation
Porta fontium	96.5	83.0
RDCL 2019	78.9	79.2
Europeana	83.1	82.2

The first line shows the results for the Porta fontium dataset. The separator detection (performed by ARU-Net) has excellent results. The evaluation on RDCL 2019 brought, as expected, worse general results, including separator detection. This is caused primarily by the large number of different types of regions that our tools did not anticipate. Nevertheless, the general results for the relatively difficult RDCL 2019 datasest, which has not been part of a training, are decent. On Europeana, we have achieved comparable result for the overall evaluation. The results on separator detection are worse in this case.

5.2   Text-Line Segmentation

This experiment describes the results of our text-line segmentation (see Sect. 3.3) being the final output of the proposed method. This experiment is realised only on the newly created Porta fontium dataset using annotation with baselines, because the other two corpora do not have this type of annotation.

We run the baseline detection in two ways. The first one is applying ARU-Net on the whole page. We will denote this approach as *page-level*. The second way is applying ARU-Net on already separated text regions. We will report it as *region-level*. We show these two approaches in order to point out the impact of our two-step algorithm in direct comparison with single-step approach.

We compare the results of our approach with Transkribus which is able to process a whole document and export a PAGE XML file containing baselines (see Table 5).

Table 5. Comparison of our text-line detection algorithm applied on the whole page (*page-level*) and results of the presented system (*region-level*); evaluation metrics are adopted from [10]; we report average values for P-value, R-value and F-value

Model	P-value	R-value	F-value
Transkribus (baseline)	0.866	0.951	0.906
page-level approach	0.921	**0.994**	0.957
region-level approach	**0.960**	0.993	**0.976**

The results of *region-level* approach are better than those of *page-level* one. The main reason is that when applying the detection algorithm on regions we can reduce some false detections near region borders and we can also merge incorrectly split baselines which are frequent in ARU-Net predictions. Merging line candidates in the whole page would often lead to baselines crossing region separators.

Table 6 shows the numbers of detected baselines and it confirms better result for the *region-level* approach since the number of detected lines is closer to the ground truth number.

Table 6. Number of detected (hypothesis) lines in Porta fontium

Model	Detected lines
Ground Truth	1,267
Transkribus	1,618
page-level	1,358
region-level	**1,293**

6 Conclusions and Future Work

In this paper, we have presented a novel approach for page segmentation into individual text lines. We have decomposed the problem into two separate steps: text-block segmentation and baseline detection. The text-block segmentation is solved by fully convolutional networks. The baseline detection is carried out using ARU-net architecture.

For evaluation, we have created a novel dataset from the data available in Porta fontium portal which is freely available for research purposes. We have compared the results of baseline detection algorithm applied directly on the whole page and that of our final system. By this approach, we managed to reduce the number of false positives in the detected baselines. We believe that by decomposing the problem into separate tasks, we can better express the document structure and not defile the reading order. Our two-step approach for baseline detection outperforms the single-step one applied directly on the whole page.

We have compared our text-line detection approach with Transkribus system and we have outperformed it with both page-level and region-level approaches. Moreover, it is evident from the results that the two-step approach performs better than the single-step one. We would like to highlight that the task of text-line segmentation is crucial for the following OCR processing that is usually the main goal of the historical document analysis.

One direction for further work would be to learn both the text segmentation and separator detection in one step. Another possibility is to train ARU-Net also for the task of x-line detection which could improve the x-height estimation process.

Acknowledgements. This work has been partly supported from ERDF "Research and Development of Intelligent Components of Advanced Technologies for the Pilsen Metropolitan Area (InteCom)" (no.: CZ.02.1.01/0.0/0.0/17_048/0007267).

References

1. Achanta, R., Shaji, A., Smith, K., Lucchi, A., Fua, P., Süsstrunk, S.: Slic superpixels. Technical report (2010)
2. Breuel, T.M., Ul-Hasan, A., Azawi, M.I.A.A., Shafait, F.: High-performance OCR for printed English and fraktur using LSTM networks. In: 2013 12th International Conference on Document Analysis and Recognition, pp. 683–687 (2013)
3. Chen, K., Seuret, M., Hennebert, J., Ingold, R.: Convolutional neural networks for page segmentation of historical document images. In: 2017 14th IAPR International Conference on Document Analysis and Recognition (ICDAR), vol. 1, pp. 965–970. IEEE (2017)
4. Chen, K., Seuret, M., Liwicki, M., Hennebert, J., Ingold, R.: Page segmentation of historical document images with convolutional autoencoders. In: 2015 13th International Conference on Document Analysis and Recognition (ICDAR), pp. 1011–1015. IEEE (2015)
5. Clausner, C., Papadopoulos, C., Pletschacher, S., Antonacopoulos, A.: The ENP image and ground truth dataset of historical newspapers. In: 2015 13th International Conference on Document Analysis and Recognition (ICDAR), pp. 931–935. IEEE (2015)
6. Clausner, C., Pletschacher, S., Antonacopoulos, A.: Aletheia-an advanced document layout and text ground-truthing system for production environments. In: 2011 International Conference on Document Analysis and Recognition, pp. 48–52. IEEE (2011)
7. Clausner, C., Pletschacher, S., Antonacopoulos, A.: Scenario driven in-depth performance evaluation of document layout analysis methods. In: 2011 International Conference on Document Analysis and Recognition, pp. 1404–1408. IEEE (2011)
8. Diem, M., Kleber, F., Fiel, S., Grüning, T., Gatos, B.: CBAD: ICDAR 2017 competition on baseline detection. In: 2017 14th IAPR International Conference on Document Analysis and Recognition (ICDAR), vol. 1, pp. 1355–1360. IEEE (2017)
9. Galibert, O., Kahn, J., Oparin, I.: The zonemap metric for page segmentation and area classification in scanned documents. In: 2014 IEEE International Conference on Image Processing (ICIP), pp. 2594–2598. IEEE (2014)
10. Grüning, T., Labahn, R., Diem, M., Kleber, F., Fiel, S.: Read-bad: a new dataset and evaluation scheme for baseline detection in archival documents. In: 2018 13th IAPR International Workshop on Document Analysis Systems (DAS), pp. 351–356. IEEE (2018)
11. Grüning, T., Leifert, G., Strauß, T., Labahn, R.: A Two-Stage Method for Text Line Detection in Historical Documents (2019). arxiv.org/abs/1802.03345
12. Krizhevsky, A., Sutskever, I., Hinton, G.E.: Imagenet classification with deep convolutional neural networks. In: Advances in Neural Information Processing Systems, pp. 1097–1105 (2012)
13. Lenc, L., Martínek, J., Král, P.: Tools for semi-automatic preparation of training data for OCR. In: MacIntyre, J., Maglogiannis, I., Iliadis, L., Pimenidis, E. (eds.) AIAI 2019. IAICT, vol. 559, pp. 351–361. Springer, Cham (2019). https://doi.org/10.1007/978-3-030-19823-7_29
14. Li, X.H., Yin, F., Xue, T., Liu, L., Ogier, J.M., Liu, C.L.: Instance aware document image segmentation using label pyramid networks and deep watershed transformation. In: 2019 International Conference on Document Analysis and Recognition (ICDAR), pp. 514–519. IEEE (2019)

15. Long, J., Shelhamer, E., Darrell, T.: Fully convolutional networks for semantic segmentation. In: Proceedings of the IEEE Conference on Computer Vision and Pattern Recognition, pp. 3431–3440 (2015)

16. Mechi, O., Mehri, M., Ingold, R., Amara, N.E.B.: Text line segmentation in historical document images using an adaptive U-net architecture. In: 2019 International Conference on Document Analysis and Recognition (ICDAR), pp. 369–374. IEEE (2019)

17. Novikov, A.A., Lenis, D., Major, D., Hladuvka, J., Wimmer, M., Bühler, K.: Fully convolutional architectures for multiclass segmentation in chest radiographs. IEEE Trans. Med. Imaging 37(8), 1865–1876 (2018)

18. Pletschacher, S., Antonacopoulos, A.: The page (page analysis and ground-truth elements) format framework. In: 2010 20th International Conference on Pattern Recognition, pp. 257–260. IEEE (2010)

19. Powers, D.: Evaluation: from precision, recall and F-measure to ROC, informedness, markedness & correlation. J. Mach. Learn. Technol. 2(1), 37–63 (2011)

20. Rezatofighi, H., Tsoi, N., Gwak, J., Sadeghian, A., Reid, I., Savarese, S.: Generalized intersection over union: a metric and a loss for bounding box regression. In: Proceedings of the IEEE Conference on Computer Vision and Pattern Recognition, pp. 658–666 (2019)

21. Ronneberger, O., Fischer, P., Brox, T.: U-Net: convolutional networks for biomedical image segmentation. In: Navab, N., Hornegger, J., Wells, W.M., Frangi, A.F. (eds.) MICCAI 2015. LNCS, vol. 9351, pp. 234–241. Springer, Cham (2015). https://doi.org/10.1007/978-3-319-24574-4_28

22. Sherrah, J.: Fully convolutional networks for dense semantic labelling of high-resolution aerial imagery. arXiv preprint arXiv:1606.02585 (2016)

23. Shi, B., Bai, X., Yao, C.: An end-to-end trainable neural network for image-based sequence recognition and its application to scene text recognition. IEEE Trans. Pattern Anal. Mach. Intell. 39(11), 2298–2304 (2016)

24. Simistira, F., Ul-Hassan, A., Papavassiliou, V., Gatos, B., Katsouros, V., Liwicki, M.: Recognition of historical greek polytonic scripts using LSTM networks. In: 2015 13th International Conference on Document Analysis and Recognition (ICDAR), pp. 766–770. IEEE (2015)

25. Simonyan, K., Zisserman, A.: Very deep convolutional networks for large-scale image recognition. arXiv preprint arXiv:1409.1556 (2014)

26. Wick, C., Puppe, F.: Fully convolutional neural networks for page segmentation of historical document images. In: 2018 13th IAPR International Workshop on Document Analysis Systems (DAS), pp. 287–292. IEEE (2018)

27. Xu, Y., He, W., Yin, F., Liu, C.L.: Page segmentation for historical handwritten documents using fully convolutional networks. In: 2017 14th IAPR International Conference on Document Analysis and Recognition (ICDAR), vol. 1, pp. 541–546. IEEE (2017)

28. Zhong, X., Tang, J., Yepes, A.J.: Publaynet: largest dataset ever for document layout analysis. In: 2019 International Conference on Document Analysis and Recognition (ICDAR), pp. 1015–1022. IEEE (2019)

Stacked Ensemble of Convolutional Neural Networks for Follicles Detection on Scalp Images

Dymitr Ruta[1(✉)], Ling Cen[1], Andrzej Ruta[2], and Quang Hieu Vu[3]

[1] EBTIC, Khalifa University, Abu Dhabi, UAE
{dymitr.ruta,cen.ling}@ku.ac.ae
[2] PLACEMAKE.IO, London, UK
andrzej.ruta@placemake.io
[3] Data Science Group, Zalora, Singapore
quanghieu.vu@zalora.com
https://www.ebtic.org/, https://www.placemake.io

Abstract. An average person's head is covered with up to 100000 hairs growing out of follicular openings on the scalp's skin. Automated hair therapy requires precise detection and localization of follicles on the scalp and still poses a significant challenge for the computer vision and pattern recognition systems. We have proposed an automated vision system for follicles detection based on the classification of digitized microscopic scalp images using an ensemble of convolutional neural networks (CNN). A pool of adapted state-of-the-art CNNs have been transfer-trained on over 700k microscopic skin image regions of 120×120 pixels and their outputs further fed to the final stacked ensemble learning layer to capture a wider context of the connected neighboring regions of the original FullHD scalp images. A high validated f1 score (0.7) of detecting regions with follicles beats the industry's benchmark and brings this technology a step closer towards automated hair treatment as well as other emerging applications such as personal identification based on follicular scalp map.

Keywords: Pattern recognition · Convolutional neural networks · Stacked ensemble learning

1 Introduction

Automated hair or scalp treatments depend on the ability of accurately detecting follicular openings on the scalp's skin and since an average head comprises as many as 100000 of them, the task remains a challenge for the computer vision in terms of both the complexity of the detection problem and its scale. The images of scalp are full of hair occlusion, light, color and contrast vary massively while the hair entering the follicle under the thin skin often appears as a regular hair or can be confused with the loose hair tip. Figure 1 shows few examples of

© The Author(s), under exclusive license to Springer Nature Switzerland AG 2023
L. Rutkowski et al. (Eds.): ICAISC 2022, LNAI 13589, pp. 49–58, 2023.
https://doi.org/10.1007/978-3-031-23480-4_4

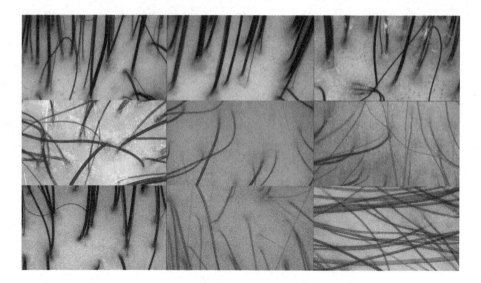

Fig. 1. 9 examples of hair follicles on microscopic scalp photos.

scalp micro-images with hair and visually highlights the challenges of follicles detection problems mentioned above.

Deep learning (DL) has achieved significant successes in computer vision [5]. Instead of defining hand-crafted, problem-specific features, such as histograms of oriented gradients (HoG) [3], deep learning models take raw image pixels as inputs and engineer hierarchically linked layers of emergent, highly predictive features and different levels of abstraction that bridge the complexity gap between the pixels and the predicted target function. Convolutional Neural Networks (CNN) are probably the most popular and successful examples of deep learning architectures for discriminative classification and regression problems [1,4]. A CNN can be seen as a feed-forward neural network with multiple characteristic convolutional and pooling layers that gradually isolate the predictive image regions from noise while reducing the data rate from the layer below. They achieve this thanks to the specific hierarchical aggregations resulted from convolving very simple receptive filters along the image pixels. The weight sharing in the convolutional layer, together with appropriately chosen pooling schemes, endows CNNs with some invariance properties (e.g., translation invariance), which have been found highly effective in image pattern recognition [1,3,4].

Detecting complex, highly versatile objects or scenes from images requires flexible and large architectures with many degrees of freedom that pushed CNNs to grow in depth and complexity often resulting in models with millions of parameters that require large image datasets to capture the diversity of the detection problem during training at an adequate level and thereby to avoid model overfitting [4].

Transfer learning is a popular approach in deep learning, which enables to reuse the core architecture and structure of a pre-trained deep network to efficiently adapt it to the new prediction problem [6]. Transfer learning starts from the pre-trained network optimized usually for large predictive problems involving hundreds or thousands of object classes. It then typically follows the replacement of the top fully-connected layers with the new problem-specific layers adapted to the number of new classes, possibly also combined with freezing top-level convolutional layers against the weights changes and retraining the whole newly reconnected structure using new labeled image examples. Transfer learning has been successfully applied in many machine learning applications [7], e.g. text sentiment classification [8], image classification [9–11], human activity classification [12], software defect classification [13], and multi-language text classification [14,15], and it marks a radical change from recent tendency of isolated purpose-built one-off models towards accumulatively improved general purpose model platforms.

This work focused on building a follicles detector from microscopic scalp skin images. Due to rather high problem complexity and ambiguity of the target objects we chose to apply an ensemble of state-of-the-art CNNs transfer-learnt and fine-tuned on a large sample of over 700k microscopic labeled skin image regions of 120×120 pixels. The outputs of retrained CNNs from the original and surrounding examples were further fed to the final stacking ensemble learning layer to capture a wider context of the connected neighboring skin regions of the original high definition scalp images and improve the detection accuracy through combining multiple diverse insights. Due to high computational cost of training a single well-performing CNN detector instead of boosting or bagging the stacking [16,17] with a simple logistic regression meta learner has been used to try to efficiently improve the performance of an ensemble of several pretrained CNN predictors.

The remainder of the paper is organized as follows. eSensei Challenge is introduced in Sect. 2, followed by a description of the proposed detection model evolving from traditional image transformation techniques through SVM trained over HOG features covered in Sect. 3.1 to the ensemble model based on adapted state-of-the-art CNNs transfer-trained on binary labeled microscopic skin image regions presented in Sect. 3.2. The concluding remarks are provided in Sect. 4.

2   Follicles Detection Challenge

The problem of automated follicles' detection has been addressed as a part of the 2018 eSensei Challenge, in which the competitors were provided with 5880 fullHD scalp images split into a grid of 144 labeled (containing/not containing follicles) regions of size 120×120 pixels each, and asked to use them as training examples to build an automated follicle detection model. The competing models were evaluated live on the preliminary leader-board set of 100 test images representing 10% of the complete testing set of 1000 fullHD images, that was used for final evaluation after the preliminary stage of model build and refinement had been closed.

3   Model Description

Our initial explorations included traditional image recognition strategies based on detecting a follicle object with Histogram of Oriented Gradient (HOG) features [3], isolating follicle color, coloring hairs or building a hair skeleton filter models. Very quickly, however, it became clear that the problem was much harder to solve than traditional object detection due to a large variety of follicles, their orientations in tandem with hairs, and foremost – overwhelming similarity between the positive and the negative class of non-follicle cases that contained hairs. In contrast, an ad-hoc attempt with a very simple convolutional neural network (CNN) resulted in still insufficient yet significantly better follicles detection rate and hence the remaining effort was put almost exclusively on designing or adapting more complex deep CNN architectures.

In Sect. 3.1 we outline two techniques of hair follicle detection that showed the most promising preliminary results. In Sect. 3.2 we focus on the CNN-based models that were finally selected to solve the 2018 eSensei Challenge problem.

3.1   Image Enhancements and Initial Models

Our initial naive assumption was that the raw RGB pixel data were fully sufficient for a human to determine the locations of hair follicles, hence we attempted several transformations of the original input images to isolate the target objects.

Observations of various scalp images gave an initial impression that the images are dominated by two objects along with their distinct colors: the hair and the skin, while the follicle appears to blend somewhat in between of the two. This triggered our first approach, in which we simply tried to isolate the color of the follicle by setting a pair of color/shade thresholds within the valley separating the skin and hair peaks of the image histogram. The depiction of a sample result is shown in Fig. 2, where the hairs are marked in gray, the skin background in black, while the potential follicles in white.

Although the predicted follicle patches appear to be co-located with the green-labeled true follicle regions, the thorough testing revealed large number of false-positives, especially for very hairy scalp images. Further refinements trying to establish the optimal fraction of the follicle-colored pixels or a combination of follicle and hair-colored pixels within a box brought rather limited improvements.

Next, we considered image contrast to be critical to distinguish dark, elongated hair patches from lighter skin background. For that reason we converted RGB images to gray-scale and computed the maps of directional gradients along with gradient magnitudes. Subsequently, each 120×120 pixels block of the above mentioned gradient maps was described with the well-known 9-bin histogram of oriented gradients (HOG), as described in [3]. However, due to low complexity of hairs compared to the complexity of human figures on natural background, for which HOGs were invented, we composed the descriptor of only 2×2 60-pixels cells giving final descriptor a dimensionality of 36. Unfortunately, a conventional linear/RBF kernel SVM classifier trained on top of such image representation

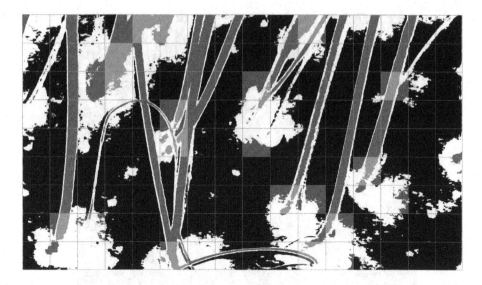

Fig. 2. Color threshold based follicle detection

did not yield promising detection accuracy. Even after expanding such a classifier into the most powerful ensemble schemes like AdaBoost [2], the f1-measured detection performance did not reach 50%.

Our third approach to hair follicle detection required image binarisation. For this purpose, adaptive local thresholding was applied, i.e. pixel intensities in local 100×100 pixels neighbourhoods were converted to 0/1 values depending on whether or not they exceeded neighbourhood average intensities multiplied by a constant. In the binarised images we eliminated the noise and filled the "holes" using combinations of morphological opening and closing operators. Further, straight line segments detection was run but due to the lack of clear concept of how to encode them for modelling, this direction was abandoned. Instead, on top of the binary images we performed so-called "skeletonisation" aimed at reducing thick hair regions to their one-pixel-wide axes. Since output of this operation inevitably contained undesired noise, we consequently implemented a recursive axis-following algorithm that enabled to "colour" individual suspected hairs. To avoid excessive false positives we considered hair follicles to be endpoints of the axes of length bigger than a threshold of 100 pixels. Despite this technique being intuitive and in agreement with how humans' visual attention seems to work when capturing hair follicles, the above detector yielded good recall but also too frequent false positives, without any fine-tuning yielding the maximum f1 score in the range of 46–50% (Figs. 3 and 4).

3.2 Convolutional Neural Network Based Detector

The detectors based on the rather simple image proprocessing techniques described in the previous section could not overcome the F1 score in excess of

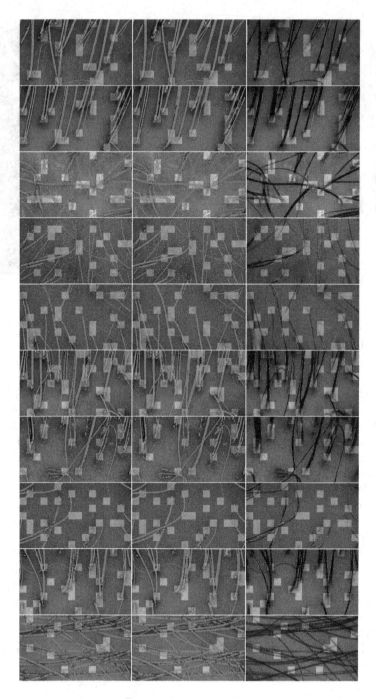

Fig. 3. Image preprocessing and enhancements focused on hair identification. Column 1: binarised images after smoothing and morphological opening applied, Column 2: binarised images after additional morphological closing applied, Column 3: straight line segments detected

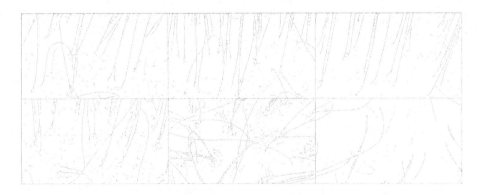

Fig. 4. Examples of skeletonization on binarized hair images

50%. However, observation of the magnified images led us to the believe that the human eye should be able to deliver more accurate detection results and hence we decided to employ several state-of-the-art CNN designs reported in the literature including: AlexNet, VGG16/19, GoogleNet, InceptionV3 and Resnet50/101, for this task. We followed standard adaptations of these networks through transfer-learning that involved replacing of the last 3 layers with our problem specific set of fully connected layers and sigmoid-activated binary classification layer.

We have experimentally established that the detection performance bene-fits independently from two critical characteristics of the training set composition: the ability to see all the available training examples and equal distribution between the positive and negative class examples. Given around 700k image patches and only 15% positives, we have composed the training set to include all negative cases and the positive class cases up-sampled to match the number of negatives. Given such a large training set and limited computing resources we trained all 7 CNNs only for 1 epoch using a fixed learning rate of 0.0001, which took over a week to complete. The trained CNNs were then applied to classify both the training and the testing set to produce vital outputs for fur-ther research and to evaluate the models on the leaderboard. The results gave AlexNet and Inception3 the f1 score in the region 0f 0.62, while all the remaining CNNs scored in the range of 0.65.

In an attempt to further improve the predictions, additional training was attempted with VGG-16 utilizing paid Amazon Cloud with 8 T K80 GPUs and GoogleNet on a local pair of desktop fitted with GTX1080Ti GPUs. Training VGG-16 beyond 5 epochs yielded some improvement, yet at a huge compute cost we could no longer afford, while GoogleNet came out as the most efficient in terms of predictive value (f1 > 0.67) achieved per unit of time/compute cost. The default Googlenet architecture, as depicted in Fig. 5, is used to transfer-learn to detect the hair follicle regions on scalp image segments.

To further boost the predictive performance, various simple ensemble meth-ods were tested upon the probabilistic outputs from 7 trained CNNs, however the best results were obtained with a lightweight 2nd stage classification with logistic regression trained upon the outputs from a subset of just 4 CNNs: GoogleNet, Inception3, VGG-16 and Resnet50.

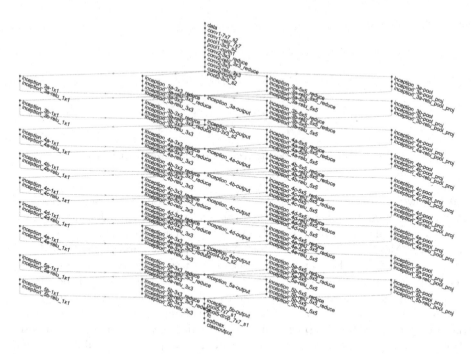

Fig. 5. GoogleNet - the most efficient individual CNN used for hair follicle detection

Further improvements of the predictive performance have been achieved from more extensive utilization of the target class predictions around the target cell. These predictions were taken as inputs for the logistic regression training to give the final classification label. The motivation for this step was to expand the perspective for the predictor by trying to learn additional insight inferred from the neighbourhood of the target cell, such as checking if the hair extended beyond the target cell or completing the patterns of follicle that extended into the neighboring cell. All these additional improvements brought only small performance gains and returned the final model with the estimated f1 score of 0.71. Table 1 summarizes incremental gains in f1-measured detection performance for our models, while Fig. 6 illustrates sample testing image with overlaid positive follicle cells detected by our top model.

Table 1. Incremental follicle detection performance results.

Model	Features	F1 score
Color thresholds	Color frequencies	0.42
SVM	HOG	0.47
SVM	Colored hair + skeletons	0.49
Alex, Inception3	Raw images	0.62
VGG16/19, Resnet50/101	Raw images	0.65
GoogleNet	Raw images	0.67
Ensemble (GNet, VGG, Res)	Raw images	0.69
LogR (GNet, VGG, Res)	Raw images	0.70
LogR (CNNs (targets+neighborhood))	Raw images	0.71

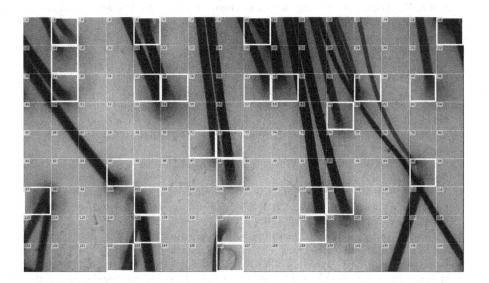

Fig. 6. Sample testing image with overlaid follicles predictions for each cells.

4 Conclusions

Automated hair follicle detection from micro-images remains a very challenging ML problem. Our proposed method involves selected transfer-learnt deep convolutional neural networks retrained on hundreds of thousands positive-class up-sampled scalp image patches of 120×120 pixels extracted from 1000s of high resolution scalp micro-images. Our initial image enhancements aimed at better discrimination of the target follicle objects yielded rather limited gains both in terms of their sole predictive power and the pre-processing step for CNN detectors. CNNs seem to unscramble the predictive power from very ambiguous pixel structures quite well, while additional improvements, pushing the f1 score beyond 0.71, were achieved by adding the second stage simple logistic regression predictor learning to correct the outputs from the selected best performing CNNs, notably VGG-16 and GoogleNet, and reusing CNNs predictive outputs around the neighborhood of the target region, thereby confirming or rejecting the initial predictions of the central region.

References

1. LeCun, Y., Haffner, P., Bottou, L., Bengio, Y.: Object recognition with gradient-based learning. In: Shape, Contour and Grouping in Computer Vision. LNCS, vol. 1681, pp. 319–345. Springer, Heidelberg (1999). https://doi.org/10.1007/3-540-46805-6_19

2. Schapire, R.E.: A brief introduction to boosting. In: Proceedings of the 16th International Joint Conference on Artificial Intelligence, vol. 2, pp. 1401–1406 (1999)

3. Dalal, N., Triggs, B.: Histograms of oriented gradients for human detection. In: Proceedings of the IEEE International Conference on Computer Vision and Pattern Recognition, pp. 886–893 (2005)

4. Krizhevsky, A., Sutskever, I., Hinton, G.E.: Imagenet classification with deep convolutional neural networks. In: Advances in Neural Information Processing Systems, pp. 1097–1105 (2012)

5. Deng, L.: Three classes of deep learning architectures and their applications: a tutorial survey. APSIPA Trans. Signal Inf. Process. **57**, 58 (2012)

6. Olivas, E., Guerrero, J., Sober, M., Benedito, J., Lopez, A.: Handbook of Research on Machine Learning Applications and Trends: Algorithms, Methods and Techniques (2009)

7. Weiss, K., Khoshgoftaar, T., Wang, D.: A survey of transfer learning. J. Big Data **3**(9) (2016)

8. Wang, C., Mahadevan, S.: Heterogeneous domain adaptation using manifold alignment. In: Proceedings of the 22nd International Joint Conference on Artificial Intelligence, vol. 2, pp. 541–546 (2011)

9. Duan, L., Xu, D., Tsang, I.: Learning with augmented features for heterogeneous domain adaptation. IEEE Trans. Pattern Anal. Mach. Intell. **36**(6) (2012)

10. Kulis, B., Saenko, K., Darrell, T.: What you saw is not what you get: domain adaptation using asymmetric kernel transforms. In: IEEE 2011 Conference on Computer Vision and Pattern Recognition (2011)

11. Zhu, Y., et al.: Heterogeneous transfer learning for image classification. In: Proceedings of the National Conference on Artificial Intelligence, vol. 2 (2011)

12. Harel, M., Mannor, S.: Learning from multiple outlooks. In: Proceedings of the 28th International Conference on Machine Learning (2011)

13. Nam, J., Kim, S.: Heterogeneous defect prediction. In: Proceedings of the 10th Joint Meeting on Foundations of Software Engineering, pp. 508–519 (2015)

14. Zhou, J., Pan, S., Tsang, I., Yan, Y.: Hybrid heterogeneous transfer learning through deep learning. In: Proceedings of the National Conference on Artificial Intelligence, vol. 3, pp. 2213–2220 (2014)

15. Zhou, J., Tsang, I., Pan, S., Tan, M.: Heterogeneous domain adaptation for multiple classes. In: Proceedings of the International Conference on Artificial Intelligence and Statistics, pp. 1095–103 (2014)

16. Wolpert, D.: Stacked generalization. Neural Netw. **5**(2), 241–259 (1992)

17. Smyth, P., Wolpert, D.: Linearly combining density estimators via stacking. Mach. Learn. **36**(1–2), 59–83 (1999)

RGB-D SLAM with Deep Depth Completion

Ali Osman Serhatoglu[1]([✉]), Oguzhan Guclu[2], and Ahmet Burak Can[1]

[1] Hacettepe University, Ankara, Turkey
{aoserhatoglu,oguzhanguclu,abc}@cs.hacettepe.edu.tr
[2] Sahibinden, Istanbul, Turkey

Abstract. RGB-D indoor mapping has been an active research topic in the last decade with the release of various depth sensors. Researchers proposed impressive SLAM systems such as ORB-SLAM2. However, the depth sensors are sensitive to illumination conditions and have limited range, which lead to missing or invalid depth data. This situation negatively affects the performance of RGB-D SLAM systems. Moreover, deep learning based approaches for estimating depth data from color frames have been proposed recently. Therefore, in this study, we aim to analyze deep depth estimation performance on SLAM. We propose a depth completion approach which merges sensor depth data and the estimated depth. To do this, we integrate a deep depth estimation method into a state-of-the-art indoor RGB-D SLAM system. The experimental results show that the proposed depth completion approach improves mapping performance.

Keywords: SLAM · Depth estimation · Depth completion

1 Introduction

Simultaneous Localization and Mapping (SLAM) has a fundamental place in robotics research. Autonomous operation such as in self-driving cars, smart drones, and manipulator robots requires effective mapping performance. The availability of inexpensive RGB-D cameras has made SLAM research more popular and thus impressive SLAM systems have been developed by researchers. SLAM research seems to retain its fundamental attention in the near future due to integration of autonomous mobility into daily life.

In the last decade, significant improvement has been made in computer vision topics such as object detection, [18,19,24], recognition [12,17,32], and semantic segmentation [3,4,20] due to the great progress in deep learning based approaches. As a positive effect of this, various parts of SLAM has been covered by deep learning techniques such as keypoint extraction and matching [25,27,29,31], camera localization [2,15], place recognition [5,11,14,21], and depth estimation [1,7,9,13].

In this work, we investigate deep learning based depth estimation performance on SLAM and propose a method for depth data integration. To do so, we

L. Rutkowski et al. (Eds.): ICAISC 2022, LNAI 13589, pp. 59–67, 2023.
https://doi.org/10.1007/978-3-031-23480-4_5

merge sensor depth data and the estimated depth by developing a depth completion approach. Four popular deep depth estimation approaches are integrated into a state-of-the-art indoor RGB-D SLAM system [10]. Firstly, performances of the depth estimation methods are compared in the monocular SLAM scenario. Then, the depth estimation approaches are used to complete missing parts of the depth frames within the RGB-D SLAM system of Guclu et al [10]. Experiments show that our depth completion approach allows promising results with improved SLAM performance.

2   Related Work

RGB-D based SLAM problem has been a big focus of research in the robotics literature, especially in the last decade. In order to solve the problem, feature based approaches have been proposed as well as direct (dense) methods. The feature based methods [6,10,22] are more well suited for real time operation since they extract salient keypoints on the frames and compute transformations by utilizing less data. The dense approaches [16,23,30] on the other hand, are able to compute more robust odometry by relying on pixel-wise error minimization.

In this work, the proposed approach is integrated on a state-of-the-art feature based RGB-D SLAM pipeline [10]. This system extracts sparse features from RGB frames, locates them in 3D by using the corresponding depth data from sensor, and computes frame-to-frame transformations using RANSAC. Loop closure detection is performed by employing a deep feature based indexing mechanism on color frames. As stated before, the depth data from the sensor is noisy and has missing parts, which hinder mapping accuracy. From this viewpoint, we change the depth data acquisition stage by employing deep learning based depth estimation to get more accurate depth information.

The depth estimation methods we utilize are; DORN [7], Monodepth2 [9], RevisitingSIDE (Single Image Depth Estimation) [1], and DenseDepth [13]. The DORN method uses deep convolutional neural networks (CNNs) and specifically focuses on image-level information and hierarchical features. It is trained with NYUv2 [26] and KITTI [8] datasets, which contain indoor and outdoor scenes respectively. Monodepth2 is self-supervised learning method. Unlike others, this method is trained on monocular video, stereo pairs and both of them together. KITTI dataset has been used for training the Monodepth2 model. Revisiting-SIDE also uses convolutional neural networks and focuses on depth maps with higher spatial resolution. This method has been trained with the NYUv2 dataset. Finally, another CNN based method is DenseDepth, which employs transfer learning and trained with both KITTI and NYUv2 datasets.

3   Method

In the study, performances of various deep depth estimation methods are compared by directly integrating into a monocular SLAM pipeline firstly. As a second work, these depth estimation approaches are utilized to complete missing parts of the depth frames within a state of the art RGB-D SLAM method.

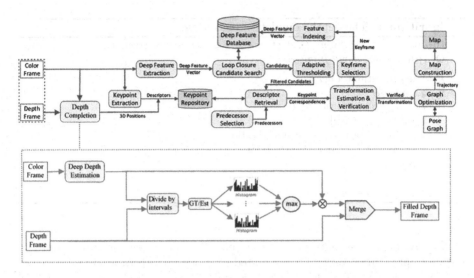

Fig. 1. Proposed improved indoor RGB-D SLAM pipeline.

3.1 Monocular SLAM with Depth Estimation

As a first study, four popular deep depth estimation approaches are integrated into the SLAM system of Guclu et al [10]. This SLAM system is based on RGB-D sensor data. However, in order to analyze performances of the related depth estimation methods, the SLAM system is used in a monocular manner by only utilizing RGB frames from the sensor. Each of the depth estimation methods take RGB frame data as input and produce pixel-wise depth data as output. Therefore, in the pipeline, RGB frames acquired from the sensor are directly fed into a depth estimation network. Then, the estimated depth data is used in a pixel-wise manner to locate the image keypoints in 3D within motion estimation.

The exact depth data estimated by a deep network has the problem of scale ambiguity, since the network is trained on a completely different scene than the SLAM dataset used. In order to overcome the scale ambiguity problem, random frames are selected from the dataset. A histogram is created by dividing per pixel sensor values by the corresponding estimation values for each frame. Mean of the values obtained from the histogram peak region is determined as the scale factor. Finally, the estimated depth values are scaled with the computed factor and used in the pipeline like acquired from a sensor.

3.2 Depth Completion in RGB-D SLAM

In the next study, the depth data estimated by the deep networks is used in a depth completion pipeline. The aim of this approach is to improve the performance of RGB-D SLAM by completing missing parts of the depth frames provided by the sensor. Figure 1 shows the depth completion pipeline.

Algorithm 1: The depth completion algorithm.

Input   : *RGB rgb frame F, depth frame D, distance intervals u*
Output: *scaled depth frame S*

$P :=$ depth_estimation(F)
for *each u* **do**
   | $i \rightarrow$ *data indexes in the u*
   | ratio $:=$ F$[i]$ / P$[i]$
   | $h :=$ hist(ratio) $\rightarrow$ histogram of ratios
   | $k :=$ max$(h) \rightarrow$ peak of histogram
   | S$[i] :=$ P$[i]$ * k
end
return *scaled depth frame*

Firstly, depth estimation for the input RGB frame is performed and the output depth frame containing estimated pixel-wise depth values is constructed (see Algorithm 1). The depth frame is divided into distance intervals which have been determined according to sensor sensitivity range (0.5 m–5 m). At these distance ranges, the ratios are calculated by dividing the pixel-wise sensor values by the corresponding estimation values. Then histograms of these ratios are computed, and the resulting scale factor is determined as the ratio belonging to the histogram peak region. The scaled depth frame is constructed in a pixel-wise approach by multiplying each estimated depth value with the computed scale factor.

In the RGB-D SLAM pipeline, the original depth frame acquired by the sensor is handled, and the missing parts of the frame are completed with related regions of the computed scaled depth frame. Then the constructed resulting depth frame is employed within the whole system for motion estimation operations.

4   Experiments

The experiments are performed on the popular TUM RGB-D dataset [28]. The *fr1* and *fr2* sequences of the dataset are employed in the experiments, which contain scenes of a middle-sized office and an industrial hall environment respectively. Both groups of sequences have important challenges such as missing depth data caused by sensor range limit, illumination conditions, and thin structures.

Two groups of experiments are carried out in order to analyze depth estimation performances on monocular SLAM and depth completion on RGB-D SLAM. The following subsections demonstrate the experimental results, contain performance comparisons, and discuss the effect of the proposed depth completion approach.

Table 1. Monocular SLAM experiment results in terms of RMS-ATE (in meters) on *fr1* and *fr2* sequences of the TUM RGB-D benchmark [28]. ('x' represents incomplete mapping failures. The letter 'N' next to the methods means trained with the NYUv2 dataset and the letter 'K' means trained with the KITTI dataset.)

System	fr1/ 360	fr1/ desk	fr1/ desk2	fr1/ floor	fr1/ plant	fr1/ room	fr1/ teddy	fr2/large_ no_loop	fr2/large_ with_loop	fr2/pioneer_ 360	fr2/pioneer_ slam	fr2/pioneer_ slam2	fr2/pioneer_ slam3
ORB-S2 [22]	**0,111**	**0,097**	0,280	x	0,477	**0,174**	0,607	x	**0,820**	x	x	x	x
DD-N [13]	0,416	0,367	0,341	0,614	0,399	0,441	0,505	0,517	0,887	1,172	2,735	1,172	1,073
DD-K [13]	0,168	0,859	0,947	0,691	0,712	0,972	0,940	2,135	1,384	1,797	1,923	2,011	1,994
DORN-N [7]	0,375	0,434	**0,269**	0,473	**0,242**	0,430	**0,388**	**0,407**	1,011	**0,936**	2,481	1,158	**0,920**
DORN-K [7]	0,313	0,404	0,595	**0,375**	0,485	1,040	0,682	0,819	1,339	1,493	1,591	1,265	1,359
R-N [1]	0,288	0,373	0,464	0,506	0,368	0,555	0,488	0,809	1,164	1,199	**1,445**	**1,081**	1,180
Md2-K [9]	0,178	0,789	0,883	0,590	0,598	0,936	0,877	2,061	1,258	1,845	1,958	2,025	1,961

4.1 Monocular SLAM Experiments

In the first group of experiments, performances of the deep depth estimation approaches are evaluated on a monocular SLAM setup. Table 1 shows the results and makes comparison with the state of the art monocular SLAM method of ORB-SLAM2 [22]. We have reproduced the results of monocular ORB-SLAM2 on the TUM dataset from the original implementation provided by the authors.

Table 1 demonstrates that DORN trained with the NYUv2 dataset produces the best results in 3 sequences on *fr1*. It also produces an acceptable result for *fr2/floor* where ORB-SLAM2 failed. On the other hand, the ORB-SLAM2 system outperforms the proposed approach in 3 sequences. Overall, it can be inferred that DORN and ORB-SLAM2 perform similarly on *fr1* sequences. On the other hand, DORN-NYUv2 produces better results than others on *fr2*.

4.2 Depth Completion Experiments

The next group of experiments have been performed to measure the effect of the proposed depth completion method on RGB-D SLAM performance. The results are represented in Table 2.

It can be seen that the trajectories estimated for the *fr1* sequences are very close to the ground-truth. DenseDepth trained with the KITTI dataset performs best for 4 sequences of *fr1*. It has greatest improvement for *fr1/room* with 6%. According to the *fr2* results, it shows achievements in less number of sequences compared to *fr1*.

Generally, it can be seen that the trajectory error values for the *fr2* sequences are much bigger than *fr1*. The main reason behind this situation is that the ratio of invalid depth data for *fr1* sequences are much lower. The invalidness of the data is mainly caused by missing parts in the depth frames and thus it mostly affects trajectory accuracy negatively. The detailed information about invalid depth ratios are represented in Table 3.

Table 2. Depth completion experiment results in terms of RMS-ATE (in meters) on *fr1* and *fr2* sequences of the TUM RGB-D benchmark [28]. The letter 'N' next to the methods means trained with the NYUv2 dataset and the letter 'K' means trained with the KITTI dataset.

System	fr1/360	fr1/desk	fr1/desk2	fr1/floor	fr1/plant	fr1/room	fr1/teddy	fr2/large_no_loop	fr2/large_with_loop	fr2/pioneer_360	fr2/pioneer_slam	fr2/pioneer_slam2	fr2/pioneer_slam3
R-SLAM [10]	0.051	**0.019**	0.029	0.027	**0.034**	0.049	**0.036**	**0.135**	**0.344**	**0.148**	0.380	0.160	0.272
DD-N [13]	0.053	0.020	0.029	**0.026**	0.040	0.049	0.040	0.552	0.384	0.308	**0.345**	0.168	0.320
DD-K [13]	**0.050**	0.019	0.028	**0.026**	0.037	**0.046**	0.042	0.169	0.429	0.625	**0.298**	0.210	**0.253**
DORN-N [7]	0.052	0.020	**0.028**	0.027	0.037	**0.047**	0.039	0.174	0.419	0.230	**0.328**	0.161	0.296
DORN-K [7]	0.053	**0.019**	**0.027**	0.027	0.038	**0.047**	0.040	0.459	0.426	0.213	**0.343**	0.166	**0.270**
R-N [1]	0.052	0.020	**0.026**	**0.026**	0.039	**0.046**	0.040	0.201	0.412	0.249	**0.313**	**0.159**	0.360
Md2-K [9]	0.054	0.020	**0.027**	**0.026**	0.039	**0.047**	0.042	0.663	0.434	0.225	**0.370**	0.177	0.283

Table 3. Invalid depth ratio of the TUM RGB-D benchmark [28]

TUM dataset	Invalid depth %	Invalid depth used %
fr1/360	5,5	10,4
fr1/desk	9,2	13,8
fr1/desk2	9,9	13,7
fr1/floor	2	6,2
fr1/plant	8,1	16,3
fr1/room	9,4	12
fr1/rpy	5,3	10,6
fr1/teddy	6,6	12,5
fr1/xyz	6,3	10,6
fr2/large_no_loop	17,5	28,7
fr2/large_with_loop	24,9	35,8
fr2/p_360	20,2	43
fr2/p_slam	16	34,4
fr2/p_slam2	9,4	23,9
fr2/p_slam3	21,3	39,9

Consequently, the proposed approach performs better in the *fr1* sequences than *fr2*. The main factor for this effect could be that the SLAM system of Guclu et al. [10] does not use invalid depth data since it masks the frames to get rid of wrong or missing depth values. Furthermore, with the proposed method, estimated depth data inherently containing errors is incorporated into the pipeline. In fact, a higher rate of prediction data (with errors) has been used for *fr2* than *fr1* because of the amount of missing frame parts.

5   Conclusion

In this paper, we investigate performance of deep depth estimation methods on SLAM and propose a depth completion approach. For this purpose, we integrate four popular deep depth estimation approaches into a state-of-the-art indoor RGB-D SLAM system. DORN-NYUv2 for Monocular SLAM and DenseDepth-KITTI in deep completion for RGB-D SLAM achieved the best performance. The experimental results are promising for further improvements such as merging sensor depth and the estimated depth. Especially the improved results obtained for the TUM *fr1* sequences indicate that the proposed approach might be prominent.

References

1. Alhashim, I., Wonka, P.: High quality monocular depth estimation via transfer learning. arXiv preprint arXiv:1812.11941 (2018)
2. Brachmann, E., Rother, C.: Learning less is more - 6D camera localization via 3D surface regression. In: 2018 IEEE/CVF Conference on Computer Vision and Pattern Recognition, pp. 4654–4662 (2018). https://doi.org/10.1109/CVPR.2018.00489
3. Chen, L.C., Papandreou, G., Kokkinos, I., Murphy, K., Yuille, A.L.: DeepLab: semantic image segmentation with deep convolutional nets, atrous convolution, and fully connected CRFs. IEEE Trans. Pattern Anal. Mach. Intell. 40(4), 834–848 (2018). https://doi.org/10.1109/TPAMI.2017.2699184
4. Chen, L.-C., Zhu, Y., Papandreou, G., Schroff, F., Adam, H.: Encoder-decoder with atrous separable convolution for semantic image segmentation. In: Ferrari, V., Hebert, M., Sminchisescu, C., Weiss, Y. (eds.) ECCV 2018. LNCS, vol. 11211, pp. 833–851. Springer, Cham (2018). https://doi.org/10.1007/978-3-030-01234-2_49
5. Chen, Z., et al.: Deep learning features at scale for visual place recognition. In: 2017 IEEE International Conference on Robotics and Automation (ICRA), pp. 3223–3230 (2017). https://doi.org/10.1109/ICRA.2017.7989366
6. Endres, F., Hess, J., Sturm, J., Cremers, D., Burgard, W.: 3-D mapping with an RGB-D camera. IEEE Trans. Rob. 30(1), 177–187 (2014). https://doi.org/10.1109/TRO.2013.2279412
7. Fu, H., Gong, M., Wang, C., Batmanghelich, K., Tao, D.: Deep ordinal regression network for monocular depth estimation. In: 2018 IEEE/CVF Conference on Computer Vision and Pattern Recognition, pp. 2002–2011 (2018). https://doi.org/10.1109/CVPR.2018.00214
8. Geiger, A., Lenz, P., Stiller, C., Urtasun, R.: Vision meets robotics: the KITTI dataset. Int. J. Robot. Res. 32(11), 1231–1237 (2013). https://doi.org/10.1177/0278364913491297
9. Godard, C., Aodha, O.M., Firman, M., Brostow, G.: Digging into self-supervised monocular depth estimation. In: 2019 IEEE/CVF International Conference on Computer Vision (ICCV), pp. 3827–3837 (2019). https://doi.org/10.1109/ICCV.2019.00393
10. Guclu, O., Caglayan, A., Can, A.B.: RGB-D indoor mapping using deep features. In: 2019 IEEE/CVF Conference on Computer Vision and Pattern Recognition Workshops (CVPRW), pp. 1248–1257 (2019). https://doi.org/10.1109/CVPRW.2019.00164
11. Hausler, S., Garg, S., Xu, M., Milford, M., Fischer, T.: Patch-NetVLAD: multi-scale fusion of locally-global descriptors for place recognition. In: 2021 IEEE/CVF Conference on Computer Vision and Pattern Recognition (CVPR), pp. 14136–14147 (2021). https://doi.org/10.1109/CVPR46437.2021.01392

12. He, K., Zhang, X., Ren, S., Sun, J.: Deep residual learning for image recognition. In: 2016 IEEE Conference on Computer Vision and Pattern Recognition (CVPR), pp. 770–778 (2016). https://doi.org/10.1109/CVPR.2016.90
13. Hu, J., Ozay, M., Zhang, Y., Okatani, T.: Revisiting single image depth estimation: toward higher resolution maps with accurate object boundaries. In: 2019 IEEE Winter Conference on Applications of Computer Vision (WACV), pp. 1043–1051 (2019). https://doi.org/10.1109/WACV.2019.00116
14. Hui, L., Cheng, M., Xie, J., Yang, J., Cheng, M.M.: Efficient 3D point cloud feature learning for large-scale place recognition. IEEE Trans. Image Process. **31**, 1258–1270 (2022). https://doi.org/10.1109/TIP.2021.3136714
15. Kendall, A., Grimes, M., Cipolla, R.: Posenet: a convolutional network for real-time 6-DoF camera relocalization. In: 2015 IEEE International Conference on Computer Vision (ICCV), pp. 2938–2946 (2015). https://doi.org/10.1109/ICCV.2015.336
16. Kerl, C., Sturm, J., Cremers, D.: Dense visual slam for RGB-D cameras. In: 2013 IEEE/RSJ International Conference on Intelligent Robots and Systems, pp. 2100–2106 (2013). https://doi.org/10.1109/IROS.2013.6696650
17. Krizhevsky, A., Sutskever, I., Hinton, G.E.: Imagenet classification with deep convolutional neural networks. In: Advances in Neural Information Processing Systems, vol. 25 (2012)
18. Law, H., Deng, J.: CornerNet: detecting objects as paired keypoints. In: Ferrari, V., Hebert, M., Sminchisescu, C., Weiss, Y. (eds.) Computer Vision – ECCV 2018. LNCS, vol. 11218, pp. 765–781. Springer, Cham (2018). https://doi.org/10.1007/978-3-030-01264-9_45
19. Liu, W., et al.: SSD: single shot MultiBox detector. In: Leibe, B., Matas, J., Sebe, N., Welling, M. (eds.) ECCV 2016. LNCS, vol. 9905, pp. 21–37. Springer, Cham (2016). https://doi.org/10.1007/978-3-319-46448-0_2
20. Long, J., Shelhamer, E., Darrell, T.: Fully convolutional networks for semantic segmentation. In: 2015 IEEE Conference on Computer Vision and Pattern Recognition (CVPR), pp. 3431–3440 (2015). https://doi.org/10.1109/CVPR.2015.7298965
21. Lopez-Antequera, M., Gomez-Ojeda, R., Petkov, N., Gonzalez-Jimenez, J.: Appearance-invariant place recognition by discriminatively training a convolutional neural network. Pattern Recogn. Lett. **92**, 89–95 (2017). https://doi.org/10.1016/j.patrec.2017.04.017
22. Mur-Artal, R., Tardós, J.D.: ORB-SLAM2: an open-source slam system for monocular, stereo, and RGB-D cameras. IEEE Trans. Robot. **33**(5), 1255–1262 (2017). https://doi.org/10.1109/TRO.2017.2705103
23. Newcombe, R.A., et al.: Kinectfusion: real-time dense surface mapping and tracking. In: 2011 10th IEEE International Symposium on Mixed and Augmented Reality, pp. 127–136 (2011). https://doi.org/10.1109/ISMAR.2011.6092378
24. Redmon, J., Divvala, S., Girshick, R., Farhadi, A.: You only look once: unified, real-time object detection. In: 2016 IEEE Conference on Computer Vision and Pattern Recognition (CVPR), pp. 779–788 (2016). https://doi.org/10.1109/CVPR.2016.91
25. Sarlin, P.E., DeTone, D., Malisiewicz, T., Rabinovich, A.: Superglue: learning feature matching with graph neural networks. In: 2020 IEEE/CVF Conference on Computer Vision and Pattern Recognition (CVPR), pp. 4937–4946 (2020). https://doi.org/10.1109/CVPR42600.2020.00499
26. Silberman, N., Hoiem, D., Kohli, P., Fergus, R.: Indoor segmentation and support inference from RGBD images. In: Fitzgibbon, A., Lazebnik, S., Perona, P., Sato, Y., Schmid, C. (eds.) ECCV 2012. LNCS, vol. 7576, pp. 746–760. Springer, Heidelberg (2012). https://doi.org/10.1007/978-3-642-33715-4_54

27. Simo-Serra, E., Trulls, E., Ferraz, L., Kokkinos, I., Fua, P., Moreno-Noguer, F.: Discriminative learning of deep convolutional feature point descriptors. In: 2015 IEEE International Conference on Computer Vision (ICCV), pp. 118–126 (2015). https://doi.org/10.1109/ICCV.2015.22

28. Sturm, J., Engelhard, N., Endres, F., Burgard, W., Cremers, D.: A benchmark for the evaluation of RGB-D slam systems. In: 2012 IEEE/RSJ International Conference on Intelligent Robots and Systems, pp. 573–580 (2012). https://doi.org/10.1109/IROS.2012.6385773

29. Sun, J., Shen, Z., Wang, Y., Bao, H., Zhou, X.: LoFTR: detector-free local feature matching with transformers. In: 2021 IEEE/CVF Conference on Computer Vision and Pattern Recognition (CVPR), pp. 8918–8927 (2021). https://doi.org/10.1109/CVPR46437.2021.00881

30. Whelan, T., Kaess, M., Johannsson, H., Fallon, M., Leonard, J.J., McDonald, J.: Real-time large-scale dense RGB-D slam with volumetric fusion. Int. J. Robot. Res. 34(4–5), 598–626 (2015). https://doi.org/10.1177/0278364914551008

31. Yi, K.M., Trulls, E., Lepetit, V., Fua, P.: LIFT: learned invariant feature transform. In: Leibe, B., Matas, J., Sebe, N., Welling, M. (eds.) ECCV 2016. LNCS, vol. 9910, pp. 467–483. Springer, Cham (2016). https://doi.org/10.1007/978-3-319-46466-4_28

32. Zoph, B., Vasudevan, V., Shlens, J., Le, Q.V.: Learning transferable architectures for scalable image recognition. In: 2018 IEEE/CVF Conference on Computer Vision and Pattern Recognition, pp. 8697–8710 (2018). https://doi.org/10.1109/CVPR.2018.00907

Semantically Consistent Sim-to-Real Image Translation with Neural Networks

Solt Skribanek[1]([⊠])[ID], Márton Szemenyei[1][ID], and Róbert Moni[2][ID]

[1] Department of Control Engineering and Information Technology,
Budapest University of Technology and Economics, Budapest, Hungary
skribanek.solt@edu.bme.hu, szemenyei@iit.bme.hu
[2] Department of Telecommunications and Media Informatics,
Budapest University of Technology and Economics, Budapest, Hungary
robertmoni@tmit.bme.hu

Abstract. Texture-swapping of images has industrial benefits besides artistic stylization and photo editing, e.g. simulated images could be modified to look like real ones to train Computer Vision methods. Autonomous driving research could largely benefit from this as its neural network-based perception systems need a large amount of labeled training data. However, the sim-to-real texture swapping is a demanding challenge because of the large gap between the two domains. Another requirement is that the semantic meaning of the photo should not change during the translation. We found that SOTA algorithms struggle with these expectations, so in this work, we improve a former method by taking advantage of the semantic labeling of the training datasets. We show that with our two improvements, we can better conserve the scene of the image during the sim-to-real translation while the photorealism of the output image does not significantly change.

Keywords: Image translation · Sim-to-real · Neural networks

1 Introduction

Autonomous driving research requires an abundant amount of labeled training data for the neural network-based image-processing methods. While these labeled datasets come with large expenses, software-simulators can reduce the costs as training data with the corresponding labels are directly accessible in them. Using such simulators comes with further advantages: one can prepare the car for dangerous or extremely rare traffic situations at no additional cost.

Despite having such benefits, one can also face difficulties when trying to use a software-simulator like this. The data gained from the program is not realistic enough, which means an agent performing well in a simulated environment won't necessarily be suitable for driving on the street. Regarding images, simulated ones usually have very homogeneous and primitive textures, and thus they are not able to represent reality.

L. Rutkowski et al. (Eds.): ICAISC 2022, LNAI 13589, pp. 68–79, 2023.
https://doi.org/10.1007/978-3-031-23480-4_6

The solution to this problem could be an image-to-image transformation that is able to convert the textures to be more realistic, while preserving the positions of the portrayed objects.

In this work, we added two new features for a state-of-the-art architecture called Swapping Autoencoder (SAE) which we found very promising for the texture-swapping task. Our new loss functions are based on the fact that we possess the semantic labels of both the sinthesized and the real datasets. We can utilize this new information in a novel way to force the translation network to keep the semantic meaning of the picture during the transformation.

We propose two novel losses: the **Inner Semantic Loss'** responsibility is to constrain the autoencoder's latent representation to resemble the original semantic meaning of the picture, while our **Outer Semantic Loss** examines the final translated image and penalizes the generator for the semantic deviation from the original synthetic image. We show that it can outperform the baseline Swapping Autoencoder using a semantic segmentation network pretrained on the target dataset.

Our contributions are summarized as follows:

- We proposed a novel architecture called **Label-Consistent Swapping Autoencoder** by introducing two new objectives.
- We evaluated our method in terms of semantic consistency (mIoU) and translation quality (FID) and found that our innovations increased the former without ruining the texture-swapping quality.
- We created a new dataset from a widely used simulator software to serve our research.

2 Related Work

2.1 Semantic Segmentation Using Neural Networks

In semantic segmentation, the goal is to classify each pixel of an image into a class. In order to get the feature maps in the input images' resolution, segmentation networks include upscaling layers after the downscaling part of a typical CNN, resulting in an encoder-decoder structure [1]. Another important architectural solution is to make skip connections between the shallow layers and the upscaled feature maps that have the same resolution. This way, the fusion of high-level low-detail information and low-level high-detail features allows the network to produce the expected outcome [13].

The so-called DeepLab [1] networks also use a special block named ASPP in the middle of the architecture and CRFs on the final output. In this work, we utilize the newest member of the family, DeepLabV3+ [1] with ResNet [7] and MobileNet [9] backbones.

2.2 Autoencoder

The classical autoencoder is an encoder-decoder structure that is able to produce a compact representation of the input in a latent space. The dimensionality of the latent space must be less than the input's, so the encoder is forced to find the best representation in the narrower space with the least information loss possible. The decoder's task is to reproduce the input from this compact latent representation as accurately as possible.

2.3   GAN

Deep Generative Adversarial Networks [6] are the basis of the SOTA image-generation methods [11]. In the basic GAN architecture, there are two networks competing with each other: a generator that aims to synthesize realistic images from noise samples, and a discriminator that distinguishes between real and generated pictures. They are trained simultaneously, and the generator learns to produce realistic images during this rivalry.

2.4   Style Transfer and Image-to-Image Translation

Neural style transfer is a widely researched area, as generating photorealistic images based on existing ones is advantageous in many situations, such as image augmentation to increase the performance of CNNs used in computer vision [5], artistic stylization [17] or editing [16] of pictures or sim-to-real image-translation.

One group of these networks uses semantic segmentation maps directly for image generation. *pix2pix* [10] uses them as condition terms for the Conditional GAN [14] architecture, where both the generator and the discriminator have access to the labels. *SPADE* [15] rather injects the semantic information into the generator at every scale using special normalization layers. These networks require the semantic labels at inference time as well.

CycleGAN [17] *and UNIT* [12] only need domain supervision rather than a paired dataset. CycleGAN's assumption is, that an image translated to the other domain should be translated back, thus introducing the so-called cycle-consistency loss. UNIT utilizes a VAE-GAN combination to map both domains to a common latent space, also implying the cycle-consistency.

Sem-GAN [2] utilizes segmentation labels only to compute loss terms, meaning that it does not need the labels during inference time. It is a symmetric architecture aiming to learn a semantically consistent translation.

Swapping Autoencoder [16] is a fully unsupervised architecture made for texture swapping and photo editing. The main idea is related to the latent space once again: the core of the model is a special autoencoder that encodes the images into two latent components. The structure code z_s is a 3D tensor with spatial dimensions, while the texture code z_t is a 1D vector. If the decomposition of the latent components and the reconstruction are correct, the reconstruction of swapped components results in a photorealistic image that represents the structure of the first input with the texture of the second input. The reconstruction of the images must be accurate and realistic. A classical L1 loss is responsible for the former, a GAN-discriminator (D) ensures the latter demand. The same discriminator enforces the photorealistic generation from mixed latent components. There is also a patch discriminator that discriminates small patches from original and mixed images. The idea is that crops of the texture-swapped image should be indistinguishable from ones that come from the source image of z_t. D_{patch} further constraints that z_s and z_t actually represent structure and texture.

3 Methodology

We picked the Swapping Autoencoder as our starting point. We also considered CycleGAN [17] and UNIT [12], but early experiments showed that the gap between our synthetic dataset and real images is too large for them, resulting only in color transformation. Swapping Autoencoder does a real style conversion, but semantic consistency is weak. Therefore, we aimed on improving this architecture by adding more constraints utilizing the semantic labels that are available for both synthetic and real datasets.

We first modified the original architecture so it could only create sim-to-real hybrids, i.e. it could only mix structure codes extracted from synthetic images with texture codes extracted from real ones. Therefore, we had to vary the swapping GAN-loss and the co-occurrence patch discriminator's loss. The reconstruction GAN-loss and the L1 loss are computed for both datasets equally. We refer to this model as our **baseline**.

As mentioned above, we want to utilize the semantic labels to further constrain the style transfer. We introduce two novel objectives to the network: the Inner Semantic Loss and the Outer Semantic Loss.

3.1   Inner Semantic Loss

Swapping Autoencoder encodes the input images into structure and texture codes, where structure code is a 3D tensor with spatial dimensions. It should only represent the semantic information of the image, as it should not carry any information about the texture (assuming disentanglement of structure and texture). However, only two facts guarantee that structure code actually represents the semantic meaning of the image: the 3D shape of the code and the D_{patch} discriminator.

We help the encoder—referred to as E—learn a more appropriate disentanglement by training the structure code using the semantic labels. We assume that if the structure code truly represents the structure, then the ground-truth semantic label maps could be computed from the structure code (or the structure path of the encoder) with a function called $I : Z^s \rightarrow Y_{/4}$ where Z^s is the space of the structure codes, $Y_{/4}$ is the space of the labels, downsampled twice. Our inner semantic loss can be termed as:

$$\mathcal{L}_{\text{sem,in}}(E) = \mathbb{E}_{\mathbf{x} \sim \mathbf{X}^{\text{sim}} \cup \mathbf{X}^{\text{real}}, \mathbf{y}_{/4} \sim \mathbf{Y}_{/4}} \left[\mathcal{NLL} \left(I(\mathbf{z_s}), \mathbf{y}_{/4} \right) \right],$$

where $\mathbf{x}$ and $\mathbf{y}_{/4}$ are corresponding image-label pairs and $\mathcal{NLL}$ stands for the negative log-likelihood function that penalizes the deviation of the prediction from the ground-truth label.

From another point of view, the combination of E and I can be considered a universal semantic segmentation network that can predict the semantic labels at a lower resolution for both $\mathbf{X}^{\mathbf{sim}}$ and $\mathbf{X}^{\mathbf{real}}$ datasets. We used a smaller resolution at this inner segmentation network because even with learned upscaling, the network can not predict higher resolution segmentation maps without skip connections from earlier layers with larger scales. The smaller resolution also increases computational efficiency.

To implement our Inner Semantic Loss, we completed the encoder with a side branch (named inner semantic branch) that branches off from the last convolutional layer just before the structure code. Because the spatial dimensions at this point are 1/16 of the original image size, the inner semantic branch is an upsampling network to reduce the gap in the scales between its output and the ground-truth semantic maps. Thus, it performs 2 learned upscaling, as it is a composition of transposed convolutional layer and classical convolutional layer from StyleGAN2 [11], repeated twice. The channel size is halved at every upscaling. In the end, there is another convolutional layer with a kernel size of 1×1 to make the final pixel-level predictions.

This way, the side branch can be considered as the decoder part of a small-scale semantic segmentation network. We did not use skip-connections from shallower layers, because that part of the main branch contains information about the texrue as well. The outputs of this part are then compared with the ground-truth label masks using the negative log-likelihood loss. Figure 1 shows this part of the architecture.

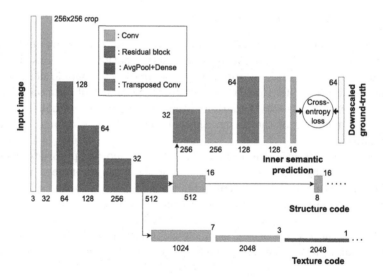

Fig. 1. Completion of the encoder part of the Swapping Autoencoder with the inner semantic branch.

3.2 Outer Semantic Loss

We also constrained our translation network from its outside: as we had access to the segmentation labels of the source dataset, we knew what the translated image should look like. We, therefore, prescribed that the result of the translation should have semantic content that corresponds to the original segmentation map.

To be able to do this, we needed a function O that segments the translated images. If the style transfer is performed correctly, then the textures of the translated image correspond to the target dataset, meaning that a segmentation network pretrained on the target dataset with frozen weights is well-suited for the task.

We note that other works such as Sem-GAN [2] also used a segmentation network similarly, but they trained their segmentation network simultaneously with the translation network. Also, at the beginning of the training, it could help the generator network create objects that are similar to the real ones. The outer semantic loss can be formulated as:

$$\mathcal{L}_{\text{sem,out}}(E, G) = \mathbb{E}_{\mathbf{x}^1 \sim \mathbf{X}^{\text{sim}}, \mathbf{x}^2 \sim \mathbf{X}^{\text{real}}, \mathbf{y} \sim \mathbf{Y}^{\text{sim}}} \left[NLL \left(O \left(G \left(\mathbf{z}_s^1, \mathbf{z}_t^2 \right) \right), \mathbf{y} \right) \right],$$

where $\mathbf{y}$ is the label for $\mathbf{x}^1$ and $\mathbf{z}_s^1, \mathbf{z}_t^2$ are the structure code of image $\mathbf{x}^1$ and the texture code of image $\mathbf{x}^2$, respectively.

In the implementation, we used a DeepLabV3+ model with a MobileNet [9] backbone as the segmentation network O on the target dataset. We chose Mobilenet because of its small size: during training, we needed to backpropagate through the full network at each iteration, so we wanted to use as small network as possible to reduce the effect of vanishing gradients, and also increase computational efficiency.

3.3 Final Objective

We combined our two new objectives with the former Swapping Autoencoder objectives, so our final objective became:

$$\mathcal{L}_{\text{total}} = \mathcal{L}_{\text{rec}} + 0.5(\mathcal{L}_{\text{GAN,rec}} + \mathcal{L}_{\text{GAN,swap}}) + \mathcal{L}_{\text{PatchGAN}} + \lambda_{in}\mathcal{L}_{\text{sem,in}} + \lambda_{out}\mathcal{L}_{\text{sem,out}},$$

from which $\mathcal{L}_{\text{rec}}$, $\mathcal{L}_{\text{GAN,rec}}$, $\mathcal{L}_{\text{GAN,swap}}$ and $\mathcal{L}_{\text{PatchGAN}}$ were the originals. We kept the original weights of the SAE objectives, while λ_{in} and λ_{out} are hyperparameters.

In summary, we added two novel objectives to the Swapping Autoencoder architecture. Both penalize deviation from semantic label maps: the inner semantic loss compares the structure codes with the downscaled labels, while outer semantic loss investigates a semantic segmentation of the translated image. The Label-Consistent Swapping Autoencoder is thus a supervised form of the Swapping Autoencoder. The full architecture can be examined in Fig. 2.

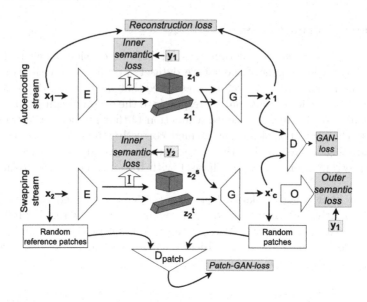

Fig. 2. Full architecture of the Label-Consistent Swapping Autoencoder with its objectives highlighted in red. (Color figure online)

4   Experiments

4.1   Datasets, Data Preparation

First, we needed a synthetic and a real dataset, both labeled. As a real dataset, we picked the widely used CityScapes Dataset [3]. It contains 3475 finely labeled pictures at 2048×1024 resolution.

To create our own synthetic set, we used the popular CARLA Simulator [4], which is an open-source software specifically created to aid autonomous driving research.

We utilized CARLA's RGB and semantic camera sensors—the latter directly provides the semantic maps—and collected 20,000 images with their corresponding labels. During data collection, we tried to create a set as similar as possible to CityScapes, so we varied the maps and weather accordingly, meaning that we collected more images from big cities and fewer from small towns, and allowed only dry weather in daylight.

After the image generation, we performed automated and manual filtering and picked 5000 images (from which 2500 comes from Town10HD, the city with the most realistic textures) for training and another 100 for test purposes.

Regarding to the labels, we had to merge some categories in both datasets in order to make sure the same label has the same meaning both in the CityScapes Dataset and our CARLA Dataset. In the end, there remained 16 classes. After that, we downsampled the images and their labels to 512×256 resolution using Lanczos and Nearest Neighbor methods respectively in order to speed up the training.

4.2   Setup

Due to the relatively small size of the real training dataset, we used random 256×256 sized crops during training, since working on images of street scenes allows us this type of augmentation. Fortunately, the fully convolutional architecture allowed us to train on crops and test on the full images.

At each training iteration, we sampled N images from both $\mathbf{X}^{sim}$ and $\mathbf{X}^{real}$. We reconstructed $N/2$ real and $N/2$ synthetic images (chosen randomly from the inputs), and computed the reconstruction loss using these. We created N hybrid images, using all of the available structure and texture codes. The image discriminator's loss is computed on the $2N$ real, the N reconstructed and the N hybrid images, where both reconstructed and hybrid images are considered fake. As for the patch discriminator, we used its basic settings (8 crops for each image, averaged features for the reference image). Note that D_{patch} in our case works only on the target domain as it discriminates between sim-to-real hybrids and their textures that come entirely from the target domain. As for the other implementation details, we also followed the basic settings of the Swapping Autoencoder. The dimensions of our texture and structure codes were 2048 and $16 \times 16 \times 8$, respectively.

For the training of our Label-Consistent Swapping Autoencoder, we used the maximum batch size that fitted into the memory of a single 32 GB Titan V100 GPU: that is 8 when training with 256×256 crops, meaning that the network works with 16 images at the same time on our hardware. We set the maximum run-time to $5M$ iterations for all our experiments.

We used an out-of-the-box DeepLabV3+ model with Mobilenet [9] backbone as our outer segmentation network[1], pretrained it on the resized and relabeled CityScapes dataset with the original train-val split. The network reached 55% on the validation set, lower than the original 72%, which can probably be attributed to the reduced resolution. As for the hyperparameters of this pretrain, we used 256×256 sized crops for training and whole 512×256 sized images for validation. The network's output stride was 16. Here we also maximized the batch size, which was 128. We ran the training until 30k iterations with an initial learning rate of 0.1 with an SGD optimizer and used a polynomial scheduling policy.

5   Results

We first trained our baseline model (i.e. $\lambda_{in} = \lambda_{out} = 0$), then trained another 6 models with different combinations of λ_{in} and λ_{out}. For evaluation, we used 100 test images from our CARLA Dataset as structure images along with 4 images from the CityScapes' test set in the full 512×256 resolution. With all the possible sim-to-real hybridizations, we get 400 result images.

As we wanted to increase semantic consistency, we used another segmentation network to evaluate it quantitatively. We picked a DeepLabV3+ with a larger backbone ResNet50 [7]. This network has circa ten times as many parameters as

---

[1] https://github.com/VainF/DeepLabV3Plus-Pytorch.

the MobileNet [9] version. We pretrained it on our modified CityScapes Dataset. For this training, we used the exact same hyperparameters as for the outer segmentation network. We segmented all the 400 translated images and compared them with the original CARLA images' ground-truth segmentation maps.

We used classical semantic segmentation metrics for quantitative investigation. The overall pixel accuracy metric is a ratio of the correctly predicted pixels. Mean pixel accuracy is computed by taking the mean of the pixel accuracies for each class. mIoU stands for mean intersection over union, it computes the intersection and the union of the correct and the predicted locations of a given class, then divides them, and takes the mean across the classes. The results can be seen in Table 1.

We also employed the Fréchet inception distance (FID) [8] to measure the similarity of two image datasets, as this metric is widely used to evaluate GAN-based image-generator networks. We computed the FID metric between the 400 result images and the 5000 images of the CityScapes Dataset.

Table 1. Validation results using semantic segmentation metrics and FID score. We **bold** the best results per column.

λ_{out}	λ_{in}	Overall acc	Mean acc	mIoU	FID
Baseline		0.530	0.167	0.112	65.26
0.0	1.0	0.500	0.170	0.109	**60.13**
1.0	0.0	0.672	0.326	0.215	61.95
1.0	1.0	0.665	0.332	0.227	68.59
1.0	5.0	0.632	0.329	0.216	64.78
2.0	2.0	0.675	0.351	0.237	66.26
3.0	0.0	**0.692**	**0.382**	**0.254**	76.08

Table 1 shows us that the use of outer segmentation loss highly increased the semantic consistency (doubled the mean pixel accuracy and the mIoU metrics) while the FID metric did not change substantially ($\lambda_{out} < 3$ cases). It seems that the inner segmentation loss did not help the effectiveness of the network, as using only this loss lowered both pixel accuracy and mIoU compared to the baseline model. This likely means that the base encoder by itself could find a better representation (regarding the generation) than the strict semantic label maps. Notably, using the inner loss resulted in a sizeable reduction in the FID metric. The best performing model based on the semantic metrics is the one with $\lambda_{out} = 3$ and $\lambda_{in} = 0$, however, its Fréchet distance is significantly higher. The model with $\lambda_{out} = \lambda_{in} = 2$ can be considered the best of both, because its semantic scores almost reach the best in this category, but there was no significant change in the FID score compared to the baseline model.

Fig. 3. Visual results. Top: original picture from CARLA and its ground-truth, Middle: transformed image with the baseline model and its segmentation result, Bottom: translation result with our $\lambda_{out} = 3$ model and its segmentation result.

Figure 3 shows the strength of our improvement: it prevents the translation from "hallucinating" cars next to the roads. CityScapes' main advantage is that it is manually filtered and thus contains a large number of traffic actors: cars and pedestrians. However, in our case, this appears as a weakness: there are too many cars in the CityScapes set, and therefore our GAN-based model collapses slightly: it can hardly imagine a road in the target domain without many cars parking aside. Our outer segmentation loss reduces this effect as it does not allow the translation network to park cars where there should not be any.

The project's GitHub Page[2] contains more images and they show other cases where our constraints resulted in more accurate translations as our model preserved the layout of the scene more accurately. There are cases that demonstrate the weakness of the CityScapes Dataset: our real dataset contains only images captured in big cities, and there are very few images that portray the open sky, therefore our model places buildings or vegetation in place of the sky.

It is worth emphasizing, however, that our model is not perfect, as errors in the generated images sometimes still persist. For instance, our model can erase bicyclists and smaller objects on the road, but still prevents adding unnecessary

---

[2] https://github.com/skribaneksolt/Label-Consistent-Swapping-Autoencoder.

cars to the scene. There are also failure cases where our constraints ruin the translation: an example can be examined on the GitHub page where the baseline places the car in the right place while our model misses the object.

It can be said, however, that overall our model improves upon the baseline in most cases, a claim evidenced by the significant improvement in the quantitative segmentation metrics. It is also worth emphasizing that we used a different segmentation architecture for training and evaluation to minimize the possibility that the generator simply learns to hide the semantic labels in the generated images.

6   Conclusions

We found that our inner semantic loss function had mixed effects on the consistency, while it improved the quality of the translation: this likely means that the Swapping Autoencoder with its original limitations is able to learn a more meaningful representation as structure code than the semantic label maps for the image-generation.

Our outer semantic loss highly increased the semantic consistency of the images as it doubled the mean pixel accuracy and the mIoU metric compared to the baseline. It is a notable outcome, but we believe it can be improved even further by additional considerations. With the unweighted cross-entropy loss, this objective forced the translation network to better preserve the layout of the scene, however, it attached greater importance to bigger objects like the road, trees, and the sky, at vehicles' and pedestrians' expense.

In our experience, the real dataset had its limitations as it was small and too homogeneous. Our GAN learned the idiosyncrasies of the target dataset— such as parking cars roadside, rich vegetation, and big buildings that cover the sky etc.—along with the real-world textures, and this caused problems in the translation.

Acknowledgment. The research presented in this work has been supported by Continental Automotive Hungary Ltd. The publication of the work reported herein has been supported by ETDB at BME.

References

1. Chen, L.-C., Zhu, Y., Papandreou, G., Schroff, F., Adam, H.: Encoder-decoder with atrous separable convolution for semantic image segmentation. In: Proceedings of the European Conference on Computer Vision (ECCV), pp. 801–818 (2018)
2. Cherian, A., Sullivan, A.: Sem-GAN: semantically-consistent image-to-image translation (2018)
3. Cordts, M., et al.: The cityscapes dataset for semantic urban scene understanding. In: Proceedings of the IEEE Conference on Computer Vision and Pattern Recognition, pp. 3213–3223 (2016)

4. Dosovitskiy, A., Ros, G., Codevilla, F., Lopez, A., Koltun, V.: CARLA: an open urban driving simulator. In: Proceedings of the 1st Annual Conference on Robot Learning, pp. 1–16 (2017)
5. Frid-Adar, M., Diamant, I., Klang, E., Amitai, M., Goldberger, J., Greenspan, H.: Gan-based synthetic medical image augmentation for increased CNN performance in liver lesion classification. Neurocomputing **321**, 321–331 (2018)
6. Goodfellow, I., et al.: Generative adversarial nets. In: Advances in Neural Information Processing Systems, vol. 27 (2014)
7. He, K., Zhang, X., Ren, S., Sun, J.: Deep residual learning for image recognition. In: Proceedings of the IEEE Conference on Computer Vision and Pattern Recognition, pp. 770–778 (2016)
8. Heusel, M., Ramsauer, H., Unterthiner, T., Nessler, B., Hochreiter, S.: GANs trained by a two time-scale update rule converge to a local nash equilibrium. In: Advances in Neural Information Processing Systems, vol. 30 (2017)
9. Howard, A.G., et al.: Mobilenets: efficient convolutional neural networks for mobile vision applications. arXiv preprint arXiv:1704.04861 (2017)
10. Isola, P., Zhu, J.-Y., Zhou, T., Efros, A.A.: Image-to-image translation with conditional adversarial networks. In: Proceedings of the IEEE Conference on Computer Vision and Pattern Recognition, pp. 1125–1134 (2017)
11. Karras, T., Laine, S., Aittala, M., Hellsten, J., Lehtinen, J., Aila, T.: Analyzing and improving the image quality of StyleGAN. In: Proceedings of the IEEE/CVF Conference on Computer Vision and Pattern Recognition, pp. 8110–8119 (2020)
12. Liu, M.-Y., Breuel, T., Kautz, J.: Unsupervised image-to-image translation networks, March 2017
13. Long, J., Shelhamer, E., Darrell, T.: Fully convolutional networks for semantic segmentation. In: Proceedings of the IEEE Conference on Computer Vision and Pattern Recognition, pp. 3431–3440 (2015)
14. Mirza, M., Osindero, S.: Conditional generative adversarial nets. arXiv preprint arXiv:1411.1784 (2014)
15. Park, T., Liu, M.-Y., Wang, T.-C., Zhu, J.-Y.: Semantic image synthesis with spatially-adaptive normalization. In: Proceedings of the IEEE/CVF Conference on Computer Vision and Pattern Recognition, pp. 2337–2346 (2019)
16. Park, T., et al.: Swapping autoencoder for deep image manipulation. In: Advances in Neural Information Processing Systems (2020)
17. Zhu, J.-Y., Park, T., Isola, P., Efros, A.A.: Unpaired image-to-image translation using cycle-consistent adversarial networks. In: 2017 IEEE International Conference on Computer Vision (ICCV) (2017)

Hand Gesture Recognition for Medical Purposes Using CNN

Jakub Sosnowski$^{(\boxtimes)}$, Piotr Pluta, and Patryk Najgebauer

Department of Intelligent Computer Systems, Czestochowa University of Technology, Armii Krajowej 36, 42-200 Czestochowa, Poland
jakub.sosnowski@pcz.pl
http://www.iisi.pcz.pl/

Abstract. The presented paper describes implementations of the gesture recognition methods based on the convolutional neural networks. For this purpose, we adopted three CNN structures. The data was obtained from a specially prepared dataset containing images with hand gestures taken on a green screen and downloaded backgrounds depicting hospital and office conditions. During experiments, the precision of recognizing individual gesture classes was measured. Experiments were carried out that showed the performance time of the gesture recognizing in images using a GPU card.

Keywords: Convolutional neural networks · Image classification · Gesture recognition

1 Introduction

Non-contact control may have a great impact on reducing the transmission of various types of bacteria and viruses. Scientific research shows that there are ten times more bacteria on the surface of devices such as keyboards or touch screens than on most toilet seats [1]. Terminals and computers in public places can be handled using gestures instead of being physically operated, reducing the number of people touching them and transmitting pathogens. Such systems can also prevent the spread of diseases and limit the extent of existing disease outbreaks.

We intend to design a control system for medical purposes that will be able to recognize gestures and function by following the commands given by these gestures. To achieve this goal, we need to build a system for recognizing images that contain gestures.

Generally, image recognition is of increasing importance in the contemporary digital world. Thanks to the field of artificial intelligence are possible, among others, face, gesture, and license plate recognition. That allows for improvement of computed tomography, securing devices and systems against unauthorized access, quick and effective recognition of people and vehicles, or recognition and extraction of text from photos and graphics (see e.g. [2–9]).

L. Rutkowski et al. (Eds.): ICAISC 2022, LNAI 13589, pp. 80–88, 2023.
https://doi.org/10.1007/978-3-031-23480-4_7

Another area where image recognition is gaining importance is gesture recognition. This technology makes it possible for non-contact operations, in which there is no need for physical contact with the device and external devices, such as a keyboard or computer mouse (see e.g. [10]).

There are two main approaches to gesture recognition using artificial intelligence. The first way is to use a glove with multiple sensors built-in, monitoring hand movement. Such systems are very accurate, color and brightness of light do not affect their performance (see e.g. [11]). Research into this kind of gesture recognition continues to improve using deep learning and big data techniques (see e.g. [12]). The second way systems are that recognize gestures like the human eye recognizes objects (see e.g. [13,14]). This approach currently uses mainly deep learning and 2D and 3D convolutional networks, which allow high precision of gesture recognition. These systems are being tested in hospitals, among other places, where doctors control advanced robots using gestures (see e.g. [13]).

In this paper, we show our first attempts to solve this challenging technical problem using three well-known CNN structures. We compare simple, small CNN with bigger networks and see the differences between them. A more detailed description of the whole experiment process in the following sections is presented, and the obtained results are shown.

2 Dataset Design

The main challenge in gesture recognition is improving the precision of the convolutional neural networks performance. There can not happen any faults during the service of a medical device controlled by gestures. To achieve such a level of precision the training dataset has to be prepared carefully, reflecting the conditions in which a piece of given medical equipment is used. It means the medical environment with lighting and typical backgrounds encountered in hospitals.

Deep learning needs a lot of data to achieve good results, but it would be difficult and time-consuming to obtain pictures manually in a hospital. For this reason, it seems rational to use other methods to get a sufficient number of training examples. In our experiments, all images which contain gestures were taken using the green screen background. Then, so prepared background was replaced by the views from the hospital environment. Unfortunately, images ware taken on a green screen usually are obtained under professional lighting without shadows, wherein natural images mostly contain them, which can entail not satisfying results during the prediction phase. These are unique situations when dealing with perfect conditions for recognizing gestures in everyday life. That is why we took photos in various lighting conditions that have to reflect natural views. Figure 1 depicts the sequence of the design of the images containing gestures.

It is an essential aspect of the work regarding the method used to extract the green color from the images. The RGB color space in color segmentation is not

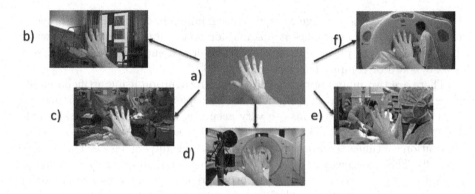

Fig. 1. Process of the design of the images containing gestures: (a) hand gesture taken on green screen; (b) (c) (d) (e) (f) training images with the same gesture on replaced backgrounds. (Color figure online)

preferred, where the information related to the color and intensity of the image has irregular characteristics. Unlike RGB, the HSV standard separates luma (or the image intensity) from chroma (the color information) [15]. Since it is possible to separate the color from its intensity and changes in lighting and shadows, it was possible to create natural images of the gestures. Hand images are made under different lighting conditions, with shadows, just like in practical use. Even if the green screen is creased and there are wrinkles on it, and the hand casts a shadow on it, the prepared algorithm can extract the area flawlessly. Figure 2 depicts the sequence of the design of the images containing gestures. Figure 1 describes differences between the RGB and the HSV standards regarding the image intensity and the color information coding.

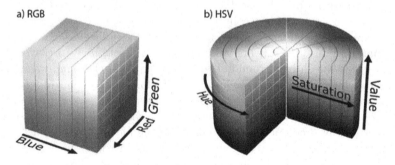

Fig. 2. The image intensity and the color information coding differences between standards: RGB (a), HSV (b) [15].

3   Convolutional Neural Networks

Deep learning is a subfield of machine learning [16]. The fundamental factor which distinguishes deep learning is the increased number of network layers (which can have even tens of them) (see e.g. [17]) in comparison to the traditional neural networks (see e.g. [18–23]). Due to the flexibility of deep learning techniques and the variety of neural network structures, deep learning has the broadest application among machine learning techniques (see e.g. [20]. They can successfully solve problems to which classic methods so far have been applied (Table 1).

Table 1. Machine learning and deep learning differences [17].

	Machine learning	Deep learning
Data	Can use small amounts of data to make predictions	Needs to use large amounts of training data to make predictions
Precision	Lower precision	Higher precision
Execution time	Takes comparatively little time to train ranging from a few minutes to a few hours	Usually takes a long time to train due to many layers
Featurization process	Requires features to be created by users and accurately identified	Learns high-level features from data and creates new features by itself
Hardware dependencies	It doesn't need a large amount of computational power	It inherently does a large number of matrix multiplication operations. GPU recommended

The most challenging step in building a network is determining the number of layers with the appropriate number of neurons. In the case of multilayer networks, the geometric pyramid method often is used. This method assumes that the number of neurons in successive layers forms a pyramidal shape and decreases from entry to exit, but it is only an approximate method. In general, learning begins with a small number of hidden layers and neurons and increases their number experimentally as the process progresses. If the network has too few neurons, it will not be able to learn proper or carry out a complex process. On the other hand, too many neurons lead to a significant extension of the calculation time and often to an overfitting effect (see e.g. [24]).

In our study, we used three convolutional network models with a different number of network layers. The first model (Table 2) has the least layers and filters to check how high precision the classes can achieve in the basic structure of the convolutional network. The second model (Table 3) is slightly more complicated. Despite the addition of just one layer and the increased number of filters, the precision of the network has visibly improved. Despite the improvement in results, the network presence is still too low to be used for recognizing gestures in practice. The third model (Table 4), by following the previously mentioned geometric pyramid rule, was increased in subsequent layers.

Table 2. First model network structure.

Layer (type)	Output shape	Params
Conv2D	(None, 94, 126, 32)	896
MaxPooling2D	(None, 47, 63, 32)	0
Conv2D	(None, 45, 61, 32)	9248
MaxPooling2D	(None, 22, 30, 32)	0
(Flatten)	(None, 21120)	0
Dense	(None, 128)	2703488
Dense	(None, 19)	2451

Table 3. Second model network structure.

Layer (type)	Output shape	Params
Conv2D	(None, 94, 126, 32)	896
MaxPooling2D	(None, 47, 63, 32)	0
Conv2D	(None, 45, 61, 32)	9248
MaxPooling2D	(None, 22, 30, 32)	0
Conv2D	(None, 20, 28, 64)	18496
MaxPooling2D	(None, 10, 14, 64)	0
(Flatten)	(None, 8960)	0
Dense	(None, 128)	1147008
Dense	(None, 19)	2451

Table 4. Third model network structure.

Layer (type)	Output shape	Params
Conv2D	(None, 94, 126, 55)	1540
MaxPooling2D	(None, 47, 63, 55)	0
Conv2D	(None, 45, 61, 89)	44144
MaxPooling2D	(None, 22, 30, 89)	0
Conv2D	(None, 20, 28, 144)	115488
MaxPooling2D	(None, 10, 14, 144)	0
Conv2D	(None, 8, 12, 233)	302201
MaxPooling2D	(None, 4, 6, 233)	0
Conv2D	(None, 2, 4, 377)	790946
MaxPooling2D	(None, 1, 2, 377)	0
(Flatten)	(None, 754)	0
Dense	(None, 144)	108720
Dense	(None, 19)	2755

4   Experimental Results

Table 5 shows the results corresponding to the time needed to train the third model. It is worth recalling that this model has the largest number of layers. The hardware platform is a notebook with processor Intel i7-7700HQ with 32GB RAM DDR4/2400MHz and graphic card Nvidia GeForce 1060 6GB. This equipment is managed by the operating system, i.e. Microsoft Windows 10 Home 64 bits. The neural network structures shortly described in Sect. 3 using the above computer system are learned.

Table 5. Time results of network training on multi-threads CPU: Intel i7-7700HQ (4-cores, 8-threads) and GPU (Nvidia GeForce 1060).

Epochs	CPU	GPU
2	384 [s]	47.3 [s]
4	769.6 [s]	67.7 [s]
6	1160 [s]	99.8 [s]
8	1552.2 [s]	132.2 [s]
10	1947.5 [s]	163.6 [s]

In Table 5, the convolutional neural network training times using CPU and GPU are shown. The use of GPU significantly speeds up learning. Moreover, GPU implementation reduces the temperature and load on the computer components [25]. When using GPU, you may find that the script is trying to use more memory than is available. If so, there are two options: computer memory size may be extended, or the resolution of images may be reduced so that it does not exceed available memory when loaded.

Table 6 summarizes the precision of each class for each of the implemented structures.

Table 6. Precision of models for 5 choosen gestures.

	Gesture 1	Gesture 2	Gesture 3	Gesture 4	Gesture 5
1st model	0.25	0.12	0.29	0.38	0.25
2nd model	0.55	0.28	0.56	0.40	0.79
3rd model	0.82	0.94	0.92	0.81	0.93

5 Conclusion

To achieve good results in the training process of neural networks, the quality of the data set is extremely important. The prepared images must have the appropriate resolution, small enough for the computer to train the neural network in a short time, and large enough not to lose important details. Taking into account the conditions in which gestures will be recognized, a network prepared to operate in a hospital setting should be learned using images with gestures on a background appropriate to these conditions. In our experiments, the image collection includes images with backgrounds depicting medical situations, computer tomographs, and office conditions. Moreover, the medical staff usually wears medical gloves. That was taken into account when preparing the images. Each gesture was taken using various types of medical gloves. Unfortunately, it may make the network unable to recognize gestures because medical gloves are often the same color as other items of staff clothing.

For better quality training images, they were converted to the HSV color model. That allowed better green color extraction.

In this paper, it has been proven that the increasing number of layers and filters in convolutional neural networks significantly improves the precision of gesture recognition. That is of great importance in systems where the controlled devices can cause harm when calling the wrong function or performing the fault action. In the worst case, such a situation may endanger the life and health of the user and the people around him.

It has been shown the duration of training for the third model to show differences between CPU and GPU usage. The crucial factor in learning neural networks, especially in deep learning using big data sets and complex structures with many hidden layers, is the learning time of the network. The shorter the learning time of such a network, the better. Of course, if the function of the network is not disturbed by it. Shorter learning time also means lower energy consumption and a possibility to perform more calculations and tasks. To speed up the learning of neural networks, one can use techniques and algorithms that can significantly affect the learning time (see e.g. [25]).

References

1. IHPI Page. https://ihpi.umich.edu/news/your-cell-phone-10-times-dirtier-toilet-seat-heres-what-do-about-it. Accessed 18 May 2022
2. Lv, Z., Chen, D., Lou, R., Alazab, A.: Artificial intelligence for securing industrial-based cyber-physical systems. Futur. Gener. Comput. Syst. **117**(1), 291–298 (2021)
3. Iwendi, C., Ur Rehman, S., Rehman Javed, A., Khan, S., Srivastava, G.: Sustainable security for the internet of things using artificial intelligence architectures. ACM Trans. Internet Technol. **21**(3), 1–22 (2021)
4. Kumar Mohanta, B., Jena, D., Satapathy, U., Patnaik, S.: Survey on IoT security: challenges and solution using machine learning, artificial intelligence and blockchain technology. Internet Things **11**(100227), 5–14 (2020)
5. Costa, M., Oliveira, D., Pinto, S., Tavares, A.: Detecting driver's fatigue, distraction and activity using a non-intrusive AI-based monitoring system. J. Artif. Intell. Soft Comput. Res. **9**(4), 247–266 (2019)
6. Gabryel, M., Grzanek, K., Hayashi, Y.: Browser fingerprint coding methods increasing the effectiveness of user identification in the web traffic. J. Artif. Intell. Soft Comput. Res. **10**(4), 243–253 (2020)
7. Korytkowski, M., Senkerik, R., Scherer, M., Angryk, R., Kordos, M., Siwocha, A.: Efficient image retrieval by fuzzy rules from boosting and metaheuristic. J. Artif. Intell. Soft Comput. Res. **10**(1), 57–69 (2020)
8. Qing, K., Zhang, R.: Position-encoding convolutional network to solving connected text captcha. J. Artif. Intell. Soft Comput. Res. **12**(2), 121–133 (2021)
9. Cierniak, R., et al.: A new statistical reconstruction method for the computed tomography using an X-ray tube with flying focal spot. J. Artif. Intell. Soft Comput. Res. **11**(4), 271–286 (2021)
10. Chalasani, T., Ondrej, J., Smolic, A.: Egocentric gesture recognition for head-mounted AR devices. Inst. Electr. Electron. Eng. **1**, 109–114 (2018)
11. Moin, A., et al.: A wearable biosensing system with in-sensor adaptive machine learning for hand gesture recognition. Nat. Electron. **4**, 54–63 (2021)
12. Duda, P., Jaworski, M., Cader, A., Wang, L.: On training deep neural networks using a streaming approach. J. Artif. Intell. Soft Comput. Res. **10**(1), 15–26 (2020)
13. Amsterdam, B., Clarkson, M., Stoyanov, D.: Gesture recognition in robotic surgery: a review. IEEE Trans. Biomed. Eng. **68**(6), 2021–2035 (2021)
14. Doroz, R., Wrobel, K., Porwik, P., Orczyk, T.: A new hand-movement-based authentication method using feature importance selection with the Hotelling's statistic. J. Artif. Intell. Soft Comput. Res. **12**(1), 41–59 (2022)
15. Oudah, M., Al-Naji, A., Chahl, J.: Hand gesture recognition based on computer vision: a review of techniques. J. Imaging **6**(8), 9–13 (2020)
16. Pérez-Pons, M., Parra-Dominguez, J., Omatu, S., Herrera-Viedma, E., Corchado, J.: Machine learning and traditional econometric models: a systematic mapping study. J. Artif. Intell. Soft Comput. Res. **12**(2), 79–100 (2021)
17. Microsoft Page. https://docs.microsoft.com/en-us/azure/machine-learning/concept-deep-learning-vs-machine-learning. Accessed 16 May 2022
18. Shi, L., Copot, C., Vanlanduit, S.: Evaluating dropout placements in Bayesian regression resnet. J. Artif. Intell. Soft Comput. Res. **12**(1), 61–73 (2022)
19. Zini, J., Rizk, Y., Awad, M.: An optimized parallel implementation of non-iteratively trained recurrent neural networks. J. Artif. Intell. Soft Comput. Res. **11**(1), 33–50 (2021)

20. Simões, D., Lau, N., Reis, L.: Multi agent deep learning with cooperative communication. J. Artif. Intell. Soft Comput. Res. **10**(3), 189–207 (2020)
21. Bilski, J., Kowalczyk, B., Marjański, A., Gandor, M., Zurada, J.: A novel fast feedforward neural networks training algorithm. J. Artif. Intell. Soft Comput. Res. **11**(4), 287–306 (2021)
22. Gabryel, M., Scherer, M., Sułkowski, Ł, Damaševičius, R.: Decision making support system for managing advertisers by ad fraud detection. J. Artif. Intell. Soft Comput. Res. **11**(4), 331–339 (2021)
23. Nowicki, R., Seliga, R., Żelasko, D., Hayashi, Y.: Performance analysis of rough set-based hybrid classification systems in the case of missing values. J. Artif. Intell. Soft Comput. Res. **11**(4), 307–318 (2021)
24. Soon Tan, Y., Ming Lim, K., Poo Lee, C.: Hand gesture recognition via enhanced densely connected convolutional neural network. Expert Syst. Appl. **175**, 114797 (2021)
25. Bilski, J., Kowalczyk, B., Marchlewska, A., Zurada, J.M.: Local Levenberg-Marquardt algorithm for learning feedforwad neural networks. J. Artif. Intell. Soft Comput. Res. **10**(4), 229–316 (2020)

Data Mining

A Streaming Approach to the Core Vector Machine

Moritz Heusinger[1]([⊠]) and Frank-Michael Schleif[2]

[1] Department of Computer Science, UAS Würzburg-Schweinfurt,
Sanderheinrichsleitenweg 20, Würzburg, Germany
`moritz.heusinger@fhws.de`
[2] School of Computer Science, University of Birmingham, Edgbaston,
Birmingham B15 2TT, UK
`frank-michael.schleif@fhws.de`

Abstract. The Support Vector Machine (SVM) is a widely used algorithm for batch classification with a run and memory efficient counterpart given by the *Core Vector Machine* (CVM). Both algorithms have nice theoretical guarantees, but are not able to handle data streams, which have to be processed instance by instance. We propose a novel approach to handle stream classification problems via an adaption of the CVM, which is also able to handle multiclass classification problems. Furthermore, we compare our Multiclass Core Vector Machine (MCCVM) approach against another existing *Minimum Enclosing Ball* (MEB)-based classification approach. Finally, we propose a real-world streaming dataset, which consists of changeover detection data and has only been analyzed in offline settings so far.

1 Introduction

In recent years, the data which is generated in industry as well as in everyday lives has increased a lot. Especially, massive amounts of sensor data is generated by *Internet of Things* (IOT) devices on the fly [5]. These continuous data streams need to be analyzed in real-time instead of performing historical batch analysis.

Hence, we need machine learning algorithms which can process data points incrementally. Furthermore, there are additional challenges in non-stationary environments, like changing data characteristics due to hardware sensors aging process, sensor pollution or sensor replacement. Such shifts in the distribution are called *Concept Drift* (CD) [4] and can occur in different variations.

In offline settings kernel based methods are very popular [24,25]. Kernel methods aim to separate non-linear data by calculating similarities in a higher dimensional space, where the data can be separated, without explicitly calculating the coordinates in this feature space [25]. One of the most popular methods is the two class SVM [25]. The SVM is a maximum-margin classifier which uses hyperplanes to separate features into classes. The SVM is normally formulated as a *Quadratic Programming* (QP) problem. The existing solutions for this problem are at least quadratic regarding time and space complexity. Thus, the CVM

L. Rutkowski et al. (Eds.): ICAISC 2022, LNAI 13589, pp. 91–101, 2023.
https://doi.org/10.1007/978-3-031-23480-4_8

has been proposed to scale up a binary class SVM, while preserving the generalization and convergence properties [24]. In CVM the quadratic problem of the SVM has been reformulated to a MEB problem. The MEB problem can be solved using an approximation as proposed by [8]. In this work we make the following contributions:

- We propose a CVM which is able to handle data points incrementally and can handle multi class problems
- We analyze a novel real-world sensor dataset, which has not been considered in non-stationary environments yet
- We compare our approach to previous MEB-based algorithms

The paper is organized as follows: Sect. 2 describes previous work in the context of stream classification and multiclass learning. Section 3 describes the challenges which arise in supervised non-stationary environments. In Sect. 4 we propose the adaptive MCCVM. In Sect. 5 we compare our algorithm against a state-of-the-art MEB-based classification algorithm and the *Adaptive Robust Soft Learning Vector Quantization* (ARSLVQ), as a prototype based classifier approach proposed in [12]. Finally, we summarize our learning and give an outlook in Sect. 6.

2   Related Work

The field of supervised data stream analysis is still an advancing topic [5]. Many popular batch learning algorithms have been modified to work in non-stationary environments [5,12].

In [16] the K-nearest neighbour (KNN) approach has been enhanced to work only on a subset of the whole data by using a sliding window approach. Furthermore, CD is handled actively by using a long short-term memory strategy and keeping track of the prediction performance of each memory.

Classification algorithms, which are used in stream environments are often prototype-based, like various versions of the *Learning Vector Quantization* (LVQ) [12,21] due to the fact, that prototype methods do not need to store large amount of datapoints, which is memory-efficient and allows fast updates.

CD is a change in the distribution which occurs between two timesteps. If CD occurs, the algorithm should react to the changing distribution, to avoid performance drops. To tackle CD in non-stationary environments different algorithms have been proposed. In sliding window approaches, CD is implicitly handled passive, by forgetting datapoints which are older than the window size [9,14]. Furthermore, there are algorithms, which additionally try to detect CD actively to react faster to changes in the distribution, which is important for abrupt drifts [4,16,19].

Coreset based algorithms are popular in batch learning environments [2,24], which has also been extended to work with multiple classes [2]. Also, there are already approaches to the SVM on data streams, however these implementations only work on append-only streams and need to be computed from scratch for a new window, which makes them impractical for real world scenarios [18].

Recently, coresets have also been maintained over a sliding window and are applied in stream situations to tackle CD detection [13] as well as classification [14]. However, this coreset based stream classifier is conceptual limited in the way how the classification task is modeled and lacks a profound statistical background as the CVM.

A MEB is useful to describe a data distribution by means of a compact model, as detailed in the following. A high effective approach to calculate a MEB for stationary data was proposed in [8]. To maintain a MEB in non-stationary environments the authors of [26] propose multiple methods to maintain a coreset for MEB over sliding windows. One of the proposed algorithms, the *Sliding Window Minimum Enclosing Ball+* (SWMEB+) is used for our proposed CVM.

3 Preliminaries

3.1 Minimum Enclosing Ball

We denote the Euclidean distance in $\mathbb{R}^m$ between two points $\mathbf{p} = (p_1, ..., p_m)$ and $\mathbf{q} = (q_1, ..., q_m)$ as $d(\mathbf{p}, \mathbf{q}) = \sqrt{\sum_{i=1}^{m}(p_i - q_i)^2}$. A ball in $\mathbb{R}^m$ with center $\mathbf{c}$ and radius r is defined as $B(\mathbf{c}, r) = \{\mathbf{p} \in \mathbb{R}^m : d(\mathbf{c}, \mathbf{p})) \leq r\}$. In this work $\mathbf{c}(B)$ and $r(B)$ are used to denote the center and radius of a ball. The μ-extension of $B(\mathbf{c}, r)$ is denoted as $\mu \cdot B$, which represents a ball centered at $\mathbf{c}$ with a radius of $\mu \cdot r$, i.e. $\mu \cdot B = B(\mathbf{c}, \mu \cdot r)$ [3,26].

Consider a set of n points $P = \{\mathbf{p_0}, ..., \mathbf{p_n}\} \subset \mathbb{R}^m$. The MEB of P, denoted as $MEB(P)$ is the smallest ball, which contains all points in P. Center and radius of $MEB(P)$ are represented by $\mathbf{c}^*(P)$ and $r^*(P)$. For a parameter $\mu > 1$, a ball B is a μ-approximate MEB of P, if $P \subset B$ and $r(B) \leq \mu \cdot r^*(P)$. A subset $S \subset P$ is a μ-coreset for $MEB(P)$, or $\mu - Coreset(P)$, if $P \subset \mu \cdot MEB(S)$. Since $S \subseteq P$ and $r^*(S) \leq r^*(P)$, $\mu \cdot MEB(S)$ is a μ-approximate MEB of P [26].

3.2 Coresets over Sliding Windows

The SWMEB+ [26] algorithm aims to match the rapidly changing characteristics and time constraints of non-stationary environments. SWMEB+ maintains a coreset for MEB over a fixed-size sliding window. It is able to return a $(9.66 + \epsilon)$-coreset of W_t at any t, however, in almost all cases, the approximation ratio of SWMEB+ improves to $3.36 + \epsilon$ [26]. The goal is to maintain a MEB, such that

$$r_t \leq r_M EB(W_t) \leq (1 + \epsilon)r_t \tag{1}$$

where r_t is the radius of all points in the window at time t. The SWMEB+ algorithm maintains a single sequence of s indices $X_t = \{x_1, ..., x_s\}$ over the sliding window W_t at timestep t. Each index x_i corresponds to an *append-only Minimum Enclosing Ball* (AOMEB) instance $A(x_i)$ that processes a substream of points from $\mathbf{p}_{x_i}$ to $\mathbf{p}_t$. $S[x_i, t]$ represents the coreset returned by $A(x_i)$ at timestep t and $B[x_i, t]$ centered at $\mathbf{c}[x_i, t]$ with radius $r[x_i, t]$ for $MEB(S[x_i, t])$.

SWMEB+ maintains the indices based on the radii of the MEBs. To also allow shrinkage of the MEB, the following technique is applied. For any $\epsilon_2 > 0$, for three neighboring indices x_i, x_{i+1}, x_{i+2}, if $r[x_i, t] \leq (1 + \epsilon_2) r[x_{i+2}, t]$ then x_{i+2} is considered a good approximation of x_i and x_{i+1} can be deleted. So the radii of MEBs gradually decrease from x_1 to x_s, with the ratios of any two neighbouring indices close to $(1+\epsilon_2)$. Any window starting between x_i and x_{i+1} is approximated by $A(x_{i+1})$. SWMEB+ keeps at most one expired index, which must be x_1 in X_t to track the upper bound for the radius $r^*(W_t)$ of $MEB(W_t)$. The AOMEB instance corresponding to the first non-expired index (x_1 or x_2 provides the coreset for $MEB(W_t)$ [26]. The number of indices in X_t is $\mathcal{O}(\frac{\log \theta}{\epsilon})$, where $\theta = \frac{d_{max}}{d_{min}}$ is the ratio of the maximum and minimum distances between any two points in the input dataset. The time complexity of SWMEB+ to update each point is $\mathcal{O}(\frac{m \log^2 \theta}{\epsilon^4})$ while the number of points stored by SWMEB+ is $\mathcal{O}(\frac{\log^2 \theta}{\epsilon^3})$, both are independent of n [26].

3.3   Core Vector Machine

The kernel method of the SVM problem can be reformulated to a MEB problem, as described in [24]. This leads to a transformed kernel $\tilde{k}$, with a corresponding feature space mapping $\tilde{\mathcal{F}}$, mapping $\tilde{\phi}$ and constant $\tilde{\kappa} = \tilde{k}(\mathbf{z}, \mathbf{z})$. For solving the kernelized MEB problem, the algorithm of [3] is slightly modified [24]. The goal is to minimize the following cost function:

$$\min_{R,c} R^2 : \|\mathbf{c} - \phi(\mathbf{x}_i)\|^2 \leq R^2, i = 1, ..., m \tag{2}$$

The corresponding dual of Eq. (2) is:

$$\max_{\alpha_i} \sum_{i=1}^{m} \alpha_i k(\mathbf{x}_i, \mathbf{x}_i) - \sum_{i,j=1}^{m} \alpha_i \alpha_j k(\mathbf{x}_i, \mathbf{x}_j)$$
$$s.t. \quad \alpha_i \geq 0, i = 1, ..., m; \quad \sum_{i=1}^{m} \alpha_i = 1 \tag{3}$$

where $\alpha = \alpha_i, ..., \alpha_m$ are the Lagrange multipliers, and $k(\mathbf{x}_i, \mathbf{x}_j)$ is the kernel function. In matrix form we obtain:

$$\max_{\alpha} \alpha' diag(\mathbf{K}) - \alpha' \mathbf{K} \alpha : \alpha \geq \mathbf{0}, \alpha' \mathbf{1} = 1 \tag{4}$$

with $\mathbf{K}$ as kernel matrix, $\mathbf{0}$ a vector where all entries are zero and $\mathbf{1}$ a vector where all entries are ones. Due to the fact, that the kernel is constant, and the alphas need to sum up to 1, we have $\alpha' diag(\mathbf{K}) = \kappa$. Dropping this constant leads to:

$$\max_{\alpha} -\alpha' \mathbf{K} \alpha : \alpha \geq 0, \alpha' \mathbf{1} = 1 \tag{5}$$

Whenever we have a constant kernel, any QP problem of the form in Eq. (5) can be viewed as a MEB problem. In [24] it is shown, that the SVM problem can be rewritten in this form. The ball is incrementally increased by including the point furthest away from the current center of the ball. We denote the coreset, the radius and center of the ball at iteration t as S_t, $\mathbf{c}_t$ and R_t. The center and the radius of a ball B are denoted as $\mathbf{c}_B$ and r_B. Choosing an $\epsilon > 0$, the CVM works as following:

1. Initialize S_0, $\mathbf{c}_0$ and R_0 using an arbitrary point $\mathbf{z} \in S$
2. If there is no training point $\mathbf{z}$ which is mapped by $\phi(\mathbf{z})$ outside of the $(1 + \epsilon)$-ball $B(\mathbf{c}_t, (1 + \epsilon)R_t)$, then terminate
3. Find $\mathbf{z}$, so that $\phi(\mathbf{z})$ is furthest away from $\mathbf{c}_t$. Update $S_{t+1} = S_t \cup \{\mathbf{z}\}$
4. Find the $MEB(S_{t+1})$ and set $\mathbf{c}_{t+1} = \mathbf{c}_{MEB(S_{t+1}}$ and $R_{t+1} = r_{MEB(S_{t+1}}$
5. Increment t and go to step 2

The points which are added to S are called core vectors. CVM provides convergence guarantees to an approximated optimality and has a space complexity of $\mathcal{O}(\epsilon^4)$ for a fixed ϵ. The time complexity is $\mathcal{O}(\frac{m}{\epsilon^2} + \frac{1}{\epsilon^4})$ and is linear in the training set size m for a fixed ϵ.

4   Adaptive Multiclass Core Vector Machine

4.1   Maintaining the MEB

As mentioned in Sect. 3.3, the challenge is to build up a kernelized MEB. We already have shown the technique to maintain a MEB over a fixed-size sliding window in the euclidean space in Sect. 3.2.

CVM only works in a kernelized space, thus we have to transfer the euclidean method in the *reproducing kernel Hilbert space* (RKHS). Instead of using euclidean distance to calculate the distance from center to x_i, the following equation is applied:

$$d(\mathbf{c}^*, \phi(\mathbf{q}))^2 = \sum_{i,j=1}^{n} \alpha_i \alpha_j k(\mathbf{c}_i^*, \mathbf{c}_j^*) + k(\mathbf{q}, \mathbf{q}) - 2 \sum_{i=1}^{n} \alpha_i k(\mathbf{c}_i^*, \mathbf{q}) \qquad (6)$$

where k is a symmetric positive definite kernel function $k(\cdot, \cdot) : \mathbb{R}^m \times \mathbb{R}^m \to \mathbb{R}$ and $\phi(\cdot)$ is its associated feature mapping, where $k(\mathbf{p}, \mathbf{q}) = \langle \phi(\mathbf{p}), \phi(\mathbf{q}) \rangle$ for any $\mathbf{p}, \mathbf{q} \in \mathbb{R}^m$. In case of symmetric non-positive definite kernel techniques from [20] can be used. The n-dimensional Lagrange multiplier vector is denoted as $\boldsymbol{\alpha} = [\alpha_1, ..., \alpha_n]'$. Theoretically, the generalized algorithm has the same approximation ratio and coreset sizes as its linear version. Only the time complexity increases by a factor of $\frac{1}{\epsilon}$, due to fact, that the time to compute the distance between $\mathbf{c}_{t-1}$ and $\phi(\mathbf{p}_t)$ via Eq. (6) is $\mathcal{O}(\frac{m}{\epsilon})$ instead of $\mathcal{O}(m)$ [26].

4.2   Adaptive Core Vector Machine

We now need to reformulate our SWMEB+ problem to a stream CVM problem. As the authors of [24] have evaluated, the feature mapping ϕ in Eq. (6) for a one class CVM needs to be adjusted by including the SVM penalty parameter C:

$$\tilde{\phi}(\mathbf{z}_i) = \begin{bmatrix} \phi(\mathbf{x}_i) \\ \frac{1}{\sqrt{C}}\mathbf{e}_i \end{bmatrix} \tag{7}$$

where $\mathbf{e}_i$ is the vector containing all zeros except that the i-th position is a 1 and C is a user-defined parameter. For the two-class CVM the kernel mapping ϕ in Eq. (6) includes y_i:

$$\tilde{\phi}(\mathbf{z}_i) = \begin{bmatrix} y_i\phi(\mathbf{x}_i) \\ y_i \\ \frac{1}{\sqrt{C}}\mathbf{e}_i \end{bmatrix} \tag{8}$$

Note, that the supervised SVM problem has been reformulated to an unsupervised one (MEB) by encoding the label information into the feature map $\tilde{\phi}$. For more details on converting the SVM into a MEB problem we refer to [24].

4.3   Multiclass Extension

To do multiclass classification, we solve a MEB problem, which is implicitly a *Multiclass Support Vector Machine* (MSVM) problem [23]. To do so, we rewrite the dual of the MSVM as proposed in [2] problem to

$$\min_{\alpha} \sum_{i,j=1}^{m} \alpha_i\alpha_j\tilde{k}(z_i, z_j)$$

$$s.t. \sum_{i=1}^{m} \alpha_i = 1 \quad \alpha_i \geq 0 \quad \forall i \tag{9}$$

where $\tilde{k}$ is the modified kernel function given by

$$\tilde{k} = (<y_i, y_j> k(x_i, x_j) + <y_i, y_j> +\delta_{i,j}vm \tag{10}$$

As far as $k(x, x) = \kappa$ is satisfied, the transformed kernel $\tilde{k}$ also satisfies $\tilde{k}(z, z) = \kappa_1$, where κ_1 is an arbitrary constant, δ is the Kronecker delta function, v is the SVM hyperparameter and m is the size of the dataset or in non-stationary settings the size of the sliding window. We use the one-hot-encoding for y_i whereas other label encodings are also possible. Note, that the kernel only works for multiclass problems [23]. Hence, the formulated problem is a MEB problem and can now be solved via our sliding window MEB algorithm (see Sect. 3.2)[1].

5   Experiments

In the following we provide experiments, which compare the adaptive MCCVM with the stream MEB classifier of [14]. Furthermore, we propose an offline changeover detection dataset for non-stationary classification.

---

[1] Code can be found at https://github.com/foxriver76/SW-MEB-Python.

5.1   Dataset: Changeover Detection

Changeover detection is a common problem in the modern industry, which aims to optimize downtimes. The authors of [17], have proposed a framework to distinguish changeover and production phases in brownfield environments on a milling machine (DMG 100 U duoBLOCK) of the company Pabst GmbH. Six different sensors are used to track different features in real-time. An overview of the features is given in Table 1. Note, that every sensor corresponds to one feature, except for the GPS sensor, which corresponds to two features (x-axis and y-axis).

Table 1. Overview of the sensor setup

Sensor type	Measuring object	Measuring type	Data type
Ifm 5D150	Door status tool holder	Distance measurement	Float
Keyence FD-Q Series	Coolant flow	Flow measurement	Float
Velleman HAA27	Door status machine main door	Contact measurement	Binary
Velleman HAA27	Door status second chamber	Contact measurement	Binary
Wago IoT-Box 9466	Machine power/performance	Power measurement	Float
Localino indoor tracking	Operator GPS data	GPS measurement	Float tuple

In [17] the data is stored in a SQL database and is then historically analyzed via batch techniques[2]. However, the scenario can also be seen as a streaming task, where you want to predict the state of a machine in near-real time. The dataset consists of 39,591 samples and contains one changeover phase. 13,818 samples are of label *Production* and 25,773 samples are of label *Changeover*. The changeover phases are then categorized in subcategories (6 labels and 21 labels).

5.2   Setup

In our experiments, we report accuracy as well as the runtime of each execution. To be not to prone to a single run of the algorithm, we use a 5-fold strategy, reporting mean as well as standard deviation.

We use real world data streams as well as synthetic data streams. The synthetic data streams are the SEA generator [22], the LED generator [7], the MIXED generator [11] and the AGRAWAL generator [1]. The LED generator contains 10 % noise while the other two generators are used twice, once with abrupt drift and once with gradual drift. All drifts appear at iteration 150,000 of 300,000. Gradual drift has a width of 60,000 samples. The AGRAWAL generator is used without noise. Before learning the data, we standardize the data via Z-transformation by maintaining the running mean and running variance

---

[2] The dataset can be found on https://github.com/ValdsteiN/OBerA-Enhanced-Changeover-Detection-in-Industry-4.0-environments-with-Machine-Learning.

over the data stream. Besides the synthetic streams, we use the following real world streams: OberA [17], Weather [10], Electricity [11], Give Me Some Credit³, Poker-Hand [6], Moving Squares [15] For a comprehensive data stream description we refer to [15]. Table 2 shows the number of labels for each data stream. In summary, we have 10 different streams, where 3 of them are multiclass problems. ARSLVQ is parameterized with $\sigma = 1$, $\gamma = 0.9$, Adadelta as gradient optimizer and one prototype per class. The MCCVM is used with $\nu = 0.05$ as a rule of thumb. The MEB classifiers use $\epsilon = 0.1$ as approximation ratio.

Table 2. Number of classes for each data stream

Stream	SEA	LED	MIXED	AGRAWAL	Weather	Electricity	GMSC	Poker	Mov. Squares	OberA
# Labels	2	10	2	2	2	2	2	10	4	2/6/21

5.3   Results

Our experiments are executed on two cores of an Intel Core i7-10710U processor with 4 GB RAM.

Table 3. Comparison of prototype based stream classifiers w.r.t. accuracy.

Algorithm	ARSLVQ	SWMEB	A(MC)CVM
LED	75.15 ± 0	61.01 ± 0	$\mathbf{98.95 \pm 0.03}$
Poker-Hand	66.73 ± 0	70.99 ± 0	$\mathbf{78.84 \pm 0}$
Moving Squares	13.34 ± 0	$\mathbf{99.84 \pm 0}$	98.43 ± 0
$MIXED_A$	88.01 ± 0	$\mathbf{90.27 \pm 0}$	$\mathbf{90.27 \pm 0}$
SEA_A	$\mathbf{98.29 \pm 11.1}$	85.67 ± 0	85.67 ± 0
$MIXED_G$	86.18 ± 0	$\mathbf{87.35 \pm 0}$	$\mathbf{87.35 \pm 0}$
SEA_G	$\mathbf{98.25 \pm 0}$	85.91 ± 11.1	85.91 ± 11.1
AGRAWAL	$\mathbf{50.53 \pm 0}$	49.9 ± 0	49.9 ± 0
Weather	$\mathbf{66.21 \pm 0}$	65.65 ± 0	65.65 ± 0
Electricity	$\mathbf{87.25 \pm 0}$	79.73 ± 0	79.73 ± 0
GMSC	$\mathbf{85.12 \pm 0}$	83.56 ± 0	83.56 ± 0
OberA$_{binary}$	89.13 ± 7.12	99.99 ± 0	$\mathbf{99.99 \pm 0}$
OberA$_{compressed}$	65.43 ± 0.07	38.59 ± 0	$\mathbf{94.74 \pm 0}$
OberA$_{high}$	34.82 ± 6.4	21.21 ± 0	$\mathbf{93.75 \pm 0}$
Mean	76.06 ± 1.76	72.83 ± 0.79	$\mathbf{89.79 \pm 0.81}$

---

³ https://www.kaggle.com/c/GiveMeSomeCredit.

Table 4. Comparison of prototype based stream classifiers w.r.t. runtime.

Algorithm	ARSLVQ	SWMEB	A(MC)CVM
LED	341.13 ± 1.03	**165.67 ± 0.48**	431.41 ± 1.12
Poker-Hand	**349.95 ± 1.61**	406.21 ± 0.92	659.42 ± 1.2
Moving Squares	**129.19 ± 0.27**	147.15 ± 0.17	308.3 ± 0.53
$MIXED_A$	**120.28 ± 0.72**	454.18 ± 0.69	454.18 ± 0.69
SEA_A	**115.35 ± 0.58**	277.92 ± 2.09	277.92 ± 2.09
$MIXED_G$	**117.2 ± 0.6**	450.75 ± 0.41	450.75 ± 0.41
SEA_G	**111.3 ± 0.4**	249.67 ± 0.33	249.67 ± 0.33
AGRAWAL	**146.79 ± 0.17**	160.73 ± 0.31	160.73 ± 0.31
Weather	**6.79 ± 0.06**	7.51 ± 0.02	7.51 ± 0.02
Electricity	**15.52 ± 0.05**	45.6 ± 0.09	45.6 ± 0.09
GMSC	**41.82 ± 0.07**	47.05 ± 0.15	47.05 ± 0.15
OberA$_{binary}$	**7.14 ± 0.63**	70.13 ± 1.42	70.13 ± 1.42
OberA$_{compressed}$	**13.4 ± 0.61**	74.76 ± 1.24	77.48 ± 1.57
OberA$_{high}$	**37.03 ± 1.87**	89.7 ± 1.27	83.7 ± 0.9
Mean	**110.92 ± 0.62**	189.07 ± 0.69	237.42 ± 0.77

Table 3 reports the accuracy on the previously described data streams. At first, we discuss the multiclass problems, namely *LED*, *Poker − Hand* and *MovingSquares*, *OberA$_{compressed}$* (6 phases problem) and *OberA$_{high}$* (21 phases problem). We can see, that the MCCVM performs well on the Poker-Hand data set. On the *MovingSquares* stream, the performance is also high in accuracy, however, this should be the case for window-based classifiers with small enough sliding windows [16]. Thus, also the SWMEB+ approach performs well. Furthermore, on the *LED* stream, the MCCVM is capable of distinguishing the 10 classes correctly in nearly all cases, which has not been the case for the other classifiers. On both multiclass OberA datasets, the MCCVM clearly outperforms the other two prototype based classifiers. In summary, on the multiclass problems, our proposed algorithm performs better than the other stream classifiers. Note, that on binary problems, the MCCVM cannot be used and the binary CVM implicitly works as the SWMEB+ classifier in such cases. Thus, the accuracy on the binary problem is equal on both classifiers. On the OberA data the CVM performed best, while the OberA data is advantageous to predict for window-based classifiers, due to the fact, that it only contains a few label transitions. Thus, windows will often contain only the present label. In Table 4 we present the runtime of our tested algorithms. We can see, that the SWMEB+ classifier is already slower than the ARSLVQ. While the SWMEB+ classifier can also be used in the euclidean space, which saves time to compute kernels, the multiclass CVM relies on kernels, which makes it slow. This has a negative impact on the runtime, which leads to slower execution on the multiclass streams. Additionally, to the more expensive kernel in the MCCVM due to the label product, we have observed, that the CVM algorithm does not converge as fast, when using the multiclass kernel, compared to the kernel without multiclass extension. Regarding runtime, the previously proposed SWMEB+ classifier is a faster MEB-based classifier for non-stationary environments, if the data is separable in the euclidean space. Especially, the runtime of kernelized classifiers is a downside in practical stream analysis scenarios. Note, that the MCCVM always needs to use a kernel, even if SWMEB+ classifier is used in the euclidean space. This leads to downsides for linear separable data.

6   Conclusion

In summary, we have shown, that the famous CVM approach can also be applied to multiclass problems in non-stationary environments, by converting it to a MEB problem and solving it in an online fashion by maintaining a MEB over a sliding window instead of the whole data set. Unfortunately, our experiments have shown, that while multiclass CVM approach is capable of distinguishing multiclass problems well, it is slower than other approaches, which does not make it the first choice for practical scenarios. In further research, computational speedup, by using a different solver than Frank-Wolfe algorithm should be considered.

Acknowledgement. We are thankful for support in the mFUND program of the BMVI, project FlowPro, grant number 19F2128B.

References

1. Agrawal, R., Imielinski, T., Swami, A.: Database mining: a performance perspective. IEEE Trans. Knowl. Data Eng. **5**, 914–925 (1993)
2. Asharaf, S., Murty, M.N., Shevade, S.K.: Multiclass core vector machine. In: Proceedings of the 24th International Conference on Machine Learning, ICML 2007, pp. 41–48. Association for Computing Machinery, USA (2007)
3. Bâdoiu, M., Clarkson, K.L.: Smaller core-sets for balls. In: Proceedings of the Fourteenth Annual ACM-SIAM Symposium on Discrete Algorithms, SODA 2003, pp. 801–802. Society for Industrial and Applied Mathematics, USA (2003)
4. Bifet, A., Gavaldà, R.: Adaptive learning from evolving data streams. In: Adams, N.M., Robardet, C., Siebes, A., Boulicaut, J.-F. (eds.) IDA 2009. LNCS, vol. 5772, pp. 249–260. Springer, Heidelberg (2009). https://doi.org/10.1007/978-3-642-03915-7_22
5. Bifet, A., Gavaldà, R., Holmes, G., Pfahringer, B.: Machine Learning for Data Streams with Practical Examples in MOA. MIT Press, Cambridge (2018)
6. Bifet, A., Pfahringer, B., Read, J., Holmes, G.: Efficient data stream classification via probabilistic adaptive windows. In: Proceedings of the 28th Annual ACM Symposium on Applied Computing, SAC 2013, pp. 801–806. ACM, USA (2013)
7. Breiman, L., Friedman, J., Stone, C.J., Olshen, R.A.: Classification and Regression Trees. CRC Press, Boca Raton (1984)
8. Bâdoiu, M., Clarkson, K.L.: Optimal core-sets for balls. Comput. Geom. **40**(1), 14–22 (2008)

9. Cohen, L., Avrahami-Bakish, G., Last, M., Kandel, A., Kipersztok, O.: Real-time data mining of non-stationary data streams from sensor networks. Inf. Fusion 9(3), 344–353 (2008)

10. Elwell, R., Polikar, R.: Incremental learning of concept drift in nonstationary environments. IEEE Trans. Neural Networks 22(10), 1517–1531 (2011)

11. Gama, J., Medas, P., Castillo, G., Rodrigues, P.: Learning with drift detection. In: Bazzan, A.L.C., Labidi, S. (eds.) SBIA 2004. LNCS (LNAI), vol. 3171, pp. 286–295. Springer, Heidelberg (2004). https://doi.org/10.1007/978-3-540-28645-5_29

12. Heusinger, M., Raab, C., Schleif, F.M.: Passive concept drift handling via variations of learning vector quantization. NCAA 1–12 (2020)

13. Heusinger, M., Schleif, F.: Reactive concept drift detection using coresets over sliding windows. In: 2020 IEEE Symposium Series on Computational Intelligence, SSCI 2020, Canberra, Australia, 1–4 December 2020, pp. 1350–1355. IEEE (2020)

14. Heusinger, M., Schleif, F.-M.: Classification in non-stationary environments using coresets over sliding windows. In: Rojas, I., Joya, G., Català, A. (eds.) IWANN 2021. LNCS, vol. 12861, pp. 126–137. Springer, Cham (2021). https://doi.org/10.1007/978-3-030-85030-2_11

15. Losing, V., Hammer, B., Wersing, H.: KNN classifier with self adjusting memory for heterogeneous concept drift. In: Proceedings of IEEE ICDM, pp. 291–300 (2017)

16. Losing, V., Hammer, B., Wersing, H.: Self-adjusting memory: how to deal with diverse drift types. In: Proceedings of IJCAI 2017, pp. 4899–4903 (2017)

17. Miller, E., Heusinger, M., Engelmann, B.: Enhanced changeover detection in industry 4.0 environments with machine learning. Sensors 21(17), 5896 (2021)

18. Nathan, V., Raghvendra, S.: Accurate streaming support vector machines. CoRR abs/1412.2485 (2014). https://arxiv.org/abs/1412.2485

19. Raab, C., Heusinger, M., Schleif, F.M.: Reactive soft prototype computing for concept drift streams. Neurocomputing 416, 340–351 (2020)

20. Schleif, F.M., Tino, P.: Indefinite proximity learning: a review. Neural Comput. 27(10), 2039–2096 (2015)

21. Straat, M., Abadi, F., Göpfert, C., Hammer, B., Biehl, M.: Statistical mechanics of on-line learning under concept drift. Entropy 20(10), 775 (2018)

22. Street, W.N., Kim, Y.: A streaming ensemble algorithm (SEA) for large-scale classification. In: Proceedings of the 7th ACM SIGKDD, KDD 2001, pp. 377–382. ACM (2001)

23. Szedmak, S., Shawe-Taylor, J.: Multiclass learning at one-class complexity. Project report (2005). https://eprints.soton.ac.uk/261157/

24. Tsang, I.W., Kwok, J.T., Cheung, P.M.: Core vector machines: fast SVM training on very large data sets. JMLR 6(13), 363–392 (2005)

25. Vapnik, V.: Statistical Learning Theory. Wiley, Hoboken (1998)

26. Wang, Y., Li, Y., Tan, K.L.: Coresets for minimum enclosing balls over sliding windows. In: Proceedings of the 25th ACM SIGKDD, KDD 2019, pp. 314–323. Association for Computing Machinery, USA (2019)

Identifying Cannabis Use Risk Through Social Media Based on Deep Learning Methods

Doaa Ibrahim[(✉)], Diana Inkpen, and Hussein Al Osman

Ottawa University, Ottawa, ON K1N 6N5, Canada
{dibra041,diana.inkpen,hussein.alosman}@uottawa.ca

Abstract. Cannabis is the most used drug around the world with the highest risks and associated criminal problems in many countries. This research describes the process of classifying online posts to identify cannabis use problems and their associated risks as early as possible. We annotated 11,008 online posts, which we used to build robust classification models. We tested classical and deep learning classifiers. Different CNN- and RNN-based models proved to be promising approaches to detect cannabis use posts. Our system can be used by authorities (such as parents or doctors) to monitor cannabis use-related posts. It could raise an alarm to the relevant authorities to take necessary interventions to analyze the cannabis use risks associated with the posts. To the best of our knowledge, this is the first study that uses deep learning methods successfully to detect cannabis use from any text or online posts. We tested our deep learning models on the SubUse-Cann unseen dataset which contains 17,099 tweets. It is an imbalanced dataset with only 6.9% positive cannabis use tweets. Our CNN-based model performed the best with an accuracy of 95.25% and an F1-score of 92.83% for classifying cannabis use.

Keywords: CNN · LSTM · Attention · Cannabis · Marijuana · Weed · Substance use · Classification · Risk behavior

1 Introduction

Cannabis is the most used drug worldwide. "It is the main drug that brings people into contact with the criminal justice system in 69 countries around the world"[1]. It is illegal in most countries and legal in a small number of them. Usually, countries that legalized cannabis consumption provide recommendations to limit its use. In Canada, the Lower-Risk Cannabis Use Guidelines contain recommendations to reduce consumption risk. Unsurprisingly, the first recommendation is "The most effective way to avoid the risks of cannabis use is to abstain

---

[1] UNODC World Drug Report 2020 https://www.unodc.org/unodc/press/releases/2020/June/media-advisory---global-launch-of-the-2020-world-drug-report.html.

Supported by the Natural Sciences and Engineering Research Council of Canada (NSERC).

from use" [7]. That is because consuming even limited amounts of cannabis over a prolonged period of time will produce adverse health outcomes. Another Canadian recommendation is to delay the start of cannabis consumption until at least after adolescence [7]. This was proven to reduce the likelihood or severity of the adverse effects. The Lower-Risk Cannabis Use Guidelines can only be effective if they are followed properly [7]. According to the same source, research has confirmed the relation between cannabis use and different psychological and behavioral problems [27]. To find the most influential factors from social media posts prediction, we must analyze a sufficient amount of information to build a strong classification model. In some cases, identifying the features that contribute to achieving an accurate model is a complex task [7,11].

As expected, most of the substance use posts on different social media platforms belong to the cannabis category, as it is the most popular drug among youth. The same result could be found when classifying the data collected by other researchers who tried to collect different substance use posts [11,15]. Many users on social media seem to share their cannabis use experiences and feelings about the substance. Also, they ask many questions regarding cannabis and its doses. Although cannabis is one of the most commonly used drugs worldwide, its consumption is often not as socially accepted as alcohol. Users ask all types of questions in their posts to obtain information about different issues related to cannabis use from other users who have better experience [15].

Our objective is to identify cannabis use risk through social media using the most effective deep neural architectures. We chose two of the most popular deep learning methods in the field of natural language processing: Convolutional Neural Network (CNN) and Recurrent Neural Network (RNN).

Our contributions can be summarized as follows:

- We investigated the performance of several classical and deep learning architectures commonly used in NLP tasks for detecting different behavioral problems from social media for cannabis use detection. To the best of our knowledge, this is the first time that cannabis use was successfully classified using deep learning methods. Our classification methods can be used to improve the classification of other types of substance use from unstructured text.
- To improve performance, we used effective generalization techniques after carefully tuning the hyperparameters given the limited size of our two unstructured data sets compared to the large datasets typically used to train most deep neural network architectures.

2   Related Work on Cannabis Use Detection

Considering the complexity of the task and the skills required for identifying the addictive cannabis use level and the necessary treatments, detecting cannabis use risk through social media using web mining techniques could be considered a preliminary step to generating awareness of cannabis addiction. Applying deep learning methods for the detection of any substance use and the associated risk is a relatively recent research field [9,11]. Although some studies succeeded in predicting alcohol use using deep learning methods, they failed to detect any other illicit drugs or cannabis (considered as one of the illicit drugs in many countries around the world, but not in all) [9].

One pilot study was done on Instagram posts by Hassanpour *et al* [9]. Unfortunately, their proposed model to predict substance use risk (other than alcohol) did not achieve statistically significant improvements over a baseline model [9]. The collected data did not have enough examples of high-risk use of tobacco, prescription drugs, or illicit drugs. The authors built a logistic regression model with semi-automatically extracted features from the texts and images of the dataset. They found that their deep learning model outperformed the logistic regression model only for the alcohol use risk detection [9]. The model extracted predictive features using CNN for image classification and long short-term memory (LSTM) for text classification. The result of this study showed that deep learning approaches applied to social media data can be used to identify substance use risk behavior [9]. The study analyzed data collected from surveys and compared it to data collected from Instagram posts. The surveys of 2,287 persons and their substance use behavior were used as a baseline for the analysis. Also, these surveys were used to assign a label of high or low risk to each substance use type. The Instagram data for those users were collected, with an average number of 183.5 posts per user. Due to memory requirements, a sample dataset of 20 of each image, captions, and comments for each user was created. The data set was divided into 80%, 10%, and 10% for training, validation, and test, respectively. The data sets were used to build different models to predict substance use risks for alcohol, tobacco, prescription drugs, and illicit drugs. A binary substance use risk of high or low levels was predicted from each model using the binary cross-entropy function. An unseen test set of 228 randomly selected users was used.

Jenhani *et al.* [12] found that using n-grams is not useful for detecting drug use, while other researchers found that it could be useful for a few conditions or to detect some related effects of the drug use [4]. The use of unigrams was effective in classifying drugs, medical conditions, and measuring units. The values of these outcomes were generally represented in the unigrams, while other types of outcomes such as the drug side effects or routes of intervention condition were more likely to be represented with n-grams [12]. This was found to be useful in the annotation process to help with the decision of using unigrams or n-grams. In general, the capacity of the automated annotator which was based on the Stanford Named Entity Recognition NER was inversely proportional to the value of n in the n-grams [12]. The Stanford model is one of the best NER models for Twitter data [6,17]. Jenhani *et al.* added an extension to it to build a drug domain NER which was used as the automatic annotator [12].

Most of the published papers in the field of individual substance use detection (such as cannabis use) relied on classical machine learning methods [12,18,23,26]. Conversely, the few published papers that used deep learning were not able to achieve any significant results except for alcohol use detection which is far away in nature from any illicit drug use detection problem [9].

3 Data

The Ensemble-2019 dataset consists of 11,008 tweets labeled as positive or negative substance use risk [11]. It consists of two batches of the data used by Hu *et al.* [11] (Note: We obtained permission to use the data but it is not collected by our team). We annotated the substance use tweets automatically comparing different cannabis keywords to label them as positive cannabis use tweets (tweets that indicate usage of cannabis) or negative cannabis tweets (tweets that indicate no sign of any kind of cannabis use). After the annotation, the Ensemble-2019 data consists of 5,504 (62%) positive and 3,372 (38%) negative cannabis use tweets. We created a balanced dataset that contains 50% of each class by randomly selecting (without replacement) 2,132 tweets of the negative cannabis use examples to be added to the original negative tweets. Eventually, we obtain a dataset that has 11,008 tweets with balanced positive and negative cannabis use tweets (5,504 tweets each). We call this set the Ensemble-Cann data.

The second dataset that we are using in this research is the SubUse-1.0 dataset. The SubUse-1.0 dataset was collected by our research team in 2018. The data was originally collected to cover seven categories (substance use, aggression, anxiety, depression, distress, sexuality, and violence). The team employed a supervised approach that uses a CNN to identify some active users on Twitter. From the active users, the team kept users who have at least 200 posts. Our group collected the posts by searching and using a well-prepared list of more than 300 convenient short phrases and words that are related to each category. We specified different Twitter hashtags (such as substanceuse, weed, marijuana, alcohol, cocaine,..) that were expected to strongly correlate with individual categories. These hashtags were used to search for Twitter posts and for the users that can be classified into the selected categories. After reviewing a user's recent tweets, if our annotators believed that the user might be classified in at least one of our categories, all of the user's tweets were downloaded using the Twitter API. The total number of tweets collected was 17,099. We re-annotated the SubUse-1.0 posts as either positive or negative cannabis use. We name the newly re-annotated set the SubUse-Cann dataset. The SubUse-Cann dataset has 6.9% positive cannabis use tweets and 93.1% negative cannabis use tweets. The SubUse-Cann data was used as a test set in our experiments. It has a total number of 17,099 tweets with 1,184 (6.9%) positive cannabis use tweets. We trained and evaluated the models using the Ensemble-Cann dataset. Then, the models were tested on the unseen SubUse-Cann data set to assess the generalization capacity of the proposed models (see Fig. 1).

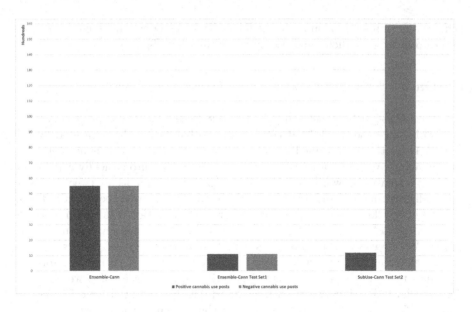

Fig. 1. Cannabis use risk data in the study.

4   Methodology

We used several text classification models to detect the risk of cannabis use in our datasets. We compared the performance of several classical and deep learning models and selected the highest performing one. We use the binary labels of {1,0} to differentiate between positive and negative posts of cannabis use. Positive posts are posts that contain the risk of cannabis use, while negative posts are posts that contain no risk of cannabis use. First, we used classical classifiers to distinguish the positive and negative tweets of cannabis use. Second, we employed deep learning classifiers to build the best possible models. The Ensemble-Cann training set was used for training the feature extraction to build the cannabis use risk identification model. The Ensemble-Cann validation set was used for the optimization of different models' hyperparameters. Finally, after training and tuning the models' hyperparameters, they were tested twice. First using the Ensemble-Cann test set then using the SubUse-Cann test set. Unlike the Ensemble-Cann test set, the SubUse-Cann test set does not follow the same distribution as the Ensemble-Cann training set. Several pre-trained word embeddings are implemented and made available by many researchers in the field. The GloVe word embedding has proven to be effective for various text processing applications [11,16]. GloVe counts accumulating word co-occurrences to produce a matrix. Then, the produced matrix is factorized to obtain lower-dimensional representations [22]. One of the commonly used GloVe word embeddings has been trained on a corpus of 6 billion tokens (Wikipedia 2014 + Gigaword 5) with a vocabulary of the top 400,000 most frequent words. It can convert the words into vectors of 50, 100, 200, or 300 dimensions [22]. We used the GloVe word embeddings as input for the deep learning methods and TF-IDF vectors for the classical methods. The GloVe vectors were shown to be efficient word embeddings for similar substance detection tasks [11,16].

For regularization, different dropout rates of 0.1 to 0.5 were applied. Also, the L2 regularization value of 0.0001 was applied at the loss function. It randomly drops neurons off the network using dropout to avoid overfitting [25].

4.1 Preprocessing

Data anonymization ensures that personal data is protected before sharing by researchers. In our case, the annotated posts have been anonymized using many tags. The tags were used to hide person names, phone numbers, places, URLs, and strings of digits mentioned in the posts. These tags were proposed to keep the confidentiality of the users' information and posts. We removed retweets, URLs, @mentions, and non-alphanumeric characters from both datasets. We also removed the stop words except for the first-, second-, and third-person pronouns. This is similar to the analysis of the expressions used in the posts for drug use and similar behavioral problems [19,21].

4.2 Word Tokenization

We used the Keras tokenizer to tokenize the posts. After tokenizing, we built a vocabulary of 8,066 unique tokens from the training Ensemble-19 dataset. This was used to encode each text as a sequence of indices. Our performance measure focuses on minimizing the loss error for the validation set. For the deep learning classifiers, we applied mini-batch gradient descent to improve the network loss function through backpropagation. We used the Adam optimizer [14] with a batch size of 30 to improve the loss function. Also, we created an embedding layer of size 300.

5   Models

We selected three classical machine learning and three deep learning algorithms to build our models. As mentioned above, we evaluated the performance of these models in classifying cannabis use tweets using two test sets.

5.1   Classical Classifiers

Multinomial Naive Bayes. This Naive Bayes (NB) algorithm is used for multinomially distributed data. It is suitable for analysis using word vectors counts. This classifier works well with TF-IDF word vectors. This is the main reason many researchers used it as the first attempt for classification. NB classifiers build basic models that can be considered as baselines, to give an idea of the difficulty of the task under study [10,11,24].

Random Forest. In a Random Forest (RF) classifier, each tree in the forest is built through bootstrap sampling. During the training phase, the best split of each node in a tree is decided using all the input features or a random subset of the features. Usually, Decision Trees (DT) have high variance and tend to overfit. However, the randomness in RF produces DTs with somewhat different prediction errors. After taking an average, some errors can cancel out. By placing all the DTs together, RF achieves a reduced variance. Eventually, variance reduction improves the classification model. The main parameters to adjust for these methods are n-estimators and max-features. The former indicates the number of trees in the forest. The larger the n-estimators values the better, but the longer it will take to compute. Moreover, results will cease to significantly improve beyond a critical number of trees. The latter hyperparameter (max-features) is the size of the random subsets of features to consider when splitting a node. The lower this number, the greater the reduction of variance, but also the greater the increase in bias[2]. As mentioned earlier, RF is fast to train but very slow for prediction on the test dataset. The denser the forest of the RF model, the more time it takes in the test stage [2].

Support Vector Machines. We built Support Vector Machines (SVM) classifiers with linear and radial basis function (RBF) kernels. As expected, the SVM classifier with an RBF kernel performed better than the classifier with a linear kernel. We tuned two parameters (C and Gamma) for both classifiers. C is called the regularization parameter. The strength of the regularization is inversely proportional to C. This parameter represents the penalty for misclassifying data. The C parameter trades off the correct classification of the training examples against the maximization of the decision function's margin. Gamma is a parameter of the RBF kernel and symbolizes the kernel spread that affects the decision region. The Gamma parameter defines how far the influence of a single training point reaches. The Gamma parameter can be seen as the inverse of the radius of points selected by the model as support vectors [20].

5.2   Deep Learning Classifiers

CNN-Based Models. After the optimization of the hyperparameters, we built the models on top of the GloVe word embeddings. For the CNN-GMax model, we used one convolutional layer followed by a global max pooling function which is known to achieve good results using the entire record representation length. Then, we added a fully connected dense layer that has 50 hidden units and uses a Rectified Linear Unit (ReLU) activation function. Finally, we applied a fully connected layer with one hidden unit. It uses a sigmoid activation to produce the binary output. For the CNN-DualChannel model, we applied two convolutional blocks each of which has 4 features and filters of lengths 5 and 3, respectively. Then, we applied a max-pooling layer to the feature map of the convolutional layer to extract more abstract information. Max-pooling extracts

---

[2] Scikit-learn 1.0.1 2021: https://scikit-learn.org/stable/supervised_learning.html.

the word features without considering the sequence order of the words [13]. Then, we applied a fully-connected layer with one hidden unit.

RNN-Based Models. RNN models are widely used in NLP as they have the advantage of recalling values over a certain time duration. RNN models maintain the sequence order. These models can suffer from exploding or vanishing the calculated weights for long sequences. This disadvantage can be mostly resolved using LSTM. LSTMs are RNN models that overcome this disadvantage using gating mechanisms. We used a bidirectional LSTM (with 100 units) instead of a vanilla unidirectional RNN which consists of two LSTMs. One LSTM processes the sequence forward from left to right, while the other processes the sequence backward from right to left. Getting more word-representative weights is the main advantage of using this bidirectional LSTM layer. For each word, its weight takes into consideration the context of the neighbors on both sides. The last layer is a dense layer that is connected to a single output node, using the sigmoid activation function [1].

6   Experiments and Results

As mentioned in the methodology section, we used the Ensemble-Cann set for the models' training. In the first group of experiments, we tested the models on the Ensemble-Cann test set, while in the second group we tested the models on the SubUse-Cann test set without re-training it. This allows us to evaluate the performance of our models on an unseen test dataset. The experiments used Python 3.6.9 and tools provided in the NLTK (v.3.2.5), Pandas (v.1.1.5), Gensim (v.3.6.0), and Keras (v.2.1.5) libraries, and the TensorFlow (v.2.4.1) platform to build the models. We extracted and used different features to create several models to classify the cannabis use posts. We calculated several evaluation measures for each model: accuracy, precision, recall, and F1-score. For the first experiments, we used accuracy as the main evaluation measure as we were dealing with a balanced binary dataset. Conversely, we used F1-score as the main evaluation measure for the second group of experiments as the SubUse-Cann dataset is highly imbalanced. The few studies that have been done on cannabis use have relied on classical classifiers to distinguish cannabis use posts from other posts [3,5,8]. However, to the best of our knowledge, no study successfully used deep learning methods for cannabis use classification [9]. Table 1 shows the results for cannabis use detection. Although the CNN-based models outperform the RNN-based models, their performance was very close to the RF model. Interestingly, the classical RF model performed higher than some deep learning models for cannabis use detection for the first group of experiments. The CNN-GMax model reached the highest performance of 86.29%. For the second group of experiments, Table 2 reports the generalization ability of our approach on the SubUse-Cann dataset. F1-score is an important evaluation measure in this case because the SubUse-Cann test set is a highly imbalanced set with only 6.9% positive cannabis use tweets. As in the first group of experiments, the CNN-GMax model reached the highest performance of 92.83% F1-score. The deep learning models outperformed the classical models. CNN-DualChannel and BiLSTM models reached a high F1-score of 87.71% and 87.34%, respectively. These are regularized models with a dropout ratio and L2 regularization of 0.2 and 0.0001, respectively. We use the Multinomial NB classifier with TF-IDF as a baseline for our binary classification task. The Multinomial NB models have the lowest performance among all other models of both groups.

Table 1. The performance results of the models using 5-fold cross-validation on Ensemble-Cann dataset.

The model	Accuracy	Precision	Recall	F1-score
NB	79.84	75.75	86.44	80.74
RF	86.19	81.44	90.77	83.73
SVM	80.38	80.33	79.29	79.81
CNN-GMax	**86.29**	91.56	78.92	84.18
CNN-DualChannel	84.22	88.01	78.58	83.02
BiLSTM	84.06	81.13	84.47	82.21

Table 2. The performance results of the models on SubUse-Cann dataset.

The Model	Accuracy	Precision	Recall	F1-score
NB	39.29	8.42	78.51	15.20
RF	89.74	35.52	59.04	44.35
SVM	76.20	17.28	64.36	27.25
CNN-GMax	95.25	88.99	97.64	**92.83**
CNN-DualChannel	87.25	87.09	89.02	87.71
BiLSTM	87.07	85.99	89.58	87.34

7 Comparison to Related Work

Given that very few studies have been done on identifying cannabis use through social media using classical classifiers, it is hard to compare our results with previous ones directly. Hassanpour et al. [9] have conducted the only study that used deep learning to identify cannabis use from Instagram posts. As they mentioned, their model failed to detect cannabis use from Instagram posts even though it is one of the most commonly used and discussed drugs on social media. Despite their model's ability to identify alcohol use risk, they specified that "results from the model for identifying tobacco, prescription drug, and illicit drug risk were not statistically significant" [9]. This could be due to the difficult task of annotating tweets of cannabis use. We overcame this obstacle by using automatic annotation with a big variety of cannabis slang terms to differentiate cannabis use posts from other substance use posts (that were annotated by two independently working annotators). Also, we checked part of the annotated posts manually to evaluate the validity of the annotation process. Hu *et al.* [11] reported high accuracy results for the Ensemble-2019 dataset. However, their results are not directly comparable to ours, as they were detecting substance use in general (they did not differentiate between alcohol, cannabis, or any other substance). Our task is more challenging as we have fewer positive cases after annotating the same data for cannabis use only. Surprisingly, we reported better results on cannabis use identification only from the same data. Hence, our models may perform better if used for identifying all substance use in the future. Our models, when testing on the Ensemble-2019 dataset, obtained a high accuracy of 86.29% using CNN-based classifier and 86.19% using the RF classifier as compared to 85.75% using the CNN-based classifier and 85.86% using RF classifier in the case of Hu *et al.* [11] results. The performance of the RF classifier is comparable to that of the deep learning classifiers using the Ensemble-2019 dataset. However, in the case of the SubUse-Cann unseen data set, it was clear that the deep learning methods outperformed classical machine learning methods. Unlike the Ensemble-2019 dataset and its cannabis use version, the SubUse-Cann dataset is highly imbalanced. This could explain why the classical classifiers did well on the analysis of the Ensemble-Cann dataset rather than the SubUse-Cann dataset. A similar result was found by Hu *et al.* [11] for the general substance use identification. In their study, they found that when the dataset is balanced, the classical machine learning model performed better than their proposed ensemble deep learning model [11].

8   Conclusion and Future Work

We have successfully implemented automated models for cannabis use detection through social media. Our best deep learning model can be used to build a system to monitor at-risk posts. A system based on this model could raise an alarm to the relevant individuals with authority to make the necessary interventions by analyzing the risks associated with the predicted posts. Based on our results of the classical and deep learning models, it was clear that social media data can be used to effectively identify cannabis use posts. The RF classical model using TF-IDF embedding performed well in the case of the balanced dataset, while the CNN-based deep learning models using GloVe embeddings performed better in the case of the imbalanced dataset. The regularization techniques helped us to find the best-generalized model for both cases. In general, the CNN-based models performed better than the RNN-based model and the classical models, including the NB baseline model. To the best of our knowledge, this is the first time that cannabis use (or any individual illicit drug use) was successfully detected by deep learning methods. In future work, we could use our deep learning model to identify other important substance use risks from social media, such as cocaine use risk.

Acknowledgment. This research was funded by the Natural Sciences and Engineering Research Council of Canada (NSERC), the Ontario Centres of Excellence (OCE), and SafeToNet.

References

1. Bahdanau, D., Cho, K., Bengio, Y.: Neural machine translation by jointly learning to align and translate. arXiv preprint arXiv:1409.0473 (2014)
2. Bansal, D., Chhikara, R., Khanna, K., Gupta, P.: Comparative analysis of various machine learning algorithms for detecting dementia. Procedia Comput. Sci. **132**, 1497–1502 (2018).. https://doi.org/10.1016/j.procs.2018.05.102
3. Bergman, B.G., Dumas, T.M., Maxwell-Smith, M.A., Davis, J.P.: Instagram participation and substance use among emerging adults: the potential perils of peer belonging. Cyberpsychol., Behav. Soc. Netw. **21**(12), 753–760 (2018)
4. Çöltekin, Ç., Rama, T.: Drug-use identification from tweets with word and character n-grams. In: Proceedings of the EMNLP Workshop SMM4H: The 3rd Social Media Mining for Health Applications Workshop & Shared Task, pp. 52–53 (2018)
5. Cox, M.J., Janssen, T., Gabrielli, J., Jackson, K.M.: Profiles of parenting in the digital age: associations with adolescent alcohol and marijuana use. J. Stud. Alcohol Drugs **82**(4), 460–469 (2021)
6. Derczynski, L., et al.: Analysis of named entity recognition and linking for tweets. Inf. Process. Manag. **51**(2), 32–49 (2015)
7. Fischer, B., Russell, C., Sabioni, P., Van Den Brink, W., Le Foll, B., Hall, W., Rehm, J., Room, R.: Lower-risk cannabis use guidelines: a comprehensive update of evidence and recommendations. Am. j. Public Health **107**(8), e1–e12 (2017)
8. George, M.J., Ehrenreich, S.E., Burnell, K., Kurup, A., Vollet, J.W., Underwood, M.K.: Emerging adults' public and private discussions of substance use on social media. Emerg. Adulthood **9**(4), 408–414 (2021)
9. Hassanpour, S., Tomita, N., DeLise, T., Crosier, B., Marsch, L.A.: Identifying substance use risk based on deep neural networks and instagram social media data. Neuropsychopharmacology **44**(3), 487–494 (2019)
10. Hu, H., et al.: An insight analysis and detection of drug-abuse risk behavior on Twitter with self-taught deep learning. Comput. Soc. Netw. **6**(1), 1–19 (2019)
11. Hu, H., et al.: An ensemble deep learning model for drug abuse detection in sparse twitter-sphere. In: MedInfo, pp. 163–167 (2019)
12. Jenhani, F., Gouider, M.S., Said, L.B.: Lexicon-based system for drug abuse entity extraction from twitter. In: BDAS, pp. 692–703 (2016)
13. Kalchbrenner, N., Grefenstette, E., Blunsom, P.: A convolutional neural network for modelling sentences. arXiv preprint arXiv:1404.2188 (2014)
14. Kingma, D.P., Ba, J.: Adam: a method for stochastic optimization. arXiv preprint arXiv:1412.6980 (2014)
15. Koratana, A., Dredze, M., Chisolm, M.S., Johnson, M.W., Paul, M.J.: Studying anonymous health issues and substance use on college campuses with YIK yak. In: AAAI Workshop: WWW and Population Health Intelligence (2016)
16. Mahata, D., Friedrichs, J., Shah, R.R., et al.: # phramacovigilance-exploring deep learning techniques for identifying mentions of medication intake from twitter. arXiv preprint arXiv:1805.06375 (2018)

17. Manning, C.D., Surdeanu, M., Bauer, J., Finkel, J.R., Bethard, S., McClosky, D.: The stanford corenlp natural language processing toolkit. In: Proceedings of 52nd annual meeting of the association for computational linguistics: system demonstrations, pp. 55–60 (2014)

18. Menon, A., Farmer, F., Whalen, T., Hua, B., Najib, K., Gerber, M.: Automatic identification of alcohol-related promotions on twitter and prediction of promotion spread. In: 2014 Systems and Information Engineering Design Symposium (SIEDS), pp. 233–238. IEEE (2014)

19. Orabi, A.H., Buddhitha, P., Orabi, M.H., Inkpen, D.: Deep learning for depression detection of twitter users. In: Proceedings of the Fifth Workshop on Computational Linguistics and Clinical Psychology: From Keyboard to Clinic, pp. 88–97 (2018)

20. Pedregosa, F., et al.: Scikit-learn: machine learning in python. J. Mach. Learn. Res. **12**, 2825–2830 (2011)

21. Pennebaker, J.W., Chung, C.K.: Expressive writing: connections to physical and mental health. In: Friedman, H.S. (ed.) The Oxford Handbook Of Health Psychology, pp. 417–437. Oxford University Press (2011)

22. Pennington, J., Socher, R., Manning, C.: Glove: global vectors for word representation. In: Proceedings of the 2014 Conference on Empirical Methods in Natural Language Processing (EMNLP), pp. 1532–1543. Association for Computational Linguistics, Doha, Qatar (October2014). https://doi.org/10.3115/v1/D14-1162,https://www.aclweb.org/anthology/D14-1162

23. Raja, B.S., Ali, A., Ahmed, M., Khan, A., Malik, A.P.: Semantics enabled role based sentiment analysis for drug abuse on social media: a framework. In: 2016 IEEE Symposium on Computer Applications & Industrial Electronics (ISCAIE), pp. 206–211. IEEE (2016)

24. Sarker, A., O'Connor, K., Ginn, R., Scotch, M., Smith, K., Malone, D., Gonzalez, G.: Social media mining for toxicovigilance: automatic monitoring of prescription medication abuse from twitter. Drug Safety **39**(3), 231–240 (2016)

25. Srivastava, N., Hinton, G., Krizhevsky, A., Sutskever, I., Salakhutdinov, R.: Dropout: a simple way to prevent neural networks from overfitting. J. Mach. Learn. Rese. **15**(1), 1929–1958 (2014)

26. Vázquez, A.L., et al.: Innovative identification of substance use predictors: machine learning in a national sample of Mexican children. Prevent. Sci. **21**(2), 171–181 (2020)

27. Yadav, S., et al.: "When they say weed causes depression, but it's your fav antidepressant": knowledge-aware attention framework for relationship extraction. PLoS one **16**(3), e0248299 (2021)

On a Combination of Clustering Methods and Isolation Forest

Michał Koziara[ID] and Paweł Karczmarek[(✉)][ID]

Department of Computer Science, Lublin University of Technology,
Nadbystrzycka 36B, 20-618 Lublin, Poland
michal.koziara@pollub.edu.pl, p.karczmarek@pollub.pl

Abstract. This study provides a comparison of the efficiency of anomaly detection in data using Isolation Forest (IF) combined with k-Means and Fuzzy C-Means algorithms. It also presents how to determine the anomaly score from the clustering results using the triangular and Gaussian membership functions. The number of clusters, the significance of the anomaly score obtained from the clustering process, and the degree of fuzziness of the clusters are additionally taken into account when testing the efficiency of anomaly detection. Moreover, we demonstrate that in most of the examined datasets, preceding IF with clustering algorithms allows obtaining significantly better results. Furthermore, combining IF with Fuzzy C-Means produces better results than combining it with k-Means. The results discussed in this paper allow one to decide which clustering method to use when combining it with IF to detect anomalies in the data. In addition, a comprehensive analysis presented in the paper sheds the light on the procedure of a choice of the parameters of the algorithms to get possibly the best results.

Keywords: Anomaly detection · Isolation forest · Clustering methods · k-Means · Fuzzy C-Means

1 Introduction

When analyzing data, it is often required to identify which observations differ from all others. These observations are known as anomalies or outliers. Anomalies can be caused by errors in the data or indicate a previously unknown process of the system under study [4]. Thus, the purpose of anomaly detection is to point out erroneous data and to reveal data resulting from processes other than those originally assumed. Obtaining this information makes it possible to fix existing errors and to describe new patterns within the system. Anomaly detection is an important problem strictly related to many areas of life. Example use cases include, but are not limited to, fraud detection, intrusion detection, event detection in sensor networks, medical anomaly detection, ecosystem disruption detection, or data cleaning.

L. Rutkowski et al. (Eds.): ICAISC 2022, LNAI 13589, pp. 114–126, 2023.
https://doi.org/10.1007/978-3-031-23480-4_10

The wide range of applications and high demand for anomaly detection have resulted in many studies on this topic. Comprehensive surveys of various approaches can be found, among others, in [1,2,12]. Most of the methods were originally designed for purposes other than anomaly detection and are based on profiling normal observations rather than directly detecting anomalies. One of the latest anomaly detection algorithms that eliminates this drawback is Isolation Forest (IF) [8,9]. This algorithm also contains many effective extensions such as Functional IF [15], CBiForest [10], k-Means-Based IF [6], Fuzzy Set-Based IF [5], and Fuzzy C-Means-Based IF [7]. However, currently known extensions of the IF method that use clustering algorithms along with the IF algorithm to detect anomalies are mostly modifications of this technique and prevent the use of proven implementations of these algorithms. The CBiForest method effectively combines k-Means and IF without modifying them. But clustering is only used to initially divide the dataset into two clusters and determine the anomaly score in the smaller cluster. Therefore, the smaller cluster is previously labeled as a group of observations containing anomalous data. For each observation located in it, the anomaly score is equal to the distance between it and the center of the larger cluster. This score represents only the distance and thus has no upper bound. The lack of an upper bound does not allow for the selection of an intuitive classification threshold from which the anomalies are determined. In the second step, an additional anomaly score is determined for all observations in the dataset by applying IF. Finally, the anomalies are determined based on both anomaly scores [10]. Observations are therefore labeled as anomalous, even if their abnormality is indicated by only one of two anomaly scores.

The main objective of this study is to propose two novel anomaly detection methods that effectively combine Isolation Forest with clustering algorithms. The first method combines IF along with k-Means. While the second method combines IF with Fuzzy C-Means. We also focus on the comparison of the efficiency of combined methods and IF without additional usage of clustering methods for various datasets. It is worth noting that our goal is not to internally change either IF or clustering methods, but only to combine them. This allows one to benefit from the use of both methods and use proven implementation of these algorithms. To achieve this, we are also interested in obtaining a single anomaly score for each observation, based on the results of the clustering methods and the IF method. Moreover, we aim to investigate the influence of clustering methods parameters such as the number of clusters or the degree of fuzziness of the clusters on the efficiency of anomaly detection in the proposed approaches.

The paper is organized as follows. In Sect. 2, we briefly recall the main properties of the IF. The details of a proposed approach are discussed in Sect. 3. The results of experiments are presented in Sect. 4, while conclusions and future work directions are detailed in Sect. 5.

2   Isolation Forest

In this section, we recall Isolation Forest [8,9]. The IF method isolates anomalies, instead of creating a profile of normal observations as density-based methods. Isolation in this case stands for separating abnormal observations from all other observations. To achieve this, the method takes advantage of the fact that anomalies are few and different, making them more susceptible to isolation than normal observations. The IF method consists of two stages. The first stage is the learning stage and the second stage is the prediction stage.

During the first stage, isolation trees are created based on samples from the dataset. Then during the second stage, an anomaly score is obtained for each observation using the isolation trees. To isolate each individual anomaly, an isolation tree structure is used. The isolation tree is built by recursively splitting a random sample from the dataset. The division of the sample is done based on a randomly selected value of one of the randomly selected features of the sample. The split values and the features are stored in nodes of the tree for use during the prediction stage. The sample is divided into two parts by comparing the selected feature value with the sample values. The left and right branches of the tree are created from these two samples. Further nodes are determined as long as the division of the sample is possible and the tree height has not reached the limit. Due to the susceptibility to isolation, anomalies are located closer to the root of the tree. Normal observations, on the other hand, are located at the further ends of the tree. The isolation of observations within the tree is what forms the basis of the algorithm. Based on the distance of the observations from the root of the tree, it is possible to determine an anomaly score that evaluates the degree to which a given observation is anomalous. Finally, the anomaly score for an observation x is calculated as $s(x, n) = 2^{-\frac{E(x)}{c(n)}}$, where $c(n) = 2H(n - 1) - (2(n - 1)/n)$ is the average path length of unsuccessful search in the binary search tree and H is the harmonic number [3], n is the number of observations in the sample, and $E(x)$ is the average length of the path that the observation x traversed from the root node to the external node.

3   Proposed Approach

The IF method is combined with k-Means and Fuzzy C-Means clustering algorithms in order to combine the advantages of both approaches and obtain methods that will allow more efficient anomaly detection. The k-Means algorithm only allows assigning observations to appropriate clusters and determining their centroids. Therefore, it is necessary to create an anomaly score that is obtained solely from these results. We propose two ways to determine the anomaly score using the k-Means algorithm. The first one determines the anomaly score c based on the degree of membership defined by a triangular function with a vertex at the centroid and side vertices at an equal distance to the cluster center, equal to the distance of the outermost point from the cluster. It is thus represented by the following formula: $c = \frac{d}{m}$, where d is the Euclidean distance between a given data point and the cluster center, and m is the Euclidean distance between the outermost point and the cluster center. To demonstrate the idea, we illustrate an example of the anomaly score utilizing the triangular function in Fig. 1. The dots indicate ten data points arranged randomly in a one-dimensional space. We assume that all points belong to one cluster. The centroid of the cluster and the anomaly score function are determined from the data points. A square indicates an outermost observation. The cross indicates the centroid of the cluster. Note that at the center of the cluster the anomaly score is 0, and at the data point farthest from the centroid the anomaly score is equal to 1. As the distance from the centroid increases, the observation anomaly score also increases. The second way is to determine the anomaly score based on the degree of membership defined by the Gaussian function. The base Gaussian function is represented as follows: $f(x) = ae^{-\frac{1}{2}\left(\frac{x-\mu}{\sigma}\right)^2}$, where $a = \frac{1}{\sigma\sqrt{2\pi}}$ corresponds to the height of the curve, μ is the expected value and corresponds to the location of the center of the curve, σ is the standard deviation, and e is the Euler number. Using the three-sigma rule, we assume that 99.7% of the observations are between $[\mu - 3\sigma, \mu + 3\sigma]$. To determine the anomaly score, we use a Gaussian function with a vertex at the centroid and side vertices at an equal distance from the cluster center, equal to the distance of the outermost point from the cluster center. Thus, assuming that $3\sigma = m$, where m is the Euclidean distance of the outermost point from the cluster center, we obtain $\sigma = \frac{m}{3}$. We additionally assume that $a = 1$ in order for the anomaly score function to return values ranging from 0 to 1. In addition, we assume the location of the center of the curve at the cluster center, hence $\mu = 0$. Finally, the anomaly score function based on the Gaussian function can be described by the following formula: $c = 1 - e^{-\frac{1}{2}\left(\frac{3d}{m}\right)^2}$, where d is the Euclidean distance between a given data point and the cluster center, m is the Euclidean distance between the outermost point and the cluster center, and e is the Euler number. As in the previous case, to show the idea, we illustrate an example of the anomaly score function obtained with the Gaussian function in Fig. 1. Again, the dots indicate ten data points arranged randomly in a one-dimensional space. We assume that all points belong to one cluster. The centroid of the cluster and the anomaly score function were determined from the data points. A square indicates an outermost observation. The cross indicates the centroid of the cluster. It is worth noting that, similarly to the triangular function-based approach, the anomaly score increases as the distance to the centroid increases. However, in the case of the Gaussian function, the increase in the score is not uniform. Additionally, at the center of the cluster, the anomaly score is 0 and at the data point furthest from the center, the anomaly score is close to 1. The output of the Fuzzy C-Means algorithm is the degree of membership the data points have to each cluster. As in the case of k-Means, we need to determine the anomaly score using the results of clustering. Therefore, the anomaly score determined by Fuzzy C-Means is based on the degree of membership. It is represented by the following formula: $c = 1 - m$, where m is the degree of membership the data point has to cluster to which it belongs the most. It should be noted that the resulting anomaly score is intuitively appealing. The smaller the degree to

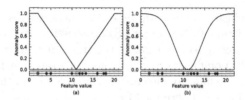

Fig. 1. Example of the anomaly score function determined by (a) the triangular function depending on the feature value and (b) the Gaussian function depending on the feature value

which an observation belongs to a cluster, the more abnormal the observation is. To join the clustering methods and the IF method, we combine the anomaly score obtained by the clustering methods and the anomaly score obtained by the IF method into a single anomaly score. The common anomaly score is determined by the following formula: $s = wc + (1 - w)f$, where w is the weight of the anomaly score obtained by clustering in the common score, c is the anomaly score obtained by clustering, and f is the anomaly score obtained by the IF method. The weight w allows one to adjust the significance of the anomaly score obtained by clustering in the common anomaly score. The smaller the weight w is, the less significant the anomaly score derived from clustering is, and thus the more significant is the anomaly score derived from the IF method. Moreover, it is obvious that the common anomaly score s satisfies $0 \leq s \leq 1$ condition. Furthermore, using the anomaly score obtained in this manner, it is possible to make an intuitive classification. If the anomaly score has a value close to 1, it means that the observation is almost certainly an anomaly. On the other hand, if the anomaly score has a value close to 0, it means that the observation is almost certainly normal. If the anomaly score is close to 0.5, it is not possible to unambiguously determine whether the observation is normal or abnormal.

4   Experimental Results

This section presents the results of numerical experiments designed to evaluate the anomaly detection effectiveness of the IF method and proposed combined methods. The conducted experiments also include the analysis of factors affecting the anomaly detection effectiveness such as the selected number of clusters, the weight of the anomaly score obtained from the clustering process, and the degree of fuzziness of the clusters.

4.1   Datasets and Measure Methods

We use seven datasets in the series of numerical experiments. Among them, there are six publicly available datasets containing real-world data from different domains and one artificial dataset that was generated for this study. The datasets containing real-world data are Annthyroid, Ionosphere, Mammography, Pima, Satellite, Smtp (KDDCUP99). These datasets are coming from the

Outlier Detection DataSets (ODDS) [13]. The selected data sets have been previously used to evaluate the anomaly detection methods [7,8]. An artificial two-dimensional dataset was created by randomly placing 20000 points inside four geometric figures, while the remaining 1050 points identified as anomalies were randomly placed outside these figures. A uniform distribution was used to randomize the positions of the points. Table 1 contains the details of all datasets sorted by the number of observations in descending order. In addition, all features of the datasets were normalized so that they equally affected the results of the clustering methods. In the case where the features of the sets had different ranges of values, the features with a larger range of values would have a greater effect on the clustering results. Min-max normalization was used to scale the values to a range from 0 to 1. Min-max normalization, unlike Z-score normalization, does not use standard deviation, so it does not eliminate the variance between clusters and thus does not significantly affect outliers, which in turn can affect the performance of anomaly detection [11]. Three quantitative measures are used to evaluate the performance of anomaly detection methods. The first one is, commonly used, accuracy. The second measure is the area under the receiver operating characteristics (ROC) curve, also known as AUROC. And, the third measure is the area under the Precision-Recall curve (AUPRC). To calculate the accuracy of anomaly detection, we set the decision threshold, which indicates observations as anomalous, to 0.5. Compared to the accuracy, the other two measures are independent of the decision threshold and provide the evaluation of predictors for all possible decision thresholds. Furthermore, we also use the AUPRC measure because AUROC may produce overly optimistic scores for highly imbalanced datasets. The parameter values of IF were determined based on the suggestions of the method authors. Thy claim that IF obtains the best results with the number of isolation trees equal to 100 and the sample size equal to 256 [8]. Moreover, in order to verify the obtained results, all experiments were performed 100 times each and the obtained results were averaged.

4.2 Analysis of Selected Parameters

First, we determine the optimal number of clusters in each dataset based on the silhouette coefficient. The silhouette coefficient measures the cohesion of clusters in relation to their separation [14]. Therefore, it allows to validate the cluster analysis results. The selected number of clusters in the dataset is considered optimal when the silhouette coefficient takes the largest value for it. For each dataset, we perform cluster analysis using the k-Means algorithm with the selected number of clusters ranging from 1 to 10. The optimal number of clusters in each dataset and the corresponding silhouette coefficient are presented in Table 2. It should be noted that there are no datasets that have less than 2 clusters and more than 4 clusters. We aim to check if the selected number of clusters during the clustering process affects the anomaly detection performance. Three proposed anomaly detection methods that combine the IF with clustering algorithms are considered in the experiment. The AUROC measure depending on the number of clusters in the artificially generated dataset is shown in Fig. 2.

Table 1. Properties of the datasets used in the experiments

Dataset	Instances	Attributes	Anomalies
Smtp (KDDCUP99)	95156	3	30 (0.03%)
Artificially generated	21050	2	1050 (4.98%)
Mammography	11183	6	260 (2.32%)
Annthyroid	7200	6	534 (7.42%)
Satellite	6435	36	2036 (32%)
Pima	768	8	268 (35%)
Ionosphere	351	33	126 (36%)

Table 2. Determined number of clusters

Dataset	Number of clusters	Silhouette coefficent
Smtp (KDDCUP99)	2	0.75
Artificially generated	4	0.68
Mammography	3	0.78
Annthyroid	2	0.44
Satellite	3	0.43
Pima	2	0.26
Ionosphere	4	0.30

Note that for all three methods, we obtain the best results for the number of clusters equal to 4. This number of clusters is also the optimal number of clusters in this dataset determined by the silhouette coefficient. In addition, notice that for Fuzzy C-Means the highest level results, where the AUROC is greater than or equal to 0.97, are obtained for any number of clusters. On the other hand, for k-Means using the triangular function and the Gaussian function top-level results are obtained only from the number of clusters equal to 4. Thus, it should be noted that the best anomaly detection results can be obtained when the optimal number of clusters is set as the selected number of clusters during the clustering

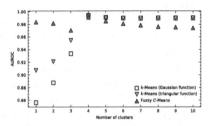

Fig. 2. AUROC depending on the number of clusters in the artificially generated dataset

process. In the further analysis for each method, the optimal number of clusters is used as the selected number of clusters, according to Table 2. In addition, the common anomaly score determined by proposed combined methods depends on the weight of the anomaly score obtained by the clustering methods, see above. We examine whether this weight affects the anomaly detection performance. The AUROC measure depending on the weight of the anomaly score obtained by the clustering in all datasets is shown in Fig. 3. It is noteworthy that in 5 out of 7 datasets, regardless of the clustering algorithm that was combined with the IF method, there is an initial increase in AUROC followed by a decrease. Furthermore, in most datasets, the best results are obtained when the weight is around 0.2. It should be emphasized that in the case of Fuzzy C-Means, the results decrease significantly as the weight approaches a value equal to 1 in most of the cases. The Fuzzy C-Means method allows to parametrize the degree of fuzziness of the prototypes. The effect of this degree on the effectiveness of anomaly detection using IF in combination with Fuzzy C-Means is also examined. This parameter is analyzed for values between 1.1 and 4. The relationship between AUROC and the degree of fuzziness in all datasets is presented in Fig. 4. In most cases, this parameter is assumed to be equal to 2. However, it is worth noting that in 6 out of 7 studied datasets, AUROC reaches the highest value for the degree of fuzziness smaller than 2. In addition, in most datasets, the best results are obtained when the degree of fuzziness is around 1.4. Thus, the proper selection of this parameter allows one to increase the effectiveness of anomaly detection using IF in combination with Fuzzy C-Means.

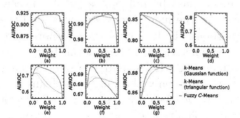

Fig. 3. AUROC depending on the weight of the anomaly score obtained by the clustering methods in the dataset: (a) Smtp, (b) artificially generated, (c) Mammography, (d) Annthyroid, (e) Satellite, (f) Pima, and (g) Ionosphere

4.3 Efficiency Analysis

Within the artificially generated dataset, the distribution of anomaly scores for each observation determined by IF and proposed methods is further investigated. In addition to the results of methods combining the IF with clustering, we also compare the anomaly detection results obtained solely by clustering methods with scoring functions such as the proposed triangular and Gaussian functions. In order to better represent the distribution of anomaly scores, the resulting range of anomaly score values in the artificial dataset is normalized for each

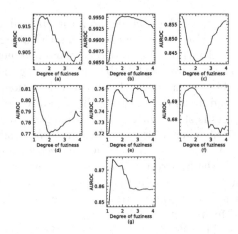

Fig. 4. AUROC depending on the degree of fuzziness of the prototypes in the dataset: (a) Smtp, (b) artificially generated, (c) Mammography, (d) Annthyroid, (e) Satellite, (f) Pima, and (g) Ionosphere

method. Min-max normalization is used to scale the value ranges. By applying the normalization, all ranges of anomaly scores are scaled to range from 0 to 1. The normalized anomaly scores for each of the observations in the artificially generated dataset are presented in six two-dimensional planes in Fig. 5. Each observation is marked with a dot. The color of the dot corresponds to the value of the anomaly score obtained by different anomaly detection methods. The more purple the dot color, the more anomalous it is, and the more orange the dot color, the more normal it is. It is important to note that for methods that combine clustering algorithms with IF, observations between squares are marked as more anomalous than when IF is used without the supporting clustering algorithms. In contrast, using only the Fuzzy C-Means method, without IF, results in only observations between squares being marked as anomalous. Paying additional attention to anomaly detection based solely on k-Means clustering, with anomaly scores determined by the Gaussian function, one can see that this method marks more points between squares as anomalous than the same method with anomaly scores determined by the triangular function. However, the method that uses the Gaussian function also labels observations at the edge of the figures as more anomalous. Finally, a performance comparison of all tested methods on all datasets is performed. We assume that the best results are obtained when the selected number of clusters is equal to the optimal number of clusters in the dataset, the weight of the anomaly score obtained from clustering is equal to 0.2, and the degree of fuzziness is equal to 1.4. The accuracy of anomaly detection methods is illustrated in Fig. 6. It is worth noting that for datasets such as Smtp, artificially generated, Mammography, and Annthyroid, that is, datasets that have a total number of anomalies less than 10%, the IF method obtains a much worse accuracy than when it is combined with the clustering methods. On

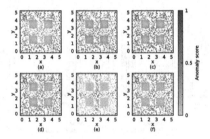

Fig. 5. Normalized anomaly score for each of the observations in the artificially generated dataset obtained by: (a) Isolation Forest, (b) k-Means (triangular function), (c) k-Means (Gaussian function), (d) IF combined with k-Means (Gaussian function), (e) Fuzzy C-Means, and (f) IF joined with Fuzzy C-Means

the other hand, for the three remaining datasets with anomaly counts greater than 30%, the accuracy obtained by using the IF method is slightly better or comparable to the accuracy obtained when the IF method is combined with the clustering approaches. The AUROC measure values for IF and the proposed combined methods in each of the datasets are in Table 3. Hereafter we use the following abbreviations: IF (Isolation Forest), IF k-M tri (IF combined with k-Means that uses the triangular function), IF k-M Gauss (IF combined with k-Means that uses the Gaussian function), and IF FCM (IF combined with Fuzzy C-Means). Paying attention to the AUROC measure, it can be seen that in 4 out of 7 datasets the IF method combined with Fuzzy C-Means produces the best results. In contrast, the IF method used without the clustering methods obtains the best result in only two datasets. The largest difference between the results obtained using the IF method and the IF method combined with Fuzzy C-Means is 0.05 of the AUROC value, which is an improvement in the result in this case by about 7%. The IF method combined with k-Means that uses the triangular function and the one that uses the Gaussian function obtain similar results to each other and outperform the use of all other methods only in the case of artificially generated dataset. In addition, the AUPRC measure values for the IF method and the proposed combined methods in each of the datasets are presented in Table 4. The IF method combined with Fuzzy C-Means yields the best results in 3 out of 7 datasets. However, it is also worth noting that in 5 out of 7 datasets this method is still better than the IF method alone. In contrast, the IF algorithm joined with k-Means that uses the triangular function outperforms the one with the Gaussian function in five datasets. However, despite it, the IF method combined with k-Means that uses the triangular function is superior to one that is merged with Fuzzy C-Means in only two of examined datasets.

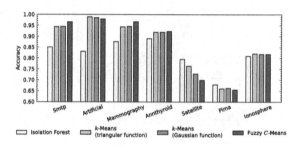

Fig. 6. Anomaly detection accuracy

Table 3. AUROC measure for each anomaly detection method. Bold text indicates the best result in a given dataset

Dataset	IF	IF k-M tri	IF k-M Gauss	IF FCM
Smtp (KDDCUP99)	0.9071	0.9146	0.9166	**0.9191**
Artificially generated	0.9873	0.9940	**0.9948**	0.9931
Mammography	**0.8666**	0.8628	0.8602	0.8567
Annthyroid	**0.8219**	0.7967	0.8045	0.7946
Satellite	0.7046	0.7110	0.6956	**0.7548**
Pima	0.6771	0.6852	0.6871	**0.6991**
Ionosphere	0.8564	0.8732	0.8789	**0.8806**

Table 4. AUPRC measure for each anomaly detection method. Bold text indicates the best result in a given dataset

Dataset	IF	IF k-M tri	IF k-M Gauss	IF FCM
Smtp (KDDCUP99)	0.0041	**0.1459**	0.0546	0.1364
Artificially generated	0.8997	0.9499	**0.9570**	0.9519
Mammography	**0.2149**	0.1906	0.1803	0.1664
Annthyroid	**0.3099**	0.2720	0.2678	0.2783
Satellite	0.6521	0.6167	0.5819	**0.6574**
Pima	0.5029	0.5031	0.5029	**0.5148**
Ionosphere	0.8105	0.8414	0.8481	**0.8543**

5   Conclusions and Future Work

In this study, we have presented two novel anomaly detection methods that effectively combine IF with Fuzzy C-Means and k-Means. Moreover, we have proposed the score function to determine the anomaly score from the results of Fuzzy C-Means and two different score functions to determine the anomaly score from the results of k-Means. The first one is based on the triangular function, and the second one is based on the Gaussian function. Based on conducted series of experiments, it was found that for most of the examined datasets, combining the IF approach with the clustering algorithms allows one to obtain better results than using the IF alone. In particular, it is worth noting that in the case of clustering algorithm selection, combining the IF method with the Fuzzy C-Means algorithm produced better results than combining the IF method with the k-Means algorithm. Considering only the IF method joined with k-Means algorithm, the choice of the triangular function or the Gaussian function to determine the anomaly score had no significant effect on the anomaly detection results obtained by this method. Finally, it was found that the best anomaly detection results for methods using clustering can be obtained when the selected number of clusters is equal to the optimal number of clusters determined by the silhouette coefficient.

Future work directions can be, among others, an application of other clustering-based methods such as k-medoids, rough clustering, or shadowed sets-based clustering. Moreover, dimensionality reduction of the dataset or an application of aggregation operators at the stage of summation of the values coming from particular trees' analyzes. Finally, other approaches including dataset granulation are worth considering in relation to IF and its modifications.

References

1. Agrawal, S., Agrawal, J.: Survey on anomaly detection using data mining techniques. Procedia Comput. Sci. **60**, 708–713 (2015)
2. Chandola, V., Banerjee, A., Kumar, V.: Anomaly detection: a survey. ACM Comput. Surv. **41**(3), article no. 15, 1–58 (2009)
3. Graham, R.L., Knuth, D.E., Patashnik, O., Liu, S.: Concrete mathematics: A foundation for computer science, pp. 480–481. Addison-Wesley, Reading, MA (1994)
4. Hawkins, D.: Identification of outliers. Monographs on Applied Probability and Statistics, Chapman and Hall, London (1980)
5. Karczmarek, P., Kiersztyn, A., Pedrycz, W.: Fuzzy set-based isolation forest. In: 2020 IEEE International Conference on Fuzzy Systems, pp. 1–6 (2020)
6. Karczmarek, P., Kiersztyn, A., Pedrycz, W., Al, E.: K-means-based isolation forest. Knowl.-Based Syst. **195**, 105659 (2020)
7. Karczmarek, P., Kiersztyn, A., Pedrycz, W., Czerwinski, D.: Fuzzy C-Means-based isolation forest. Appl. Soft Comput. **106**, 107354 (2021)
8. Liu, F.T., Ting, K.M., Zhou, Z.-H.: Isolation forest. In: 2008 Eighth IEEE International Conference on Data Mining, pp. 413–422 (2008)
9. Liu, F.T., Ting, K.M., Zhou, Z.-H.: Isolation-based anomaly detection. ACM Trans. Knowl. Discov. Data **6**(1), 1–39 (2012)
10. Liu, J.M., Tian, J., Cai, Z.X., Zhou, Y., Luo, R.H., Wang, R.R.: A hybrid semi-supervised approach for financial fraud detection. In: 2017 International Conference on Machine Learning and Cybernetics, pp. 217–222 (2017)
11. Milligan, G.W., Cooper, M.C.: A study of standardization of variables in cluster analysis. J. Classif. **5**(2), 181–204 (1988)
12. Pang, G., Shen, C., Cao, L., Van Den Hengel, A.: Deep learning for anomaly detection: a review. ACM Comput. Surv. **54**(2), article no. 38, 1–38 (2022)

13. Rayana, S.: ODDS Library. Stony Brook, NY: Stony Brook University, Department of Computer Science. https://odds.cs.stonybrook.edu (2016). Accessed 09 Jun 2021
14. Rousseeuw, P.J.: Silhouettes: a graphical aid to the interpretation and validation of cluster analysis. J. Comput. Appl. Math. **20**, 53–65 (1987)
15. Staerman, G., Mozharovskyi, P., Clémençon, S., d'Alché-Buc, F.: Functional isolation forest. In: Asian Conference on Machine Learning, pp. 332–347 (2019)

K-Medoids-Surv: A Patients Risk Stratification Algorithm Considering Censored Data

George Marinos$^{(\boxtimes)}$ ⓘ, Chrysostomos Symvoulidis ⓘ,
and Dimosthenis Kyriazis ⓘ

University of Piraeus, Piraeus, Greece
{gmar,simvoul,dimos}@unipi.gr

Abstract. Traditional survival analysis estimates the instantaneous failure rate of an event and predicts survival probabilities distributions. In fact, in a set of censored data there may exist several sub-populations with various risk profiles or survival distributions, for which regular survival analysis approaches do not take into consideration. Consequently, there is a need for discovering such sub-populations with unambiguous risk profiles and survival distributions. In this work, we propose a modified version of the K-Medoids algorithm which can be used to efficiently cluster censored data and identify diverse groups with distinct lifetime distributions.

Keywords: Survival analysis · Lifetime clustering · Risk stratification

1 Introduction

Survival analysis is a widely known family of techniques for the estimation of the remaining time until an event of interest occurs. Although it was initially created in terms of medical research and the purpose was to model a patient's survival, it can also be applied to several other application domains and this highlights its usefulness. Survival analysis can be used to estimate the probability of failure for a wide range of events such as failure of the manufacturing equipment based on the hours of operations of the death of the patient in a clinical trial. Even though time to event prediction is a very important process that assists analysts in deeply understanding the dynamics of the analyzing subjects and facilitating the prediction of future events, lifetime clustering is a relatively under-explored field. "Lifetime clustering" or "Clustering in terms of survivability" is the process of identification of underlying sub-populations with distinct lifetime distributions (and risk scores) in a set of censored data. In other words, this is a way of risk stratification when dealing with censored data. It has been proven that such a tool can be very crucial for survival analysis and generally in the process of risk group identification. Despite its importance, there are not many research studies in the literature dealing with the problem of effectively discovering groups of subjects with similar lifetime distributions.

L. Rutkowski et al. (Eds.): ICAISC 2022, LNAI 13589, pp. 127–140, 2023.
https://doi.org/10.1007/978-3-031-23480-4_11

In this paper, we propose a modified version of the K-Medoids algorithm which takes as input the estimated survival probability distribution of each subject along with the original features and aims to produce data clusters that are not only spatially, but also biologically meaningful. Finally, the identified groups not only significantly differ from each other in terms of survivability which we evaluate using a well established statistical test (log-rank), but they also seem to have clear spatial separation when visualizing them in the 2-dimensional space.

The remainder of this paper is organized as follows: Sect. 2 presents the relevant studies that have been identified in the literature, Sect. 3 describes the basic functions and notations of Survival Analysis, Sect. 4 presents the proposed approach, and Sect. 5 illustrates the evaluation of our methodology using publicly available data sets, while Sect. 6 concludes the paper.

2   Related Work

Clustering in terms of survivability cannot be performed by simply using conventional clustering techniques like K-means or DBSCAN because, in that way, it cannot be ensured that groups will always have distinct lifetime distributions which is the desired outcome. In that section, we quote the most relevant studies found in the literature regarding the "lifetime clustering" as well as the clustering of probability distributions.

There exist several studies [13] that appear in literature trying to solve the problem of survival clustering utilizing various data modalities like medical images [12,14] as well as genes [1] and tabular data. Starting with [11], Li and Gui proposed an extension of partial least squares (PLS) regression in the framework of the Cox model by providing a parallel algorithm for constructing the latent components. The proposed algorithm involves constructing predictive components by iterated least square fitting of residuals and Cox regression fitting. These components can then be used in the Cox model for building a useful predictive model for survival. Since this process has already been done, principal components can then be used for discovering various risk sub-populations in the data as well.

In [12] the authors proposed a semi-supervised probabilistic approach that aims at clustering censored data by leveraging stochastic gradient descent variational inference. Their proposed method maps input data into a latent representation using a variational autoencoder with a Gaussian mixture prior and since they aim for time to event prediction, the survival density function in this approach is given by a mixture of Weibull distributions. Results of the study showed that the proposed approach can retrieve clusters considering explanatory variables and survival information.

In [1] the authors highlighted the importance of discovering cancer sub-types using gene expression data and clinical data jointly. Discovered cancer sub-types appeared to have significant differences in terms of patients' survival when the semi-supervised technique was used. The two-step process which is proposed in this study can be described as follows: Firstly a Cox regression model is used on genes expression data in order to assign each of them a "Cox" (importance) score and then select only those genes with high "Cox" score. After that procedure, only significant genes have been left. In the next phase, after having chosen a subset of gene expression, they apply traditional clustering techniques e.g., K-means only on gene expression data in order to identify the desirable number of clusters. In the second part of the proposed approach, the authors test the cluster assignment using only the clinical data. By utilizing clinical data they set cluster assignment as the dependent variable and apply classification. Finally, it turns out that the classification returns good results which means that cluster assignments have been accurately identified.

Decision trees have been broadly utilized in several studies as a mean to perform clustering. The authors in [16] proposed a novel decision tree-based approach aimed at survival clustering. The initial step of this method is to break the data set into sub-populations and based on attribute - values test to discover the identified populations' discrepancies in survival distribution which is done using the Kaplan-Meier estimates. Kuiper [9] statistics is then used in order to measure the significance of the difference across survival distributions. The proposed method results in a tree where each leaf node has an associated population of users and thus one can observe clusters at leaf nodes. However, this approach faced the shortcoming that the degree of dissimilarity between identified clusters may not necessarily be significant when subjects with similar survival distributions will be placed closer in the tree diagram. To face that, the authors propose the usage of a complete graph that consists of leaf nodes as vertices and p-values as edge weights to ensure that the degree of dissimilarity between the clusters will be significant. In the aforementioned graph, each node will be connected to the other with the edges which will denote the significance of the relationship while finally, Markov clustering algorithm [22] is used to obtain the final graph.

A recent research study introduced DeepCLife [17], an inductive neural network-based clustering model architecture that aims to the observation of empirical lifetime distribution of underlying clusters. The purpose of this framework is to discover clusters that have different lifetime distributions whereas the subjects of the same cluster share the same lifetime distribution. A major advantage of this approach is that the model does not assume proportional hazards and hence it is not restricted only to data sets that satisfy this assumption. Furthermore, this research addresses the issue of unobservability of termination signals which is very common in real-world cases, meaning that it can be applied to data sets that termination signals have not been tracked. The main contribution of this work is the proposal of a novel clustering loss function that is used in the introduced neural network architecture and is based on the Kuiper two-sample test.

The authors in [2] focus on the characterization of time-to-event predictive distributions from a clustered latent space conditioned on covariates. Specifically, the authors proposed a mixed-type solution that aims for time-to-event modeling, as well as survival clustering. They utilize a (Bayesian non-parametric) deterministic encoder to perform risk profile-based clustering and map covariates into a latent representation, followed by a stochastic survival predictor which feeds from the latent representation. The proposed approach performed well in learning clustering structure in a latent representation for which the number of clusters is unknown. Results show that the usage of the proposed Bayesian non-parametric stick-breaking representation of the Dirichlet Process is essential for the identification of meaningful clusters in terms of survivability. In this research, a Bayesian non-parametric approach was used for the clustering process. The Bayesian approach emboldened the latent representation to act as a mixture of distributions while the distribution matching approach (in this study) follows a Dirichlet Process.

Clustering probability densities is not so commonly used compared to traditional clustering. There are few studies found in the literature that aim to perform clustering on probability densities. The authors in [15], study the notion of affinity between several densities. More recently, in [19] it is proposed the use of the maximum of k functions defined on R^n to define the joint L^1-distance of two (or more) probability density functions and this was proven to be an effective tool both for classification as well as discriminant analyses. Not only that, the proposed tool could be used for Bayesian applications and it could also be useful for the computation of Bayes error.

Following the same approach, the authors in [23] have proposed a hierarchical clustering approach using the same distance to accurately identify groups of similar distributions while maximizing the inter-cluster distance. The criterion used in this study was the cluster width defined from the L^1-distance, a special property of which is that it can be applied to a set of two or more densities, unlike in the classic case where the distance between an element and a cluster must be defined separately.

In [18], the authors proposed an extension of the batched k-means algorithm for clustering histograms using mixed divergences by associating two dual centroids per cluster. They also achieved to generalize the probabilistically guaranteed good seeding of k-means++ to mixed α-divergences. In this paper, due to the usage of the mixed α-seedings, probabilistic clustering bounds by picking up seeds from the data can be provided, and most importantly the exact (or precise) computation of centroids is not necessary.

More recently, K-Medoids has been used for similar but not identical purposes in [6] where authors aim to cluster probability density functions using the L^1-distance metric as it has been defined in [19]. Their recommended approach allows for a quick convergence based on the distance matrix which is computed exactly on time. Authors not only provide the related proof of convergence for the proposed method but also present numerical examples to demonstrate the robustness and effectiveness of their method with regard to both accuracy and computational time.

Even though that the aforementioned studies propose methods for clustering probability densities, indeed, those methods do not account for censored data. To the best of our knowledge, this is the first time that probability densities clustering is used to solve the problem of the identification of similar survival populations with regard to lifetime distribution.

3 Survival Analysis Basics

3.1 Time to Event Analysis

The triplet (X_i, y_i, z_i), represents the i-th data point in a given data set, where $X_i \in R^{1 \times P}$ is the feature vector, and z_i is the binary event indicator which is marked as 1 when the subject has experienced the event of interest, and marked 0 otherwise. The observed time is denoted by y_i and is equal to the survival time T_i if the given observation is uncensored, or C_i, if the given observation is censored.

$$y_i = \begin{cases} T_i, & \text{if } z_i = 1 \\ C_i, & \text{if } z_i = 0 \end{cases} \tag{1}$$

Traditional survival analysis aims at estimating the time until the event of interest happens for a new instance k with feature predictors denoted by a feature vector X_k. Mathematically the survival function [8,10] represents the probability that the time to the event of interest is not earlier than a specified time t. Survival function is represented as follows:

$$S(t) = P(T > t) \tag{2}$$

The equation above denotes an individual that survives longer than t and it is true that the survival function decreases while the t increases. Its starting value is 1 for $t = 0$ which implies that in the beginning of the observation all subjects have very strong probability of surviving which is decreases in the fullness of time. From the definition of cumulative death distribution function $F(t)$,

$$S(t) = 1 - F(t) \tag{3}$$

3.2 Censored Data

Censorship is a kind of missing data during the data collection period and is the main reason why a traditional regression model cannot be fitted for making predictions in data with such peculiarities. Collecting data in order to use them for modeling a survival analysis problem, is often challenging. In a censored data set, it is possible that the events of interest may not be observed for some instances. The concept of having a data set with incomplete data for the event of interest is called censoring [20]. There exist three main categories of censorship, as identified by the authors in [3]; (i) right censoring, which is the most common type of censoring and it occurs when the observed survival time is less than or equal to the true survival time, (ii) left-censoring, for which the observed survival time is greater than or equal to the true survival time, and (iii) interval censoring, for which it is only known that the event occurs during a given time interval.

4   Proposed Approach

As already mentioned, the proposed approach regards a modification of the traditional clustering algorithm K-Medoids, which can be visually depicted in Fig. 1. Of course, the K-Medoids algorithm steps are utilized, yet there exist some major differences. Bellow a list of the differences between the proposed approach and the traditional K-Medoids are presented.

- *A major difference between the classic K-Medoids algorithm and the current work's approach version is that, unlike K-Medoids which can be used on only conventional tabular data, the proposed algorithm receives as input censored data which means that apart from the conventional features there are also lifetime information and a binary event indicator.*

  In essence, in order to identify meaningful clusters in terms of survivability, we need an intermediate step which is the survival probability distribution estimation for each subject. This can be done by using a predictive model, i.e., the Cox Proportional Hazard. Consequently, we estimate the survival distribution concerning the lifetime of each subject using a kernel function (either Cox Proportional Hazards, or DeepSurv). The proposed algorithm uses those distributions in conjunction with the initial features as input to finally discover the clusters.

- *Since data partly consist of probability distributions it would not make sense to use only a distance metric like Euclidean or Manhattan distances as it is traditionally used in K-Medoids.*

  To solve this, we introduce a *Joint Distance* which not only counts the spatial distance but also the lifetime distribution dissimilarity. As stated earlier, the input data consist of the probability distributions of each subject and the original data. We compute two different distance metrics for each subject, (i) the Euclidean distance using the original data, and (ii) Jensen-Shannon distance for the probability distributions of each subject. We standardize the Euclidean distance to ensure that all values per feature are in the same range as the values in the Jensen-Shannon distance and we finally compute the summation of the scaled Euclidean and the Jensen-Shannon distances and this results in the Joint Distance measure.

- *The proposed approach utilizes the p-value of the log-rank test as a tolerance for the centroids for the change of the centroids.* Normally, in order for the optimal clusters to be discovered, K-Medoids algorithm tries to minimize the total cost which is the sum of all the costs in the cluster where each cost is defined as the distance between each subject and each selected centroid (based on the chosen distance metric). In this step, our algorithm computes the total cost of the introduced Joint Distance and in parallel uses the p-value of the log-rank test between the K selected clusters in order to decide if the centroids need to be altered or not. It iteratively computes the total cost and the log-rank test for each cluster combination and it only keeps the combination of clusters which returns the lower total cost along with a p-value lower from the tolerance level that the user has defined. The predefined tolerance level is 0.01 but it can be modified according to the user's needs. Thus, the K chosen clusters include subjects that have similar survival probability distributions, and hence the difference between the clusters in terms of survival distributions is maximized.

In Algorithm 1, the K-Medoids modification is described, which receives censored data as input and computes the survival probability distribution predicted from a Survival Analysis model. This variation can effectively divide the subjects into groups that have a statistically significant difference in their survival distributions.

Algorithm 1: Modified K-Medoids using Joint Distance

Input: Data: X_{i_1}, X_{i_2}, ..., X_{i_n}, T_i, E_i // where X_{i_1} ... X_{i_n} are the data features, T_i is time and E_i is the binary event indicator
  $Probs$ = Estimated survival probability distributions for each i
Parameters: k = Number of desired clusters,
  n = Number of observations,
  EU$[n,k]$ = Euclidean Distance matrix,
  NormEU$[n,k]$ = Normalized Euclidean Distance matrix per feature,
  JS$[n,k]$ = Jensen-Shannon Distance matrix
  JD$[n,k]$ = Joint Distance matrix
  logrank_tol = the significance level of log rank score // usually is 0.01
  max_iterations // maximum number of iterations is defined by the user
Output: k clusters of similar data points

1 **Function** Modified K-Medoids:
2   Randomly choose initial k centroids from data
3   Estimate the survival probability distributions $p*$ for each i
4   EU$[n,k]$ $\leftarrow$ Initialize empty $n \times k$ matrix for the Euclidean distances
5   JS$[n,k]$ $\leftarrow$ Initialize empty $n \times k$ matrix for the Jensen-Shannon distances
6   **while** $cc \leq max\_iterations$ **do**
7   **for** $i \leftarrow 1$ to n in Data **do**
8   **if** i is not a centroid **then**
9   **for** $j \leftarrow 1$ to k in centroids **do**
10            $JS(i, j) = \sqrt{\frac{D(p_i*||m)+D(p_j*||m)}{2}}$
11   Append $JS(i, j)$ in JS[n,k]
12            $EU(i, j) = \sqrt{\sum_{i=1}^{n}(p_i - p_j)^2)}$
13   Append $EU(i, j)$ in EU$[$ n,k $]$

14   NormEU$[n,k]$ $\leftarrow$ Normalize the EU$[n,k]$ in the range of $(0,1)$ per feature
15   JD$[n,k]$ $\leftarrow$ NormEU$[n,k]$ + JS$[n,k]$
16   **for** $i \leftarrow 1$ to n in $JD[n,k]$ **do**
17   **for** $j \leftarrow 1$ to k in $JD[n,k]$ **do**
18   cluster(i) $\leftarrow$ argmin(JD$[n,k]$) //assign i to the nearest centroid

19   Compute the log_rank test's p_value between the data points of the different clusters
20   **for** $j \leftarrow 1$ to k **do**
21   **for** $i \leftarrow 1$ to n **do**
22   total_costs $\leftarrow \sum_{j=1}^{k} \sum_{i=1}^{n}(JD[n,k])$// summation of the total cost distances

23   **if** $total\_costs < best\_total\_cost$ and log_rank test's $p\_value < logrank\_tol$
24   **then**
25   best_total_cost $\leftarrow$ total_cost
26   SWAP centroids

27   **else**
28   continue

29      $cc = +1$
30   **return** k clusters

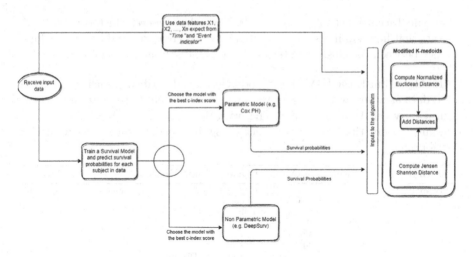

Fig. 1. Proposed approach steps

This variation of K-Medoids algorithm, allows for differences between features as well as the lifetime distribution of the subject and therefore the identified groups include subjects that are similar to each other. Such an approach satisfies our goal, in which we aim to discover groups of data that have statistically significantly different (distinct) lifetime distributions from each other, where each group will contain subjects that have similar survival distributions as well as similar features.

4.1   Custom Distance Function

Many distance measures have been proposed in the literature to accurately divide data points into groups. Most of them are based on the spatial distance of the data point with the most widely known of them being Euclidean, Manhattan, Chebyshev, Mahalanobis, etc., and others are used to measure the difference between probability distributions. The most popular of the them is the Kullback-Leibler divergence (KL divergence) which is also called relative entropy. This statistical distance measures how much a probability distribution differs from a second one.

To the best of our knowledge, this is the first time that a mixed type distance is used to cluster data points within the K-Medoids algorithm. In this paper we introduce the *Joint Distance (JD)* which is the summation of Euclidean and Jensen-Shannon distance:

$$JD = EU(p_i - q_i) + JS(p*_i, q*_i) \tag{4}$$

$$\sqrt{\sum_{i=1}^{n}(p_i - q_i)^2} + \sqrt{\frac{D(p_i * \| m) + D(q_i * \| m)}{2}} \tag{5}$$

where p and q are two randomly chosen data points from the original data $X_1, X_2, ..., X_n$, p^* and q^* are the estimated survival probability distribution of the relative data points p and q respectively, obtained from either a semi-parametric model like the Cox Proportional Hazard estimator or from a non-linear deep learning model like DeepSurv, m is the pointwise mean of p and q and D is the Kullback-Leibler divergence.

An very important part of the proposed distance measure is that since the Jensen Shannon distance is the square root of the Jensen-Shannon divergence return values in the range of (0,1). However, Euclidean distance's results are not bounded in the range of (0,1) and could practically return results in any range. To solve that, we perform the Euclidean distance matrix standardization in the range of (0,1) per feature in order for the summation of those two distance measures to be meaningful.

In essence, the Jensen-Shannon distance is a symmetric version of the Kullback-Leibler divergence. In the proposed distance we use the Jensen Shannon distance metric which is the square root of Jensen-Shannon divergence. We chose this metric because:

- Jensen-Shannon divergence is a bounded symmetrization of the KL divergence and therefore returns values in the range of [0,1],
- It does not require the condition of absolute continuity.

As in the KL divergence values near zero indicate similarity between distributions and positive values indicate divergence in distribution. Consequently, the bigger the returned value from the JS (Jensen-Shannon) distance, the larger the divergence.

4.2   Optimal Number of Clusters

Our method aims at identifying clusters that are robust with regards to the lifetime distribution, as well as spatial differences between features. The proposed method requires the desired number of clusters to be defined by the user and thus the user needs to experimentally search and obtain the various results that each choice of the number of clusters will bring. There are several methods proposed in the literature that can be used for optimal number of clusters discovery like the Silhouette method, Gap statistics, elbow rule etc. We have chosen to utilize the Silhouette method in order to provide a way for the optimal number of cluster identification. However, Silhouette method works well for conventional clustering algorithms like K-means, K-Medoids, DBSCAN, etc., but could not apply to our algorithm since the Silhouette method inherently computes the Euclidean distance, unlike our approach where we use a custom distance function. Therefore, we propose a modified version of the Silhouette method where the Joint Distance introduced in this paper is used, instead of the Euclidean distance. The Silhouette score $s(i)$ is traditionally calculated according to the following formulas:

$$s(i) = \frac{b(i) - a(i)}{max(a(i), b(i))}, if |C_I| > 1 \tag{6}$$

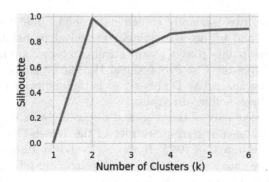

Fig. 2. Silhouette score for different numbers of k for the NKI data set.

where a(i) and b(i) are defined as follows:

$$a(i) = \frac{1}{|C_I| - 1} \sum_{j \in C_I, i \neq j} d(i,j) \tag{7}$$

$$b(i) = min_{J \neq I} \frac{1}{|C_J|} \sum_{j \in C_J} d(i,j) \tag{8}$$

a(i) is the average distance between i and all other data points in the same cluster (intra-cluster distance), $|C_I|$ is the number of points belonging to cluster i, b(i) is the smallest mean distance of i to all points in any other cluster of which i is not a member (inter-cluster distance). Finally, d(i,j) is the distance between i and j where instead of using Euclidean distance as traditionally used in Silhouette method we use our Joint Distance metric which was introduced earlier. For example, Fig. 2 demonstrates the optimal number of clusters found after calculating the silhouette score for a range of the different number of clusters.

4.3   Clustering Evaluation

The validation of our results is performed through:

- the log-rank p-value of the final clusters
- visually by reducing the dimensions of the data set in 2D and visualize the clusters to obtain if the resulted groups are well separated
- visually by Kaplan-Meier curves

Dimensionality reduction is being performed using the t-distributed Stochastic Neighbor Embedding (tSNE) algorithm which traditionally internally uses the Euclidean distance to transform the data, but in our case instead of Euclidean distance, the proposed custom Joint Distance is used, for the dimensionality reduction and the visualization to be meaningful.

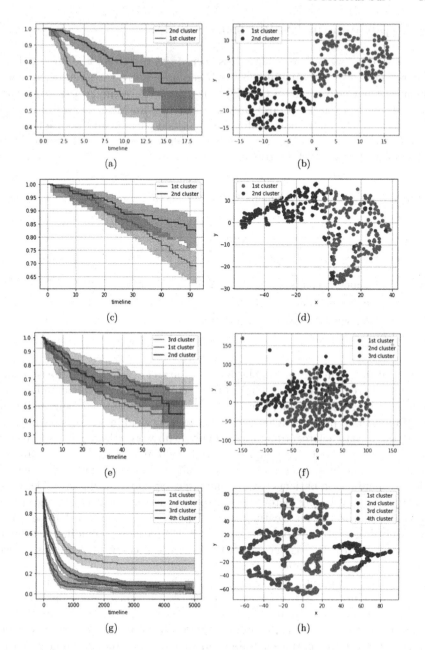

Fig. 3. Kaplan-Meier curves for the NKI data set and the tSNE plot are depicted in sub-figures (a) and (b) respectively with log-rank test's p-value = 0.000292. For the Rossi data set in sub-figures (c) and (d) with log-rank test's p-value = 0.003436, for the Diabetes data set the Kaplan-Meier curves (sub-figure (e)) and tSNE plot (sub-figure (f)) are also show, with log-rank test's p-value = 0.009410, while for the LeukSurv data set the same plots are presented in sub-figures (g) and (h) with log-rank test's p-value = 0.001245.

5   Experimental Evaluation

In this section, we utilize open access (censored) data sets and the final purpose is to discover meaningful clusters based on data features and their spatial similarities but also based on their lifetime distributions similarities. To visually evaluate our approach, we also present both Kaplan-Meier curves along with the p-value of the log-rank score for the identified clusters and we use the tSNE method to reduce the dimensions and be able to visualize the groups in the 2-dimensional space with scatter plots.

Table 1. This table presents summary statistics about the below-mentioned datasets.

	NKI	Rossi	Diabetes	LeukSurv
Number of observations	272	432	394	1,043
Number of missing values	-	-	-	-
Number of events	29%	26%	39%	84%
Number of numerical features	6	6	2	6
Number of categorical features	8	1	4	1

The utilized publicly available data sets are (*i*) NKI: a 272 breast cancer patients cohort which includes clinical risk factors, and gene expression measurements, histologic type grade, and diameter, as well as time and survival which were found to be prognostic for metastasis-free survival in an earlier study [21], (*ii*) Rossi: a data set regarding 432 convicts who were released from Maryland state prisons in the 1970s s and who were followed up [4] for one year after release, (*iii*) diabetes: a data set that was collected to test a laser treatment for delaying blindness in patients with diabetic retinopathy. The data set [7] consists of 394 patients including factors related to their age, survival status, the specific eye that was selected for treatment, etc., and (*iv*) LeukSurv: a data set [5] on the survival of acute myeloid leukemia for 1,043 patients, including factors and patient attributes such as age, sex, white blood cell count at diagnosis, as well as the Townsend score, among others.

Figure 3 demonstrates the performance of the proposed approach in the identification of clusters for 4 different data sets. As previously mentioned, we have identified the optimal number of clusters for each data set and we present the results in this section. It is apparent that for all data sets, the assigned labels reveal that are both spatial separable and they differ in terms of lifetime distributions. In the aforementioned figure, it is obvious that the algorithm achieved to identify clusters that, as depicted, have a very good spatial separation, and in parallel a very clear distinction of the Kaplan-Meier curves which is something that the log-rank tests' p-values of all the data sets also denotes. In more detail, as far as the NKI (p-value = 0.000292) and Rossi (p-value = 0.003436) data sets are concerned, we observe a clear spatial separation in the tSNE plots (see also Fig. 2(b) and 2(d)), which is also depicted in the Kaplan-Meier curves (2(a) and 2(c) respectively). Results for the Diabetes data set (p-value=0.009410) also show that there is a clear separation of the Kaplan-Meier curves (Fig. 2(e)) despite the existence of a little overlapping between the clusters as depicted in the related figure (Fig. 2(f)). Good distinctive performance has also been achieved in the LeukSurv data set with regard to the statistical difference of the log-rank test (p-value=0.001245) between the 4 clusters and Kaplan-Meier curves (2(g)) and the tSNE visualization (2(h)).

6   Conclusions

This paper introduced a modified version of the K-Medoids algorithm, which aims to solve the problem of clustering not only in terms of spatial differences between data features of a data set but also in terms of survival probability distributions between subjects. To solve that, the *Joint Distance* measure was also introduced, which is the summation of the (previously normalized) Euclidean and Jensen-Shannon distances. The above-mentioned method has been proven to be effective for survival clusters identification based on (i) the (log-rank) statistical test results, which showed statistically significant differences between the identified clusters, and (ii) the clusters' visual representation.

Acknowledgements. The research leading to the results presented in this paper has received funding from the European Union's funded Project iHelp under grant agreement no 101017441.

References

1. Bair, E., Tibshirani, R., Golub, T.: Semi-supervised methods to predict patient survival from gene expression data. PLoS Biol. **2**(4), e108 (2004)
2. Chapfuwa, P., Li, C., Mehta, N., Carin, L., Henao, R.: Survival cluster analysis. In: Proceedings of the ACM Conference on Health, Inference, and Learning, pp. 60–68 (2020)
3. Clark, T.G., Bradburn, M.J., Love, S.B., Altman, D.G.: Survival analysis part i: basic concepts and first analyses. Br. J. Cancer **89**(2), 232–238 (2003)
4. Fox, J., Carvalho, M.S.: The rcmdrplugin. survival package: extending the r commander interface to survival analysis. J. Statist. Softw. **49**, 1–32 (2012)
5. Henderson, R., Shimakura, S., Gorst, D.: Modeling spatial variation in leukemia survival data. J. Am. Stat. Assoc. **97**(460), 965–972 (2002)
6. Ho-Kieu, D., Vo-Van, T., Nguyen-Trang, T.: Clustering for probability density functions by new-medoids method. Sci. Program. **2018**, 1–7 (2018)
7. Huster, W.J., Brookmeyer, R., Self, S.G.: Modelling paired survival data with covariates. Biometrics **45**(1), 145-56 (1989)
8. Klein, J.P., Moeschberger, M.L.: Survival Analysis: Techniques for Censored and Truncated Data. Springer, New York (2006). https://doi.org/10.1007/978-1-4757-2728-9
9. Kuiper, N.H.: Tests concerning random points on a circle. In: Nederlandse Akademie van Wetenschappen. roceedings. Series A.,vol. 63, pp. 38–47 (1960)

10. Lee, E.T., Wang, J.: Statistical Methods for Survival Data Analysis, vol. 476. John Wiley & Sons (2003)
11. Li, H., Gui, J.: Partial cox regression analysis for high-dimensional microarray gene expression data. Bioinformatics **20**(suppl_1), i208–i215 (2004)
12. Manduchi, L., et al.: A deep variational approach to clustering survival data. arXiv preprint arXiv:2106.05763 (2021)
13. Marinos, G., Kyriazis, D.: A survey of survival analysis techniques. In: HEALTH-INF, pp. 716–723 (2021)
14. Marinos, G., Symvoulidis, C., Kyriazis, D.: Micsurv: medical image clustering for survival risk group identification. In: 2021 4th International Conference on Bio-Engineering for Smart Technologies (BioSMART), pp. 1–4. IEEE (2021)
15. Matusita, K.: On the notion of affinity of several distributions and some of its applications. Ann. Inst. Stat. Math. **19**(1), 181–192 (1967)
16. Mouli, S.C., Naik, A., Ribeiro, B., Neville, J.: Identifying user survival types via clustering of censored social network data. arXiv preprint arXiv:1703.03401 (2017)
17. Mouli, S.C., Teixeira, L., Neville, J., Ribeiro, B.: Deep lifetime clustering. arXiv preprint arXiv:1910.00547 (2019)
18. Nielsen, F., Nock, R., Amari, S.i.: On clustering histograms with k-means by using mixed α-divergences. Entropy **16**(6), 3273–3301 (2014)
19. Pham-Gia, T., Turkkan, N., Vovan, T.: Statistical discrimination analysis using the maximum function. Commun. Stati. Simul. Comput. ® **37**(2), 320–336 (2008)
20. Prinja, S., Gupta, N., Verma, R.: Censoring in clinical trials: review of survival analysis techniques. Indian j. Commun. Med. **35**(2), 217 (2010)
21. Vasn de Vijver, M.J., et al.: A gene-expression signature as a predictor of survival in breast cancer. N. Engl. J. Med. **347**(25), 1999–2009 (2002)
22. Van Dongen, S.: Graph Clustering by Flow Simulation. 2000. University of Utrecht (2001)
23. Van Vo, T., Pham-Gia, T.: Clustering probability distributions. J. Appl. Stat. **37**(11), 1891–1910 (2010)

A Benchmark of Process Drift Detection Tools: Experimental Protocol and Statistical Analysis

Caio Raduy[1] , Denise M. V. Sato[1,2(✉)] , Eduardo A. Franciscon[1] , and Edson E. Scalabrin[1]

[1] Pontifical Catholic University of Paraná (PUCPR), Imac. Conceição. 1155, Curitiba 80215-901, Brazil
{caio.raduy,denise.maria,eduardo.franciscon}@pucpr.edu.br,
edson.scalabrin@pucpr.br
[2] Federal Institute of Paraná (IFPR), João Negrão. 1285, Curitiba 80230-150, Brazil
denise.sato@ifpr.edu.br

Abstract. Business processes are sequences of activities performed to achieve a specific goal, e.g., applying a clinical protocol to a patient. Process mining provides tools and techniques for analyzing and enhancing business processes. However, these processes are usually dynamic and can change because of new regulations, emergencies, or other reasons; and these changes are named concept or process drifts. Detecting drifts allows managers to improve the process analysis and act proactively upon these changes. We benchmarked three process drift detection tools and compared them based on an experimental protocol designed to evaluate the accuracy of the detected drifts. The selected tools detect sudden drifts in event logs. The experimental protocol generated a dataset containing the accuracy metric and the parameter configuration applied in each scenario. We applied statistical tests to verify significant differences in the accuracy between the tools when performed using distinct parameter configurations. The findings indicate that the parameter configuration affects the accuracy of the detected drifts and the dataset configuration. Another contribution of this paper is the designed experimental protocol, which can be applied to objectively evaluate the process drift detection tools and the dataset containing the results of the accuracy calculated over the performed experiments.

Keywords: Process drift · Process drift detection · Changing processes · Process drift accuracy

1 Introduction

Business processes represent a sequence of tasks to achieve a final goal [2]. Companies store the data related to the execution of such tasks in the information

Supported by CAPES (Coordenação de Aperfeiçoamento de Pessoal de Nível Superior-Brasil)-Finance Code 001, Grant Nos.: 88887.321450/2019-00, 88887.607090/2021-00; and PUCPR PIBIC (Programa Institucional de Bolsas de Iniciação Científica).

systems, and it is possible to extract event logs from such records. An event log contains information about a business process: when it was performed, by whom, the performed activities and their order, and information about the context of the process. Process mining techniques provide different approaches and tools for analyzing the event logs by extracting valuable information about the business process [2].

However, the processes are constantly changing due to, e.g., improvements, changes in regulations, emergencies, and new protocols. The authors in [12] named the situation where the process is modified while being analyzed as concept drift. Process analysts can act upon these changes, also named process drifts, if they have tools to detect them as early as possible [7]. Besides, current process mining techniques tend to analyze these processes as if they do not change [4]. In this context, process drift detection can improve the application of such techniques by applying drift detection as a pre-processing step on the process mining pipeline, enhancing the analysis.

The process drifts are classified into sudden, gradual, incremental, or recurring, based on how the change affects the process model [11]. A sudden drift represents an abrupt change, occurring when all the current processes start following differently, e.g., an emergency protocol is established. In a gradual drift, the new behavior and the current one co-exist for a period, e.g., a new curriculum applied for the new students. In the incremental drifts, minor process changes are implemented, e.g., a company applying a new governance model department by department. Moreover, the recurring drifts represent seasonality, where the behavior of the process instances changes, but after a while, they return to act as before the change, e.g., a logistic company that changes the delivery process to attend to the high demand in the Christmas time.

The process mining community proposed different approaches and tools for process drift detection. However, comparing the accuracy of the distinct methods is still an open challenge. There are few artificial event logs containing drifts public available, and the experimental protocol applied sometimes does not include objective metrics, or the metric applied is not thoroughly explained. The authors in [11] suggested using the F-score metric with clearly defined parameters to test these algorithms with artificial logs.

To fulfill this gap, we conducted a benchmarking of process drift detection tools to evaluate the accuracy of the drift detectors. We selected only process drift detection tools that (i) report the change point, i.e., the exact point where the change occurred, (ii) have the source code publicly available, and (iii) conduct an offline drift analysis, i.e., uses the event log as input of the analysis. We have also designed an experimental protocol that can be applied to compare process drift detection approaches using statistical analysis and an objective metric.

We organized this paper as follows. In Sect. 2, we described the selected process drift detection methods. The experimental protocol applied, and the datasets are described in Sect. 3. Section 4 presents the experimental results, and Sect. 5 discusses the results and concludes the paper.

2   Related Work

We selected three process drift detection tools for the benchmarking that attend to (i), (ii), and (iii) criteria raised in Sect. 1. The ProM - Concept Drift plugin was introduced in [3] and enhanced with gradual drift analysis [4]. The approach applies statistical hypothesis tests in a stream of features derived from the traces using a sliding window. The main idea is that if a process drift occurred, we would detect a significant difference in the features from adjacent windows. In [8], the authors include an adaptive window approach to avoid specifying a window size *a priori*. The artificial datasets applied in the evaluation are not publicly available.

Another tool is the Apromore - ProDrift plugin from [6,7], which detects sudden and gradual drifts in event logs or streams[1]. We only included the approach for detecting drifts in event logs (*runs* approach), which applies statistical hypothesis tests in an abstraction derived from the traces. It uses a sliding window with fixed and adaptive size over the stream of runs to define the samples for the statistical test, as the ProM - Concept Drift. The dataset used in the experiments reported in [7] is publicly available; unfortunately, the dataset from [6] is not. The authors compared the fixed and adaptive windowing strategies for sudden drift detection against the ProM - Concept Drift. However, the authors do not explicitly specify which parameter configuration they defined in the ProM - Concept Drift for each scenario, and they do not explicitly define how they calculate the components of the F-score metric reported.

The Visual Drift Detection (VDD) system [13] is a more recent process drift detection tool able to analyze the four types of drifts using different visualizations (drift maps, drift plots, autocorrelation plots) and metrics (spread of constraints, erratic measure, Augmented Dickey-Fuller test). The VDD system differs from previous tools because it uses a change point detection algorithm instead of statistical hypothesis testing over features. Firstly, the VDD system uses a sliding window approach to calculate a measure (e.g., confidence) using the declare constraints[2] of each window (sub-log) and the declare constraints derived from the complete event log, resulting in a multivariate time series. Then, the VDD system clustered these multivariate time series and applied a change point detection algorithm named PELT (Pruned Exact Linear Time). The metrics and visualizations are generated based on the constraint clustered measures. The authors evaluate VDD system accuracy using a dataset provided by [10]. However, the comparison lacks details: the names of the datasets between the two papers are different, and the F-score metric reported is not clearly defined.

We have identified two previous papers comparing tools for process drift detection [5,9]. The authors in [9] compared the accuracy of the ProM - Concept Drift and the Apromore - ProDrift using a real-world dataset. The accuracy comparison cannot be performed using objective metrics (e.g., F-score) because we do not *a priori* know the actual drifts in real datasets. In the second paper [5],

---

[1] Contains information about events performed in a process, similar to event logs. However, the events are stored as they occurred, allowing online process mining.

[2] Declare constraints are obtained from DECLARE process models [1].

the authors compare the detection accuracy of five process drift detection tools using the public dataset from [7] and a new dataset created for the paper. In this comparison, the authors combine different event logs with distinct parameter configurations and apply statistical analysis to the results. The main difference between our work is the chosen accuracy metric. The authors apply two traditional regression metrics: Mean Squared Error (MSE) and Root Mean Squared Logarithmic Error (RMSLE) [5] because proposing metrics was out of the scope of the work. Using these two metrics makes it possible to compare if the total number of detected drifts is next to the real number of drifts; however, it is not possible to evaluate how "far" from the exact change point the detected drift is. Neither of these papers applies an objective metric to identify if the detected drifts really represent a real drift, such as the F-score clearly defined as the proposed in [11]. The experimental protocol proposed in this paper objectively compares the accuracy of the process drift detection tools, considering if the reported change point is close to the real one. Furthermore, the method applied for the benchmarking can be replayed using different datasets and tools.

3   Method

For the benchmark, we define the experimental protocol in Fig. 1, which compares the accuracy of the detected drifts between the selected tools using the F-score.

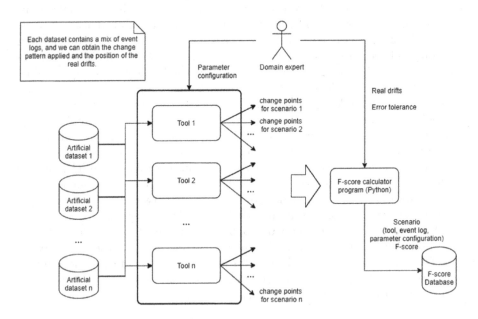

Fig. 1. Experimental protocol for comparing the accuracy of detected process drifts.

The F-score metric (Eq. 1) is known in the data mining fields and represents the harmonic mean of recall and precision, which are calculated based on: the true positives (TP), the false positives (FP), and the false negatives (FN). For process drifts, a TP indicates a detected drift that happened in the process, a FP indicates a detected drift that did not occur, and a FN is a real drift not detected by the tool [8]. Any detection mechanism is reactive, i.e., it needs to read the information after the drift to detect it. So, it is essential to consider a TP only if the reported drift position is after the real one and within a tolerance period. We adapted the error tolerance (et) proposed in [14]. A TP is counted if the detected drift occurred in the interval $[RD, RD + et]$, where RD indicates the position of the real drift and et indicates the error tolerance.

$$F - score = \frac{TP}{TP + \frac{FP+FN}{2}} \tag{1}$$

We have selected three process drift detection tools: Apromore - ProDrift, ProM - Concept Drift, and the VDD system. We evaluated the two approaches from the Apromore - ProDrift (fixed and adaptive) and the adaptive approach in the ProM - Concept Drift. The VDD system only provided a fixed approach. By the time we had performed the experiments, we could not execute the fixed approach in ProM - Concept Drift because the tools raised errors.

We selected two synthetic datasets which are publicly available [5,7]. The first dataset[3] [8] contains events from a loan application process with sudden changes over time. The authors applied 12 simple and 6 complex changes, deriving 18 types of event logs. Each change pattern contains 4 event logs with 2,500, 5,000, 7,500, and 10,000 traces. All the event logs contain 9 drifts, with intervals between the drifts at 10% of the total size (250, 500, 750, and 1,000 traces). We performed the tests using the event logs containing 5,000 traces because it was the smallest size considered in all tests reported in [7]. For the change patterns *lp* and *re* we used the event logs *lp2.5k* and *re2.5k*, because these are the ones with 5,000 traces.

The second dataset[4] [5] contains event logs from the same process model from [7] and part of the reported change patterns (15 in total). We selected the sudden drift event logs with 0% noise and 1,000 traces for our benchmark. The event logs selected contain only one drift reported in the middle of the log, i.e., at trace index 500.

The combination of tool, dataset, and parameters (Fig. 1) is related to the scenarios we want to evaluate in the analysis: (a) influence of window size on the accuracy for the fixed approaches; (b) influence of the initial window size on the accuracy of the Apromore - ProDrift adaptive, and (c) comparing the accuracy between all selected methods. We have performed 825 executions using Apromore - ProDrift fixed and adaptive, VDD system, and ProM - Concept Drift adaptive to verify the defined scenarios[5]. We developed a Python program

---

[3] https://data.4tu.nl/articles/dataset/Business_Process_Drift/12712436.

[4] http://dx.doi.org/10.21227/2kxd-m509.

[5] https://www.kaggle.com/caioraduy/f-score-results-comparing-process-drift-tools.

to automatically extract the outputted information of the tools (when provided as text) and calculate the F-score metric[6]. After calculating the F-score, we propose two types of analysis (Fig. 2).

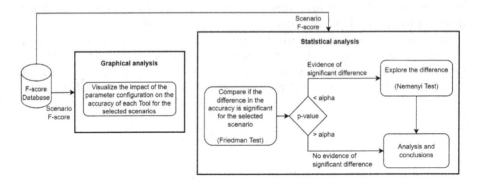

Fig. 2. Suggested analysis for the calculated F-scores.

4   Experiments and Results

4.1   Influence of Window Size on the Accuracy of the Fixed Approaches

Both Apromore - ProDrift fixed and VDD detects the drifts by applying a sliding window of fixed size on the input derived for each method (runs or constraint measures). We tested both methods using window sizes from 50 to 150 in increments of 25 and the window size of 200, the default initial window size of Apromore - ProDrift Adaptive. We included the window sizes 32 and 162, which are the default values suggested for VDD [13] from both datasets based on their size ($dataset1_{size} = 5,000; dataset2_{size} = 1,000$). For the VDD system [13], we defined the slide size parameter as half of the window size (rounding down) and the *driftAll* option set to *True*, to detect overall changes. We calculated the average of the F-scores for each tool and dataset using the window size as the error tolerance.

We can observe in Fig. 3 that the Apromore - ProDrift shows promising results with dataset1; the F-score starts with a value higher than 0.55 and increases until it stabilizes at 0.9, with 125 as the window size. The last window size (200) reaches an F-score of 0.93. In dataset2, the Apromore - ProDrift presented a similar behavior; the F-score increases until stabilizing at window size 125, but with lower values. It does not surpass the F-score of dataset1 with any tested window sizes.

We can observe in Fig. 3 that the VDD system's accuracy is more sensitive to the selected window size. For dataset1, the F-score started with almost 0.4

---

[6] https://github.com/caioraduy/F-score-calculating.

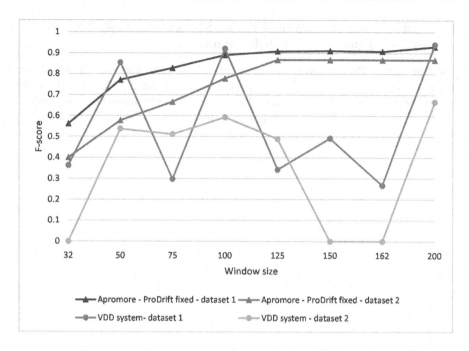

Fig. 3. Average of the F-score from Apromore - ProDrift fixed and VDD system.

and had an unstable behavior, reaching the highest average (0.940) at window size 200 and the lowest average with window size 32 (0.2683) - which is the window size suggested by the tool. In dataset2, we had three windows with an average F-score of 0: 32, 150, and 162. However, the last window size (200) had an average F-score of almost 0.7.

To confirm the insights of the visual analysis, we applied Friedman's test with $\alpha = 0.05$ to verify whether this difference between the F-scores calculated for each window size using the same tool and dataset is statistically significant, i.e., verify if the chosen window size impacts the accuracy of the tools (Table 1). There is a statistically significant difference in the accuracy for both datasets depending on the window size set.

Table 1. Friedman's test results for the influence of the window size on the fixed approaches.

Tool	Dataset1 ($n = 18$)	Dataset2 ($n = 15$)
Apromore - ProDrift	$[X^2(7) = 67.078; p < 0.001]$	$[X^2(7) = 33.050; p < 0.001]$
VDD system	$[X^2(7) = 100.931; p < 0.001]$	$[X^2(7) = 88.539; p < 0.001]$

4.2 Influence of the Initial Window Size on the Accuracy of the Apromore - ProDrift Adaptive

The Apromore - ProDrift provides an adaptive approach; however, the user must define the initial window size. In this scenario, we evaluated the impact of the initial window size on the accuracy of the detected drifts using the same values of the first analysis. The F-score is calculated using an $et = 100$ because the window size adapts over time.

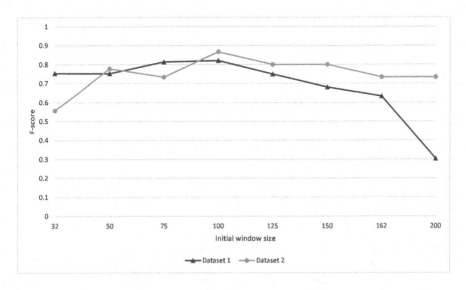

Fig. 4. Average of the F-score from Apromore - ProDrift adaptive.

We can observe in Fig. 4 that the F-scores from dataset1 show a slight variation until the initial window size of 125. Then the value considerably drops, ending with an average F-score of 0.3 using the default initial window size (200). In dataset2, we identify a slight variation in all values, with the highest average F-score with initial window size 100 (0.87) and the lowest at initial window size 32 (0.56). We applied Friedman's test ($\alpha = 0.05$), and there is a statistically significant difference in the accuracy depending on which initial window size is set for dataset1, $[X^2(7) = 42.975; p < 0.001]$. However, for dataset2 there is no evidence of a statistically significant difference $[X^2(7) = 11.288; p = 0.127]$.

4.3 Comparing the Accuracy of All Selected Methods

In this scenario, we aim to compare the accuracy of all tested methods. We selected the parameter configuration from the previous experiments, resulting in the highest F-score for each dataset and approach. In the ProM - Concept Drift, by the time we performed the tests, we could only get results using the configuration: do not split the log, local features (all activities selected), follows relation, numeric value, J-measure = 10, Kolmogorov-Smirnov, ADWIN, sudden drift search, trace amount, population minimum size = 50, p-value threshold = 0.4, step size =1 and population maximum size = 500.

Table 2. Average of the F-scores for the best parameter configuration.

Dataset	Apromore ProDrift fixed	VDD system	Apromore ProDrift adaptive	ProM Concept Drift adaptive
1	0.930	0.940	0.821	0.128
2	0.867	0.667	0.867	0.333

We calculated the average of the F-score (Table 2) for each dataset using the best configuration from previous experiments; then, we applied Friedman's test ($\alpha = 0.05$) pairwise. For the fixed approaches the error tolerance is set to the window size, and for the adaptive approaches, we used the value 100. There is a significant difference in the accuracy between the tools for both datasets. Thus, we plot the Nemenyi post-hoc critical distance diagrams to explore the differences. For dataset1 (Fig. 5), there is no statistically significant difference between both approaches of Apromore - ProDrift, and VDD; however, there is a statistically significant difference in the ProM - Concept Drift accuracy. Furthermore, in dataset2 (Fig. 6, there is no statistically significant difference between all the tools.

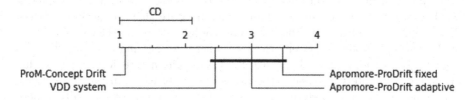

Fig. 5. Nemenyi *post-hoc* critical distance diagrams for dataset1.

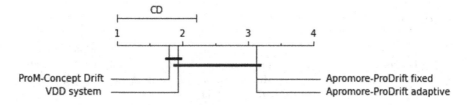

Fig. 6. Nemenyi *post-hoc* critical distance diagrams for dataset2.

5   Discussion and Conclusions

The Apromore - ProDrift shows promising results for both datasets. The experiments with dataset1 show that the initial window's size influenced the Apromore - ProDrift adaptive accuracy. The ProM - Concept Drift presented low accuracy for both datasets, with the highest F-score of 0.56. However, we could only perform the experiments with a specific parameter configuration, which can explain these results. The VDD system obtained the highest average F-score for the first dataset (0.940) and the third average F-score for dataset2 (0.667). Still, our experiments show that the accuracy is very sensitive to the window size configuration, contradicting [13], where the authors stated that the window size marginally influenced VDD's detected drifts.

Therefore, analyzing the best parameter configuration of each tool, we have no significant differences between the tools for dataset2, and only ProM - Concept Drift presented evidence of a difference for dataset1. We believe this reinforces that we must tune the parameter configuration of the methods for the dataset's characteristics to improve the accuracy of the drift detection. The ProM - Concept Drift is the only tool we cannot perform this tuning because we could only evaluate it with a specific parameter configuration. We also highlight that applying any process drift detection tools in real datasets is still challenging because we do not *a priori* know the drifts and their characteristics for performing a parameter configuration tuning. Therefore, the lack of publicly available synthetic datasets with process drifts also challenges performing such tuning in a controlled environment.

We aimed to compare the accuracy of process drift detection tools and design a generic experimental protocol that can be replayed for evaluating other tools, datasets, or scenarios. The designed protocol provided an objective comparison between the selected tools. The findings indicate that the accuracy of the detected drifts is sensitive to the parameter configuration and the characteristics of the dataset applied as input. However, there are not many public datasets available, and the ones available still miss some characteristics of the real-life drifts, e.g., the interval between the drifts is usually fixed. In future work, we plan to apply the same experimental protocol in a synthetic dataset with a more diverse drift configuration.

References

1. van der Aalst, W.M.P., Pesic, M., Schonenberg, H.: Declarative workflows: balancing between flexibility and support. Comput. Sci. Res. Dev. **23**(2), 99–113 (2009)
2. van der Aalst, W.M.: Process Mining: Data Science in Action. Springer, Berlin (2016). https://doi.org/10.1007/978-3-662-49851-4
3. Bose, R.P.J.C., van der Aalst, W.M.P., Žliobaitė, I., Pechenizkiy, M.: Handling concept drift in process mining. In: Mouratidis, H., Rolland, C. (eds.) CAiSE 2011. LNCS, vol. 6741, pp. 391–405. Springer, Heidelberg (2011). https://doi.org/10.1007/978-3-642-21640-4_30

4. Bose, R.P.C., Van Der Aalst, W.M., Zliobaite, I., Pechenizkiy, M.: Dealing with concept drifts in process mining. IEEE Trans. Neural Netw. Learn. Syst. **25**(1), 154–171 (2014)

5. Ceravolo, P., Marques Tavares, G., Junior, S.B., Damiani, E.: Evaluation goals for online process mining: a concept drift perspective. IEEE Transa. Servi Computi. **15**, 2473–2489 (2020)

6. Maaradji, A., Dumas, M., Rosa, M.L., Ostovar, A.: Detecting sudden and gradual drifts in business processes from execution traces. IEEE Trans. Knowl. Data Eng. **29**(10), 2140–2154 (2017)

7. Maaradji, A., Dumas, M., La Rosa, M., Ostovar, A.: Fast and accurate business process drift detection. In: Motahari-Nezhad, H.R., Recker, J., Weidlich, M. (eds.) BPM 2015. LNCS, vol. 9253, pp. 406–422. Springer, Cham (2015). https://doi.org/10.1007/978-3-319-23063-4_27

8. Martjushev, J., Bose, R.P.J.C., van der Aalst, W.M.P.: Change point detection and dealing with gradual and multi-order dynamics in process mining. In: International Conference on Business Informatics Research, pp. 1–15 (2015)

9. Omori, N.J., Tavares, G.M., Ceravolo, P., Barbon, S.: Comparing concept drift detection with process mining tools. In: SBSI'19: Proceedings of the XV Brazilian Symposium on Information Systems, pp. 1–8. ACM, Aracaju, Brazil (2019)

10. Ostovar, A., Maaradji, A., La Rosa, M., ter Hofstede, A.H.M., van Dongen, B.F.V.: Detecting drift from event streams of unpredictable business processes. In: Comyn-Wattiau, I., Tanaka, K., Song, I.-Y., Yamamoto, S., Saeki, M. (eds.) ER 2016. LNCS, vol. 9974, pp. 330–346. Springer, Cham (2016). https://doi.org/10.1007/978-3-319-46397-1_26

11. Sato, D.M.V., De Freitas, S.C., Barddal, J.P., Scalabrin, E.E.: A Survey on concept drift in process mining. ACM Comput. Surv. **54**(9), 1–38 (2022), https://dl.acm.org/doi/abs/10.1145/3472752

12. Vander Aalst, W.S., et al.: Process Mining manifesto. In: Daniel, F., Barkaoui, K., Dustdar, S. (eds.) BPM 2011. LNBIP, vol. 99, pp. 169–194. Springer, Heidelberg (2012). https://doi.org/10.1007/978-3-642-28108-2_19

13. Yeshchenko, A., Di Ciccio, C., Mendling, J., Polyvyanyy, A.: Visual drift detection for sequence data analysis of business processes. IEEE Trans. Visual. Comput Graphics pp. 1–1 (2021). https://eeexplore.ieee.org/document/9316994/

14. Zheng, C., Wen, L., Wang, J.: Detecting process concept drifts from event logs. In: Panetto, H., et al. (eds.) OTM 2017. LNCS, vol. 10573, pp. 524–542. Springer, Cham (2017). https://doi.org/10.1007/978-3-319-69462-7_33

Improving Solar Flare Prediction by Time Series Outlier Detection

Junzhi Wen[(✉)] [iD], Md Reazul Islam, Azim Ahmadzadeh[iD],
and Rafal A. Angryk[iD]

Georgia State University, Atlanta, GA 30302, USA
jwen6@student.gsu.edu

Abstract. Solar flares not only pose risks to outer space technologies and astronauts' well being, but also cause disruptions on earth to our high-tech, interconnected infrastructure our lives highly depend on. While a number of machine-learning methods have been proposed to improve flare prediction, none of them, to the best of our knowledge, have investigated the impact of outliers on the reliability and robustness of those models' performance. In this study, we investigate the impact of outliers in a multivariate time series benchmark dataset, namely SWAN-SF, on flare prediction models, and test our hypothesis. That is, there exist outliers in SWAN-SF, removal of which enhances the performance of the prediction models on unseen datasets. We employ Isolation Forest to detect the outliers among the weaker flare instances. Several experiments are carried out using a large range of contamination rates which determine the percentage of present outliers. We assess the quality of each dataset in terms of its actual contamination using TimeSeriesSVC. In our best findings, we achieve a 279% increase in True Skill Statistic and 68% increase in Heidke Skill Score. The results show that overall a significant improvement can be achieved for flare prediction if outliers are detected and removed properly.

Keywords: Solar flare prediction · Time series classification · Outlier detection · Multivariate time series · Isolation forest

1 Introduction

Solar flares are abrupt bursts of energy from the Sun that emit large amounts of electromagnetic radiation. They are frequently accompanied by a coronal mass ejection (CME), which is a huge bubble of radiation from the Sun. While most of the radiation and particles from a solar flare are filtered by the earth's atmosphere, intense solar flares still can release radiation that may penetrate and interfere with the radio communications, cause power outages, and pose irreversible health risks to astronauts engaging in extravehicular activities. Based on the peak flux of soft X-ray with a range of wavelengths from 0.1 to 0.8 nm detected by National Oceanic and Atmospheric Administration (NOAA)'s GOES

© The Author(s), under exclusive license to Springer Nature Switzerland AG 2023
L. Rutkowski et al. (Eds.): ICAISC 2022, LNAI 13589, pp. 152–164, 2023.
https://doi.org/10.1007/978-3-031-23480-4_13

satellites, solar flares are logarithmically classified into five classes as follows, namely A, B, C, M, and X, from weakest to strongest. An X-class flare is ten times stronger than an M-class flare, a hundred times stronger than a C-class flares, and so on. Among the five classes, M- and X-class flares are often targeted in space-weather prediction because they are much more likely to cause adverse effects to the earth.

Due to the potential threats that solar flares pose to human society, flare prediction has been receiving a lot of attention during the past two decades. As machine learning techniques have achieved remarkable success in multiple fields over the recent years, they have been employed to predict flares as well [21]. Since solar flares are a spatiotemporal phenomenon with a pre-flare phase [20], it has been suggested that to achieve a higher and more robust performance, time series of predictive parameters for flare forecasting should be used rather than point-in-time values. While a number of machine learning methods using time series have been introduced to improve flaring prediction [1,10,17], none of them, to our best knowledge, has systematically investigated the impact of outliers in time series data for flare prediction.

Outlier detection is an important task for many data mining and machine learning applications, and it has been applied broadly in various domains such as economy, biology, and astronomy [5]. Outliers are the data instances that differ substantially from the rest of the data. In classification tasks, outliers can mislead the classifier resulting in poor performance. Therefore, the discovery of outliers is crucial to better understand the underlying nature of the data and develop more efficient methods. In this study, we investigate the possibility of improving the performance of flare prediction algorithms by removing outliers that are detected by an outlier detection algorithm, named Isolation Forest (iForest) [19], from a multivariate time series benchmark dataset, Space Weather ANalytics for Solar Flares (SWAN-SF) [2]. We hypothesize that there exist outliers that negatively impact flare prediction in SWAN-SF.

The rest of the paper is organized as follows: In Sect. 2, we introduce some advanced and popular outlier detection algorithms for time series data. In Sect. 3, we briefly describe the SWAN-SF dataset that we use in this study. In Sect. 4, we discuss our selection of the outlier detection algorithm. We also introduce the experiment design, methodology for tackling the class-imbalance issue, the classifier chosen, and the hyperparameter tuning process, as well as the metrics for evaluation. In Sect. 5, we talk about our experiments and discuss the results. Finally, we conclude and propose future work in Sect. 6.

2   Related Work

Generally, outlier detection can be done locally and globally for time series data. The former means the detection of outliers within time series and the latter concerns the outliers among a set of time series data. In global outlier detection, different algorithms have been proposed based on the requirements in different fields and they could be supervised or unsupervised depending on the availability of labels in the data [14].

Unsupervised outlier detection can be achieved by discriminative methods, which rely on a similarity function that measures the similarity between two time series sequences. Once the similarity function is defined, a clustering mechanism is applied to cluster the data instances such that within-cluster similarity is maximized and between-cluster similarity is minimized. An outlier score is then assigned to each testing instance based on the distance to its closest cluster's centroid (or medoid). SequenceMiner [9] uses longest common subsequence (LCS) as the similarity measure to handle time series with different lengths, but the time series has to be discretized, which will cause the loss of information. In [5], dynamic time warping (DTW) is used to address the distortion in the time axis and calculate more accurate similarity between two time series. However, the high computational complexity makes it impossible to train a model on a large dataset in a reasonable time.

More recently, unsupervised outlier detection methods based on deep learning have received a lot of attention. The GGM-VAE [13] employs Gated Recurrent Unit (GRU) cells under a variational autoencoder (VAE) framework to discover the correlations among multivariate time series data. The Robust Deep Autoencoders (RDA) [26] combines deep autoencoders and robust principal component analysis (RPCA) to isolate noise and outliers in the input data. The Multi-Scale Convolutional Recursive EncoderDecoder (MSCRED) [25] jointly considers time dependence, noise robustness, and interpretation of anomaly severity. Although these methods obtain good results, they do not take into account the training time (or energy consumption) on large datasets.

Isolation Forest (iForest) [19], different from other outlier detection methods, is a tree-based algorithm that explicitly isolates outliers. Because of its capability of running fast on large and high-dimensional datasets, iForest has been utilized for multivariate time series data in a variety of domains [3,11]. More details about iForest are explained in Sect. 4.1.

3   SWAN-SF Dataset

Space Weather ANalytics for Solar Flares (SWAN-SF) is a benchmark dataset introduced by [2], which entirely consists of multivariate time series (MVTS) data. The development of SWAN-SF provides a unified testbed for solar flare prediction algorithms. The dataset contains 4,075 MVTS data instances from active regions of 10,000 flare reports spanning over 8 years of solar active-region data from Solar Cycle 24 (May 2010–December 2018). Each MVTS data instance of SWAN-SF represents a 12-h observation window of 51 flare-predictive parameters. Each time series has 60 records with a 12-min cadence, and corresponds to a reported active region.

The data instances in SWAN-SF are collected through a sliding-window methodology with a 1-h step size. A MVTS data instance is labeled by the class of the strongest flare reported within a 24-h prediction window right after the observation window. If no flare happens or only A-class flares are reported within an observation window, the data instance is labeled as a flare-quiet instance, denoted by N. In this work, we run a few experiments on a dichotomous version of SWAN-SF, as well as its original 5-class version. That is, we group the X- and M-class flares into one group and treat them as *flaring* instances, and group the other classes (including the N class) and treat them as *non-flaring* instances. The instances of the former group are also denoted by the XM class, and correspondingly, the instances of the latter group are denoted by the CBN class.

Because of the sliding-window methodology used for the creation of SWAN-SF, caution must be taken when dealing with the *temporal coherence* of data [1], which can be briefly described as follows: since temporally adjacent time series have over 91% of overlap, random sampling of data in order to create non-overlapping training, validation, and test sets, will fail. This introduces bias to the learner and also obscures possible overfitting. To properly deal with the temporal coherence, we take advantage of the fact that the dataset is already split into five non-overlapping partitions, and each partition has approximately the same number of X- and M-class flares. Therefore, the training and testing datasets in our experiments are selected from different partitions to prevent the effect of temporal coherence. The details of the sample sizes for each partition of SWAN-SF are listed in Table 1.

SWAN-SF also exhibits extreme class imbalance. In each partition, the number of the flaring instances, which are referred to as the minority class, is significantly less than the instances of the non-flaring instances, which are referred to as the majority class. For instance, there is a 1:364 imbalance ratio between X and N and a 1:58 imbalance ratio between XM and BCN in partition 1. The imbalance ratio of each partition of SWAN-SF is listed in Table 1.

Table 1. The sample sizes and imbalance ratios of each partition in SWAN-SF

Partition	Class					Imbalance ratio	
	X	M	C	B	N	X:N	XM:CBN
Parition 1	165	1089	6416	5692	60130	1:364	1:58
Parition 2	72	1392	8810	4978	73368	1:1019	1:62
Parition 3	136	1288	5639	685	34762	1:256	1:29
Parition 4	153	1012	5956	846	43294	1:283	1:43
Parition 5	19	971	5763	5924	62688	1:3299	1:75

4   Methodology

4.1   Outlier Detection Algorithm

Isolation Forest (iForest) [19] is a tree-based algorithm that detects outliers (anomalies) efficiently and effectively. Unlike most existing model-based outlier detection approaches, which construct a profile of normal instances and then identify instances that do not conform to the profile as outliers, iForest explicitly isolates outliers by random partitioning [19]. Since anomalies consist of fewer instances and are assumed to have attribute values that are very different from normal instances, they are more likely to be isolated earlier during random partitioning (i.e., shorter paths in a tree structure).

An iForest consists of a number Isolation Trees (iTrees), and each iTree is built on a random sample of data by recursively dividing it with a randomly selected attribute and a randomly selected split value until a terminating condition is satisfied. Each data instance is then passed through the iForest and receives an outlier score. A pre-defined *contamination rate*, i.e., the proportion of outliers in the dataset, is required to provide the iForest algorithm with a halting criterion. Given a contamination rate r, the first $r\%$ of the instances with higher outlier scores will be flagged as the outliers. Because of the effectiveness of iForest on large collections of high-dimensional data [19], it fits very well to our outlier detection investigation on SWAN-SF.

4.2 Experiment Design

The experiments in this study are generally designed to test the hypothesis that there exist outliers in the SWAN-SF dataset, which can negatively affect flare prediction. As mentioned in Sect. 3, there are 51 flare-predictive parameters in SWAN-SF. In this proof-of-concept study, we limit our experiments to a subset of five flare-predictive parameters only. These parameters are chosen from the top ranked features previously discovered in [8], namely (as named in the dataset) TOTUSJH, TOTBSQ, TOTPOT, TOTUSJZ, and ABSNJZH (see [2] for the list of all parameters and their definitions). To minimize the learning bias, we use Partition 1 as the training set in all experiments and report the performance on the remaining partitions.

Using iForest, given a contamination rate, we first detect outliers among the non-flaring instances in Partition 1. We then remove the detected outliers and combine the rest of the non-flaring instances with the flaring instances to build a flare forecasting model. We repeat this by gradually increasing the contamination rate (from 0.0 to 50.0) and producing new training datasets. In each case, we apply the *climatology-preserving undersampling* (as discussed in Sect. 4.3) to mitigate the effect of the extreme class-imbalance issue on training our classifiers. Min-max normalization is followed to avoid the influence of different scales between parameters [15]. To show the robustness of the model, we carry out 10-fold cross validation, i.e., we repeat the undersampling and normalization processes 10 times and report the average and the variance of the performance of the classifiers trained on those different datasets for each contamination rate.

The trained models are tested on Partitions 2 through 5 separately. The test sets are kept unchanged except for normalization; no outlier detection, outlier removal, or undersampling is applied. This strategy provides a series of unbiased experiments, which is closest to the operational setting. The design of our experiments is shown in Fig. 1.

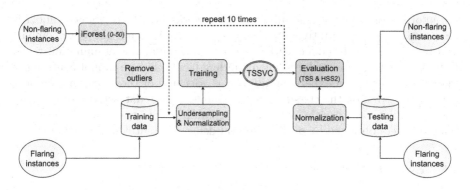

Fig. 1. Experiment Design: the outlier detection operation is only applied to the non-flaring instances in Partition 1. Only normalization is applied to the test sets. 10-fold cross validation is applied for a robust evaluation. Note that the flaring instances and non-flaring instances used in Sect. 5.1 and Sect. 5.2 are different.

4.3 Tackling the Class-Imbalance Issue

The class-imbalance issue can affect the performance of any classifier resulting in superficially good classification/prediction scores, as thoroughly discussed in [1]. We use the so-called climatology-preserving undersampling, as suggested in [1], which achieves a 1:1 ratio between the minority and majority instances while preserving the distribution of flare classes. Preserving the distribution of flare classes is important during undersampling because it produces more realistic data for training the model, which will hence generate a reliable model performance on testing data. Although more advanced strategies, like synthetic data generation, can be used to handle the class-imbalance issue, to avoid confounding variables that we cannot control, climatology-preserving undersampling is preferred in our study.

4.4 Evaluation Model and Hyperparameter Tuning

To evaluate the performance of our binary classification task, we choose Support Vector Machines (SVM) as the classifier. However, since we use multivariate time series data, regular SVM is not appropriate for our experiments. TimeSeriesSVC, from the tslearn machine learning toolkit [24], is the SVM classifier designed specifically for time series data and is employed as the classifier in this study. TimeSeriesSVC operates support vector classification by casting DTW distances measure as definite kernels for time series [12].

In order to achieve an optimal performance for TimeSeriesSVC on SWAN-SF, we use the exhaustive grid-search method to tune models' hyperparameters. Ideally a grid search should be carried out separately using each dataset to find one optimal model per contamination rate. However, although such an independent optimization may result in higher classification performance, it may not necessarily bring robustness; in operational settings (i.e., real-time flare forecasting) the information about which subset of data works best for a trained model is unknown. More importantly, note that our objective is to present a fair comparison between all such models with respect to their unique contamination rates. Data-specific tuning introduces a confounding factor that we cannot control and therefore, it results in experimentation bias. Because of these reasons, we only apply grid search on the classifier that is trained on the dataset without removing any outliers, and use the optimal hyperparameters to train models across all other datasets obtained by outlier removal using different contamination rates.

4.5   Evaluation Metrics

Many measures have been developed for evaluation of the deterministic performance of classifiers using the four quantities of the confusion matrix [23]: true positives (TP), true negatives (TN), false positives (FP), and false negatives (FN). In flare prediction studies, the True Skill Statistic (TSS) [16] and the updated Heidke Skill Score (HSS2) [4] are typically used for performance evaluation (e.g., in [6–8,18,22]), and they are used in this study as well. Next, we briefly review these measures and the reasons justifying their appropriateness for a rare-event classification problem such as flare forecasting.

TSS, as shown in Eq. 1, measures the difference between the probability of true prediction (i.e., true positive rate) and the probability of false alarm (i.e., false positive rate). TSS ranges from –1 to 1, where –1 indicates that every prediction the classifier makes is incorrect, and 1 indicates a perfect performance meaning the classifier is correct for all of its predictions.

$$TSS = \frac{TP}{TP + FN} - \frac{FP}{FP + TN} \tag{1}$$

However, the drawback of TSS is that it equates all models for which the difference between the true-positive and false-positive rates is the same. This is not a universally sound assumption (see the numerical examples in [1]). Therefore, it might be misleading to use TSS alone. This is why it is coupled with HSS2. As shown in Eq. 2, HSS2 measures the fractional improvement of prediction that the classifier has over a random guess (no-skill) model. Similar to TSS, HSS2 ranges from –1 to 1, with 1 indicating a perfect performance, –1 indicating reverse assignment of labels to all instances, and 0 indicating no skill (i.e., as same as a random guess).

$$HSS2 = \frac{2((TP \cdot TN) - (FN \cdot FP))}{P(FN + TN) + N(TP + FP)} \tag{2}$$

5   Experiments, Results, and Discussion

5.1   Experiment A: Impact of Outliers on X-N Classification

We start by simplifying the flare prediction problem where only X-class and N-class instances are used. Since they are the most extreme classes (i.e., X is the strongest and N is the weakest), the data points are far apart, and it is easy for the classifier to distinguish between them. Through this experiment, we investigate the impact of outliers on the simplest case. Outlier detection is applied to N-class instances, which are the non-flaring instances used in this case. Random undersampling is applied in this experiment since there is only one class in the majority. The hyperparameters tuned to be used for the TimeSeriesSVC classifier are RBF kernel with the coefficient γ being 0.01 and a soft margin constant C of 100.

As we can see in Fig. 2, a significant improvement is achieved by the removal of outliers. Both TSS and HSS2 increase as the contamination rate increases in the early phase, and the classifier becomes more and more robust (i.e., smaller variance). After a certain contamination rate which is unique to each partition of SWAN-SF, HSS2 starts dropping while TSS remains on the same level. Our empirical investigation shows that because at earlier stages, when the contamination rate is low, the outliers that confuse the classifier are detected and removed from the training set, hence resulting in an improvement on TSS and HSS2. However, after a certain contamination rate, iForest is forced to detect normal instances as outliers. This makes the decision boundary of the TimeSeriesSVC classifier move further towards the majority instances, so there are more FP and fewer TN, hence a smaller HSS2.

5.2   Experiment B: Impact of Outliers on XM-CBN Classification

In this experiment, we have a more complex and realistic case where the data instances are composed of all five classes of SWAN-SF, as opposed to the X-versus N-class flares that we investigated before. The non-flaring instances in this experiment consist of all instances from C, B, and N classes. Outlier detection is then applied to the group of these three classes. In addition, climatology-preserving undersampling is applied to preserve the portions of subclasses in the majority. The hyperparameters used for the TimeSeriesSVC classifier in this experiment are the same as those in Experiment A.

As Fig. 3 shows, there is a significant improvement of TSS on Partitions 3 and 5 while the same level of HSS2 is preserved. A minor improvement is also achieved on Partitions 2 and 4. There is a drop in terms of HSS2 at last several contamination rates on each testing partition, which is expected because too many normal data instances are forced to be removed, and this makes the decision boundary move to negative class, causing more FP, hence a smaller HSS2. Furthermore, the best TSS remains around 0.8 on each testing partition, and this may be because there also exist some outliers in flaring instances in our training data. Since, in our experiments, we only apply outlier detection to non-flaring instances (in order to preserve as many flaring instances as possible), the outliers in flaring instances are not removed. This causes more FN, hence a smaller TSS. Nevertheless, we still see that improvement can be achieved even in this simplified case.

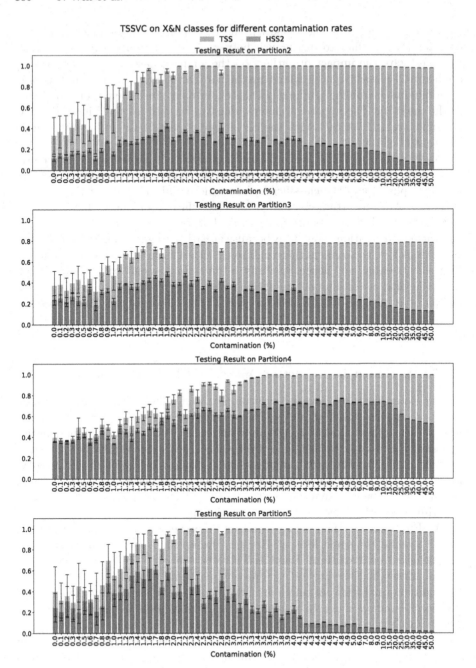

Fig. 2. Result of Experiment A. The blue bar represents TSS value and the green bar represents HSS2 value at each contamination rate. The height and the black error bar of each bar represents the mean value and the variance, respectively, of the corresponding measure over 10-fold cross validation. The contamination of 0.0% means no outlier detection applied and is the baseline. (Color figure online)

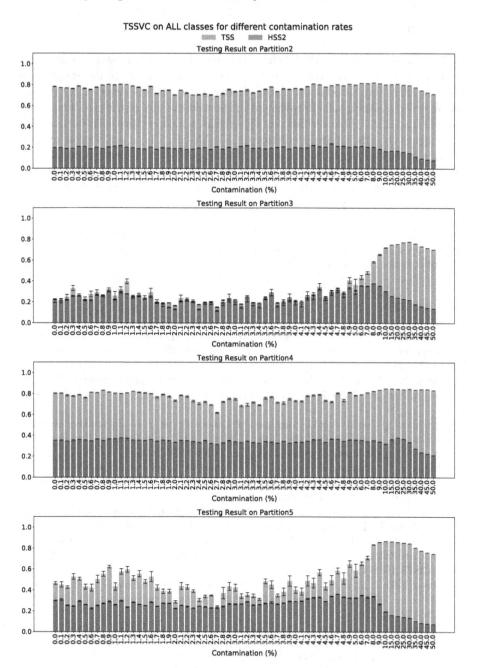

Fig. 3. Result of Experiment B. The blue bar represents TSS value and the green bar represents HSS2 value at each contamination rate. The height and the black error bar of each bar represents the mean value and the variance, respectively, of the corresponding measure over 10-fold cross validation. The contamination of 0.0% means no outlier detection applied and is the baseline. (Color figure online)

6   Conclusion and Future Work

In this study, we used the SWAN-SF benchmark dataset to investigate the impact of outliers in time series data. We designed two experiments to investigate how outliers affect flare prediction. After removing the outliers detected by iForest, we observed a significant improvement in the performance of the classifiers based on TSS and HSS2. This verified our hypothesis, which is there exist outliers that can negatively affect flare prediction in SWAN-SF. There are several avenues we can further explore. For example, if we can achieve a larger improvement, in Experiment B, by doing outlier detection on each individual class of non-flaring instances. Additionally, a more advanced outlier detection algorithm for multivariate time series can be utilized.

Acknowledgement. This project has been supported in part by funding from CISE, MPS and GEO Directorates under NSF award #1931555, and by funding from the LWS Program, under NASA award #80NSSC20K1352.

References

1. Ahmadzadeh, A., et al.: How to train your flare prediction model: revisiting robust sampling of rare events. Astrophys. J. Suppl. Ser. **254**(2), 23 (2021). https://doi.org/10.3847/1538-4365/abec88
2. Angryk, R.A., et al.: Multivariate time series dataset for space weather data analytics. Nat. Sci. Data **7**(1), 227 (2020). https://doi.org/10.1038/s41597-020-0548-x
3. Audibert, J., et al.: USAD: unsupervised anomaly detection on multivariate time series. In: Proceedings of the 26th ACM SIGKDD International Conference on Knowledge Discovery & Data Mining pp. 3395–3404 (2020)
4. Balch, C.C.: Updated verification of the Space Weather Prediction Center's solar energetic particle prediction model. Space Weather **6**(1), S01001 (2008). https://doi.org/10.1029/2007SW000337
5. Benkabou, S.E., et al.: Unsupervised outlier detection for time series by entropy and dynamic time warping. Knowl. Inf. Syst. **54**(2), 463–486 (2018)
6. Benvenuto, F., et al.: A Hybrid supervised/unsupervised machine learning approach to solar flare prediction. Astrophys. J. **853**(1), 90 (2018). https://doi.org/10.3847/1538-4357/aaa23c
7. Bloomfield, D.S., et al.: Toward Reliable Benchmarking of Solar Flare Forecasting Methods. Astrophys. J. Lett. **747**(2), L41 (2012). https://doi.org/10.1088/2041-8205/747/2/L41
8. Bobra, M.G., Couvidat, S.: Solar flare prediction usingsdo/HMI vector magnetic field data with a machine-learning algorithm. Astrophys. J. **798**(2), 135 (2015). https://doi.org/10.1088/0004-637x/798/2/135

9. Budalakoti, S., et al.: Anomaly detection and diagnosis algorithms for discrete symbol sequences with applications to airline safety. IEEE Trans. Syst. Man Cybernet. Part C (Appl. Rev.) **39**(1), 101–113 (2009). https://doi.org/10.1109/TSMCC.2008. 2007248

10. Chen, Y., Kempton, D.J., Ahmadzadeh, A., Angryk, R.A.: Towards synthetic multivariate time series generation for flare forecasting. In: Rutkowski, L., Scherer, R., Korytkowski, M., Pedrycz, W., Tadeusiewicz, R., Zurada, J.M. (eds.) ICAISC 2021. LNCS (LNAI), vol. 12854, pp. 296–307. Springer, Cham (2021). https://doi.org/10.1007/978-3-030-87986-0_26

11. Cook, A.A., et al.: Anomaly detection for IoT time-series data: a survey. IEEE Internet Things J. **7**(7), 6481–6494 (2020). https://doi.org/10.1109/JIOT.2019. 2958185

12. Cuturi, M.: Fast global alignment kernels. In: Proceedings of the 28th International Conference on Machine Learning (ICML-2011), pp. 929–936 (2011)

13. Guo, Y., et al.: Multidimensional time series anomaly detection: a GRU-based gaussian mixture variational autoencoder approach. In: Proceedings of The 10th Asian Conference on Machine Learning. Proceedings of Machine Learning Research, vol. 95, pp. 97–112. PMLR (14–16 November 2018). https://proceedings.mlr.press/v95/guo18a.html

14. Gupta, M., et al.: Outlier detection for temporal data: a survey. IEEE Trans. Knowl. Data Eng. **26**(9), 2250–2267 (2013)

15. Han, J., et al.: 3 - data preprocessing. In: Data Mining, 3rd edn., pp. 83–124. The Morgan Kaufmann Series in Data Management Systems,3rd edn. Morgan Kaufmann, Boston, (2012). https://doi.org/10.1016/B978-0-12-381479-1.00003-4

16. Hanssen, A., Kuipers, W.: On the relationship between the frequency of rain and various meteorological parameters: (with reference to the problem ob objective forecasting). Koninkl. Nederlands Meterologisch Institut. Mededelingen en Verhandelingen, Staatsdrukkerij- en Uitgeverijbedrijf (1965). https://books.google.com/books?id=nTZ8OgAACAAJ

17. Hostetter, M., et al.: Understanding the impact of statistical time series features for flare prediction analysis. In: 2019 IEEE International Conference on Big Data (Big Data), pp. 4960–4966 (2019). https://doi.org/10.1109/BigData47090.2019.9006116

18. Jolliffe, I.T., et al.: Forecast Verification: A Practitioner's Guide in Atmospheric Science. John Wiley & Sons (2012). https://doi.org/10.1002/9781119960003

19. Liu, F.T., et al.: Isolation forest. In: 2008 Eighth IEEE International Conference on Data Mining. pp. 413–422 (2008). https://doi.org/10.1109/ICDM.2008.17

20. Martens, P.: Solar flares: preflare phase. In: Murdin, P. (ed.) Encyclopedia of Astronomy and Astrophysics, p. 2288 (2000). https://doi.org/10.1888/0333750888/2288

21. Massone, A.M., et al.: Chapter 14 - machine learning for flare forecasting. In: Machine Learning Techniques for Space Weather, pp. 355–364. Elsevier (2018). https://doi.org/10.1016/B978-0-12-811788-0.00014-7

22. Sadykov, V.M., Kosovichev, A.G.: Relationships between characteristics of the line-of-sight magnetic field and solar flare forecasts. Astrophys. J. **849**(2), 148 (2017). https://doi.org/10.3847/1538-4357/aa9119

23. Stehman, S.V.: Selecting and interpreting measures of thematic classification accuracy. Remote Sens. Environ. **62**(1), 77–89 (1997). https://doi.org/10.1016/S0034-4257(97)00083-7

24. Tavenard, R., et al.: Tslearn, a machine learning toolkit for time series data. J. Mach. Learn. Res. **21**(118), 1–6 (2020). http://jmlr.org/papers/v21/20-091.html

25. Zhang, C., et al.: A deep neural network for unsupervised anomaly detection and diagnosis in multivariate time series data. In: Proceedings of the AAAI Conference on Artificial Intelligence, vol. 33, pp. 1409–1416 (2019)
26. Zhou, C., Paffenroth, R.C.: Anomaly detection with robust deep autoencoders. In: Proceedings of the 23rd ACM SIGKDD International Conference on Knowledge Discovery and Data Mining, pp. 665–674 (2017)

Various Problems of Artificial
Intelligence

Tourism Stock Prices, Systemic Risk and Tourism Growth: A Kalman Filter with Prior Update DSGE-VAR Model

David Alaminos[1]([⊠]) [iD] and M. Belén Salas[2,3]

[1] Department of Business, Universitat de Barcelona, Barcelona, Spain
alaminos@ub.edu
[2] Department of Finance and Accounting, Universidad de Málaga, Málaga, Spain
belensalas@uma.es
[3] Cátedra de Economía Y Finanzas Sostenibles, Universidad de Málaga, Málaga, Spain

Abstract. Dynamic Stochastic General Equilibrium (DSGE) and Vector Autoregressive (VAR) models allow for probabilistic estimations to formulate macroeconomic policies and monitor them. One of the objectives of creating these models is to explain and understand financial fluctuations through a consistent theoretical framework. In the tourism sector, stock price and systemic risk are key financial variables in the international transmission of business cycles. Advances in Bayesian theory are providing an increasing range of tools that researchers can employ to estimate and evaluate DSGE and VAR models. One area of interest in previous literature has been to design a Bayesian robust filter, that performs well concerning an uncertainty class of possible models compatible with prior knowledge. In this study, we propose to apply the Bayesian Kalman Filter with Prior Update (BKPU) in a tourism field to increase the robustness of DSGE and VAR models built for small samples and with irregular data. Our results indicate that BKPU improves the estimation of these models in two aspects. Firstly, the accuracy levels of the computing of the Markov Chain Monte Carlo model are increased, and secondly, the cost of the resources used is reduced due to the need for a shorter run time. Our model can play an essential role in the monetary policy process, as central bankers could use it to investigate the relative importance of different macroeconomic shocks and the effects of tourism stock prices and achieve a country´s international competitiveness and trade balance for this sector.

Keywords: Dynamic stochastic general equilibrium · Bayesian kalman filter · Prior update · Markov chain monte carlo · Tourism stock prices · Systemic risk · Volatility

1 Introduction

The relevance of Dynamic Stochastic General Equilibrium (DSGE) and Vector Autoregressive (VAR) models has recently become increasingly important for their application in the analysis of business cycles and the detection of recessions [1]. These models

© The Author(s), under exclusive license to Springer Nature Switzerland AG 2023
L. Rutkowski et al. (Eds.): ICAISC 2022, LNAI 13589, pp. 167–181, 2023.
https://doi.org/10.1007/978-3-031-23480-4_14

are often estimated using the Kalman filter built into the Markov Chain Monte Carlo algorithm (in its Metropolis-Hastings version) to predict the posterior distributions of the parameters considered in the model. The evaluation of these models is being carried out by the result obtained in the standard deviation, the sample size, and its complexity. For these models, the literature shows the different fitting results once various Kalman filter algorithms have been applied. For instance, models built with the classical Kalman filter have offered a fit of 0.38–1.43 standard deviation with small samples and irregular data [2], however, their fits vary to 0.27–0.98 in large and regular samples [3]. On the other hand, models using non-classical Kalman filters have achieved even better results than those described above. With small samples and irregular data, the results of these models have been 0.29–0.41 [4], in contrast to 0.14–0.38 with large samples and regular data [5]. Hence, it is noticed that the classical Kalman Filter presents an accuracy in the interval 0.82–0.43 after simulations with samples larger than 100 observations, not giving a deviation sufficiently low in a small sample (a deviation of 0.64 when using a sample size of under 100 observations) [6]. Other investigations have also used DSGE models to analyse the link between monetary policy and stock price and exchange rate volatility using the VAR method for that connection [7]. In this way, [8] analyzed the connection between American tourism companies for the period 2018–2020, including the scenario of the Covid-19 pandemic. The analysis of the risk of contagion showed a significant increase during the Covid-19 pandemic. Small businesses become more systemically important in the pandemic, while the level of serious risk contagion harms the stock performance of American tourism companies. Since the tourism sector, stock price, and systemic risk are financial variables that are crucial for the transmission of economic cycles across borders, and most of the existing research in tourism is based on the hedonic method or the traditional single regression econometric model, it is essential to analyse these macroeconomic variables in this sector with DSGE models. Recent research has applied the DSGE models in other different areas of the tourism field [8, 9]. [9] incorporated the VAR model into a DSGE framework for the analysis of tourism development and sustainable economic growth. The results show that a 10% rise in tourism productivity can improve the value-added of the tourism sector by more than 4.11% and boost about 0.5% of GDP growth. [8] suggested a DSGE model to understand the effect of an infectious disease outbreak on tourism, They concluded the suitability of DSGE model to address the impact of the health crisis in this scenario given that the duration and severity of the outbreak are uncertain.

In summary, all this research has applied DSGE models, which have an infinite-order VAR representation. Hence, VARs have been widely used in the forecasting literature evaluating DSGE models. However, due to many parameters and short time series, classical estimates of the coefficients of the unconstrained VAR are often imprecise and forecasts are of low quality due to large estimation errors. A common method to address this problem is to apply Bayesian techniques. For addressing these precision problems of the current DSGE and VAR models, this paper builds on the Bayesian Kalman Filter with Prior Updating (BKPU), having already established its methodological supremacy in other domains for accurate sampling with only a few observations and with non-regular distributions of data [10]. Compared to the classical Kalman Filter used, our results show, in terms of accuracy, a more robust estimation, especially in out-of-sample estimations, a better performance with small and irregular samples. Therefore, the misspecification of previous literature is reduced, as our results also show a better estimation of posterior distributions [11]. These results can be very valuable when applied in DSGE and VAR models, as well as in other macroeconomic models that guide policymakers and other related interest groups in performing estimations. To fill the gap in the existing literature, our investigation evaluates the tourism stock prices volatility and systemic risk applying a BKPU DSGE model for Spain, increasing the robustness of DSGE and VAR models built for small samples and with irregular data.

2 Methods

2.1 Bayesian Kalman Filter with Prior Update

This algorithm framework involves similar recursive equations to those of the classical Kalman filter with the posterior effective noise statistics in place of the ordinary noise statistics. Posterior effective noise statistics represent the posterior distribution of the noise second-order statistics, namely, the covariance matrix, where the posterior distribution is obtained by incorporating observations into the prior distribution of unknown noise parameters. Now assume that the covariance matrices of the process and observation noise are not known and parameterized as $E\left[u_k^{\theta_1}(u_l^{\theta_1})^T\right] = Q^{\theta_1}\delta_{kl}$ and $E\left[v_k^{\theta_2}(v_l^{\theta_2})^T\right] = R^{\theta_2}\delta_{kl}$, $\theta = [\theta_1, \theta_2]$ being the set of unknown parameters governed by the prior distribution $\pi(\theta)$. The state-space model belongs to an uncertainty class Θ ($\theta \in \Theta$) of possible state-space models. If θ_1 and θ_2 are statistically independent, then the state-space model can be parameterized as $x_{k+1}^{\theta_1} = \Phi_k x_k^{\theta_1} + \Gamma_k u_k^{\theta_1}$ (3) and $y_k^\theta = H_k x_k^{\theta_1} + v_k^{\theta_2}$. The intrinsically the Bayesian robust Kalman filter that provides optimal performance on average concerning a prior distribution has been developed using the notions of Bayesian orthogonality principle and Bayesian innovation process in [10], and its structure is completely similar to that of the classical Kalman filtering with the noise covariances and the Kalman gain matrix replaced by the expected noise covariances and the effective Kalman gain matrix, respectively. Being $\psi_{IBR}(y; k) = \arg\min_{\psi \in \Psi} E_\theta\left[C_\theta(x_k, \psi(y; k))\right]$ where the expectation is taken relative to the prior distribution $\pi(\theta)$ governing Θ, $C_\theta(.)$ characterizes the filter cost relative to θ, and ψ_{IBR} is called an intrinsically Bayesian robust filter [10]. Considering the state-space model in y_k^θ and $\psi_{IBR}(y; k)$, let $\Upsilon_{k-1} = \{y_0, \ldots, y_{k-1}\}$ and $X_k = \{x_0, \ldots, x_k\}$ be the sequences of observations and states up to times k − 1 and k, respectively, with $f(\theta, \Upsilon_{k-1}, X_k)$ being the joint probability distribution of the uncertainty class Θ and observations and states. In the context of optimal Bayesian filtering theory, we seek a linear filter of the form $\hat{x}_k^\theta = \sum_{l \le k-1} G_{k,l}^\Theta y_l^\theta$ $G_{k,l}^\Theta =$ $\arg\min_{G_{k,l \in G}} E_\theta\left[E\left[\left(x_k^{\theta_1} - \sum_{l \le k-1} G_{k,l} y_l^\theta\right)^T \times \left(x_k^{\theta_1} - \sum_{l \le k-1} G_{k,l} y_l^\theta\right)\right]|\Upsilon_{k-1}\right]$, where G is the vector space of all n × m matrix-valued functions, $G_{k,l \in G}$ is a mapping $G_{k,l}$: N × N→R^{nxm} such that $\sum_{k=1}^{\infty}\sum_{l=1}^{\infty}\|G_{k,l}\|_2 < \infty$, $\|\bullet\|_2$ being the L_2 norm and $\hat{x}_k^\theta$ is called the optimal Bayesian least-squares estimate of x_k^θ. The following theorem, definition, and lemma are essential for the derivation of the OBKF framework and are restatements of their counterparts in [10] concerning the posterior distribution. The linear estimate $\hat{x}_k^\theta$

obtained in the last equation of the prior paragraph, is an optimal Bayesian least-squares estimate of x_k^θ, if and only if $E_\theta\left[E\left[\left(x_k^{\theta_1} - \hat{x}_k^\theta\right)\left(y_l^\theta\right)^T\right]|Y_{k-1}\right] = O_{n \times m} \forall\, l \leq k - 1$.

Consider this state-space model and let $\hat{x}_k^\theta$ be a linear estimate of x_k^θ that satisfies this space, then the random process $\tilde{z}_k^\theta = y_k^\theta - H_k \hat{x}_k^\theta$ is a zero-mean process, called the Bayesian innovation process, and $\forall\, l, l' \leq k - 1,$, we have $E_\theta\left[E\left[\tilde{z}_l^\theta (\tilde{z}_{l'}^\theta)^T\right]|Y_{k-1}\right] = E_\theta\left[H_l P_l^{x,\theta} H_l^T + R^{\theta_2} |Y_{k-1}\right]\delta_{ll'},$where $P_l^{x,\theta} = E\left[(x_l^{\theta_1} - \hat{x}_l^\theta)(x_l^{\theta_1} - \hat{x}_l^\theta)^T\right]$ is the estimation error covariance matrix of the OBKF at time l relative to θ. Let $\check{x}_k^\theta = \sum_{l \leq k-1} G_{k,l}\, \tilde{z}_l^\theta$ be an estimate of x_k^θ obtained using the information in $\tilde{z}_l^\theta = y_l^\theta - H_k \check{x}_k^\theta$, such that, $E_\theta\left[E\left[(x_k^{\theta_1} - \check{x}_k^\theta)\,(\tilde{z}_l^\theta)^T\right]|Y_{k-1}\right] = O_{n \times m}$. Then $E_\theta\left[E\left[(x_k^{\theta_1} - \check{x}_k^\theta)\,(y_l^\theta)^T\right]|Y_{k-1}\right] = O_{n \times m}$. Using the Bayesian orthogonality principle and the Bayesian innovation process, the recursive equations constituting the OBKF can be found similar to those for the Kalman filter in [10]. According to this, we can write $\hat{x}_k^\theta$ that satisfies the previous equations as: $\hat{x}_k^\theta = \sum_{l \leq k-1} G_{k,l}^\Theta\, \tilde{z}_l^\theta$.

Using $\hat{x}_k^\theta = \sum_{l \leq k-1} E_\theta\left[E\left[x_k^{\theta_1}\left(\tilde{z}_l^\theta\right)^T\right]|Y_{k-1}\right] E_\theta^{-1}\left[H_l P_l^{x,\theta} H_l^T + R^{\theta^2} |Y_{k-1}\right]\tilde{z}_l^\theta$, an update equation for $\hat{x}_k^\theta$ can be found as $\hat{x}_{k+1}^\theta = \Phi_k \hat{x}_k^\theta + \Phi_k K_k^{\Theta*}\tilde{z}_k^\theta$, where $K_k^{\Theta*} = E_\theta\left[P_k^{x,\theta}|Y_{k-1}\right]H_k^T E_\theta^{-1}\left[H_k P_k^{x,\theta} H_k^T + R^{\theta^2}|Y_{k-1}\right]$ is the posterior effective Kalman gain matrix. Note that we use $K_k^{\Theta*}$ and K_k^Θ to distinguish between the effective Kalman gain matrix obtained relative to the posterior distribution in this paper and the one obtained relative to the prior distribution in [10]. Letting $x_k^{e,\theta} = \hat{x}_k^{\theta_1} - \hat{x}_k^\theta$ be the Bayesian least-squares estimation error at time k, the update equation for $x_k^{e,\theta}$ is $x_{k+1}^{e,\theta} = \Phi_k\left(I - K_k^{\Theta*}H_k\right)x_k^{e,\theta} + \Gamma_k u_k^{\theta_1} - \Phi_k K_k^{\Theta*}v_k^{\theta_2}$. Letting $P_{k+1}^{x,\theta} = E\left[x_{k+1}^{e,\theta}\left(x_{k+1}^{e,\theta}\right)^T\right]$ and after some mathematical manipulations, $E_\theta\left[P_{k+1}^{x,\theta}|Y_k\right] = \Phi_k(I - K_k^{\Theta*}H_k)E_\theta\left[P_k^{x,\theta}|Y_k\right]\Phi_k^T + \Gamma_k E_\theta\left[Q^{\theta_1}|Y_k\right]\Gamma_k^T$.

To implement an OBKF, we need to compute the conditional expectations $E_\theta\left[Q^{\theta_1}|Y_k\right]$ and $E_\theta\left[R^{\theta^2}|Y_k\right]$ concerning the posterior distribution $\pi(\theta|Y_k)$ $\propto f(y_k|\theta)\pi(\theta)$, where $f(y_k|\theta)$ is the likelihood function of θ given the sequence of observations y_k. As there is no closed-form solution for $\pi(\theta|Y_k)$ for many prior distributions, we employ a Markov Chain Monte Carlo (MCMC) method to generate samples from the posterior distribution $\pi(\theta|Y_k)$ and then approximate $E_\theta\left[Q^{\theta_1}|Y_k\right]$ and $E_\theta\left[R^{\theta^2}|Y_k\right]$ as sample means of the generated MCMC samples. First, we need to compute the likelihood function $f(y_k|\theta)$. Assume that node α_i has received message $\mu_{\beta_i \to \alpha_i} = (S_i, M_i, \mathcal{E}_i)$. Now we aim to compute the outgoing message $\mu_{\beta_{i+1} \to \alpha_{i+1}}$ from node β_{i+1} to node α_{i+1}. Computing $\mu_{\beta_{i+1} \to \alpha_{i+1}}$ corresponds to the computation of the following integral: $\int_{x_i} \mathcal{N}\left(x_{i+1}; \Phi_i x_i, \tilde{Q}_i^{\theta_1}\right)\mathcal{N}\left(y_i, H_i x_i, R^{\theta_2}\right) \times S_i \mathcal{N}(x_i; M_i, \mathcal{E}_i)dx_i$. The solution of the integral given in (20) is a scaled multivariate Gaussian function $S_{i+1}\mathcal{N}(x_{i+1}, M_{i+1}, \mathcal{E}_{i+1})$, whose parameters S_{i+1}, M_{i+1}, and $\mathcal{E}_{i+1}$ are given by $\mathcal{E}_{i+1}^{-1} = \left(\tilde{Q}_i^{\theta_1}\right)^{-1} - \left(\tilde{Q}_i^{\theta_1}\right)^{-1}\Phi_i\Lambda_i\Phi_i^T\left(\tilde{Q}_i^{\theta_1}\right)^{-1}$, $M_{i+1} = \mathcal{E}_{i+1}\left(\tilde{Q}_i^{\theta_1}\right)^{-1}\Phi_i\Lambda_i\left(H_i^T\left(R^{\theta_2}\right)^{-1}y_i + \mathcal{E}_i^{-1}M_i\right)$

$$\text{and} S_{i+1} = S_i \sqrt{\frac{|\Lambda_i||\mathcal{E}_{i+1}|}{|\tilde{Q}_i^{\theta_1}||\mathcal{E}_i|}} \mathcal{N}\left(y_i; O_{m\times 1}, R^{\theta_2}\right) \times \exp\left(\frac{M_{i+1}^T \mathcal{E}_{i+1}^{-1} M_{i+1} + W_i^T \Lambda_i W_i - M_i^T \mathcal{E}_i^{-1} M_i}{2}\right),$$

where $W_i = H_i^T \left(R^{\theta_2}\right)^{-1} y_i + \mathcal{E}_i^{-1} M_i$ and $\Lambda_i = (\Phi_i^T \left(\tilde{Q}_i^{\theta_1}\right)^{-1} \Phi_i + H_i^T \left(R^{\theta_2}\right)^{-1} H_i + \mathcal{E}_i^{-1})^{-1}$.

The update rules given should be iterated for $0 \leq i \leq k - 1$ to finally obtain the message $\mu_{\beta_k \to \alpha_k} = (S_k, M_k, \mathcal{E}_k)$. Then the likelihood function is obtained as $f(y_k|\theta) = \int_{x_k} \mathcal{N}\left(y_k; H_k x_k, R^{\theta_2}\right) S_k \mathcal{N}(x_k, M_k, \mathcal{E}_k) dx_k = \int_{x_k} \frac{S_k}{\sqrt{(2\pi)^m |R^{\theta_2}|} \sqrt{(2\pi)^n |\mathcal{E}_k|}} \times \exp\left(\frac{-1}{2}(y_k - H_k x_k)^T \left(R^{\theta_2}\right)^{-1}(y_k - H_k x_k) + (x_k - M_k)^T \mathcal{E}_k^{-1}(x_k - M_k)\right) dx_k$. To estimate the posterior effective noise statistics $E_\theta\left[Q^{\theta_1}|Y_k\right]$ and $E_\theta\left[R^{\theta_2}|Y_k\right]$, we employ the Metropolis-Hastings MCMC [4]. Let the last accepted MCMC sample in the sequence of samples be $\theta^{(j)}$ generated at the j-th iteration. A candidate MCMC sample θ^{candid} will be drawn according to a proposal distribution $f(\theta^{candid}|\theta^{(j)})$. The candidate MCMC sample θ^{candid} will be either accepted or rejected according to an acceptance ratio r defined as $r = min\left\{1, \frac{f(\theta^{(j)}|\theta^{candid})f(y_k|\theta^{candid})\pi(\theta^{candid})}{f(\theta^{candid}|\theta^{(j)})f(y_k|\theta^{(j)})\pi(\theta^{(j)})}\right\} = min\left\{1, \frac{f(y_k|\theta^{candid})\pi(\theta^{candid})}{f(y_k|\theta^{(j)})\pi(\theta^{(j)})}\right\}$, where the second formula is used when the proposal distribution is symmetric, $f(\theta^{candid}|\theta^{(j)}) = f(\theta^{(j)}|\theta^{candid})$. The (j + 1)-th MCMC sample is: $\theta^{(j+1)} = \begin{cases} \theta^{candid} with\ probability\ r \\ \theta^{(j)} otherwise \end{cases}$.

The positivity of the proposal distribution $(f(\theta^{candid}|\theta^{(j)}) > 0$ for any $\theta^{(j)})$ is a sufficient condition for having an ergodic Markov chain of MCMC samples whose steady-state distribution is the target distribution $\pi(\theta|Y_k)$ [12, 13].

2.2 DSGE Model

According to [13], a DSGE model can be used to examine tourism with the general balance of the economy (see Table 1).

Against this backdrop, we develop a DSGE-VAR model (see Table 2). First, we have determined a vector of endogenous variables to express the model VAR. Then, we have defined the vector of VAR variables, where is it established the trade-weighted nominal exchange rate in the United States. Therefore, growth in the trade-weighted nominal exchange rate causes the U.S. dollar to depreciate. Nevertheless, the DSGE-VAR estimation requires a hierarchical prior, for this reason, we have carried out the DSGE parameter vector. First, we use the DSGE model to generate artificial data according to the prior distributions of the DSGE-VAR estimation. Second, these data are subsequently taken as priors for the Bayesian VAR estimation [13]. Finally, is necessary to stipulate a posterior distribution: $p(\Phi, \sum_u, \theta/Y) = p(\Phi, \sum_u/\theta, Y) p(\theta/Y)$ for correctly estimating the model.

Table 1. Dynamic Stochastic General Equilibrium (DSGE)

Functions	Variables
The utility function of households	
$U = E_0 \sum_{t=0}^{\infty} \beta^t \dfrac{\left[(C_t - hC_{t-1}) + \dfrac{u_t^{1+v_1}}{1+v_1} \dfrac{(La_{l,t}\varsigma_{la,t})^{1+v_2}}{1+v_2}\right]^{1-\sigma}}{1-\sigma}$	E_0: expected utility function hypothesis β: discounted rate h: typifies the habit persistence of consumption C_t: (using a CES function) is composed by: · $C_{T,t}$: Tourism goods · $C_{NT,t}$: Non-tourism goods · $C_{P,t}$: Public services u_t: Unemployment rate $La_{l,t}$: Private land supply shock $\varsigma_{la,t}$: The exogenous variable that is estimated by an auto-regression process to represent the result of private land inputs on the economy σ, v_1, and v_2: the parameters of the constant elasticity of substitution (CES) $C_{M,t}$: is composed by: · $C_{MT,t}$: Imports of tourism products · $C_{MNT,t}$: Non-tourism products
The production functions of the tourism and non-tourism activities	
$Y_{T,t} = \Omega_{T,t} K_{T,t}^{\alpha_1} N_{T,t}^{\alpha_2} La_{T,t}^{1-\alpha_1-\alpha_2}$ $Y_{NT,t} = \Omega_{NT,t} K_{NT,t}^{\alpha_3} N_{NT,t}^{1-\alpha_3}$	$Y_{i,t}$ ($i = T, NT$): The value-added of the given sector $\Omega_{i,t}$ ($i = T, NT$): The productivity function connected to the effects of physical capital and public sector $K_{i,t}$ ($i = T, NT$): The physical capital and is calculated by the process: $K_{i,t+1} = I_{i,t} + (1 - \delta)K_{i,t}$ ($i = T, NT$) · $I_{i,t}$: The physical capital investment in every sector · δ: The depreciation rate $N_{i,t}$: Human capital enhancement: $N_{i,t} = H_t n_{i,t}$ ($i = T, NT, P$) · $n_{i,t}$: points out the labor force for the sectors · H_t: The spill-over effects of capital and the accumulation of human capital $La_{T,t}$: The private land rentals to the tourism sector
The productivity function connected to the effects of physical capital and public sector	
$\Omega_{i,t} = A_t A_{i,t} (\varsigma_{P,t} Y_{P,t})^{\varphi_{P,i}} K_{i,t}^{\varphi_i} \left(\dfrac{K_{P,t}}{K_{T,t} + K_{NT,t}}\right)^{\varphi_{c,i}}$ $i = (T, NT)$	A_t: The auto-regression processes of the total productivity shocks $A_{i,t}$: The auto-regression processes of the sector that point out sector and total productivity shocks $\varsigma_{P,t}$: The exogenous shock to the spill-over effects of public sector $Y_{P,t}$ the effect of public sector $K_{i,t}$: The effect of physical capital $\left(\dfrac{K_{P,t}}{K_{T,t} + K_{NT,t}}\right)^{\varphi_{c,i}}$: The spillover effect of $K_{p,t}$ $\varphi_{P,i}$: The effect of the public sector $\varphi_{c,i}$: The effect of the private sector ($i = T, NT$): The parameters
The spill-over effects of capital and the accumulation of human capital	
$H_t = \dfrac{EX_{T,t}^{\alpha_T}\left(Y_{T,t} - EX_{T,t}\right)^{b_T}\varsigma_{H,t}}{H_t^{\pi_T}} + \dfrac{EX_{NT,t}^{\alpha_{NT}}\left(Y_{NT,t} - EX_{NT,t}\right)^{b_{NT}}}{H_t^{\pi_{NT}}} - \delta_H H_{t-1}$	$EX_{T,t}$: The exports of tourism $\varsigma_{H,t}$: The shock to human capital accumulation $EX_{NT,t}$: The non-tourism products $E_{i,t}^{a_i}$ and $\left(Y_{i,t} - EX_{i,t}\right)^{b_i}$: The effect of the tourism product on human capital a_i, b_i and π_i: The parameters δ_H: The depreciation rate of human capital $H_t^{\pi_i}$: The externality of experience

(*continued*)

Table 1. (*continued*)

Functions	Variables
Systemic Risk (Risk Contagion)	
$X_{j,t} = g\left(\beta_{J|R_j}^T R_{j,t}\right) + \epsilon_{j,t}$	$R_{j,t} \equiv \{X_{-j,t}, M_{t-1}, B_{j,t-1}\}$
$\overline{CoVaR}_{j|\tilde{R}_j,t}^{TENET} \equiv \hat{g}\left(\hat{\beta}_{j|\tilde{R}_j}^T \tilde{R}_{j,t}\right)$	$X_{-j,t} \equiv \{X_{1,t}, X_{2,t}, \ldots, X_{k,t}\}$ is the set of (k-1) independent variables such as the log-returns of tourism stocks, except tourism stock j, and k number o tourism stocks which is 95 in our case
$\hat{D}_{j|\tilde{R}_j} \equiv \left.\frac{\partial \hat{g}\left(\hat{\beta}_{j|R_j}^T R_{j,t}\right)}{\partial R_{j,t}}\right\|_{R_{j,t} = \tilde{R}_{j,t}} = \hat{g}'\left(\hat{\beta}_{j|\tilde{R}_j}^T \tilde{R}_{j,t}\right)\hat{\beta}_{j|\tilde{R}_j}$	$\beta_{j|R_j} \equiv \{\beta_{j|-j}, \beta_{j|M}, \beta_{j|B_j}\}^T$
	$\overline{CoVaR}_{j|\tilde{R}_j,t}^{TENET}$ is the TENET risk that contains the impacts of all other tourism stocks on tourism stock j and integrates the non-linearity displayed in the shape of a link function (g.)
The exports	
$EX_{i,t} = \left(\frac{P_{i,t}}{RER_t}\right)^{\theta_{EX_i}} Y_{ROW,t}^{\omega_i}$ (i = T, NT)	$\left(\frac{P_{i,t}}{RER_t}\right)$: The real exchange rate in USD
$P_{H,t} = \overline{P_{H,t}} + \eta P_{N,t}$	RER_t: The exchange rate
$P_{T,t} = \left[a_H P_{H,t}^{\frac{\rho}{\rho-1}} + a_F P_{F,t}^{\frac{\rho}{\rho-1}}\right]^{\frac{\rho}{\rho-1}}$	$Y_{ROW,t}$: The world income level
	$P_{H,t}$: Consumer price if the Home traded goods
$P_t = \left[a_T P_{T,t}^{\frac{\phi}{\phi-1}} + a_N P_{N,t}^{\frac{\phi}{\phi-1}}\right]^{\frac{\phi}{\phi-1}}$	$\overline{P_{H,t}}$: Price of Home traded goods at producer level
	$\eta P_{N,t}$: Value of the nontraded goods that are necessary to distribute to consumers
$AC_{H,t}^p(h) = \frac{k_H^p}{2}\left(\frac{\overline{P}_t(h)}{\overline{P}_{t-1}(h)} - \pi\right)^2 D_{H,t}$	$AC_{H,t}^p(h)$ and $AC_{H,t}^{p*}(h)$: The price adjustment costs faced by firms in the traded goods according to the destination market
$AC_{H,t}^{p*}(h) = \frac{k_H^{*p}}{2}\left(\frac{\overline{P}_t^*(h)}{\overline{P}_{t-1}^*(h)} - \pi\right)^2 D_{H,t}$	$AC_t^p(n)$: The price adjustment costs faced by firms in the non-traded goods
$AC_t^p(n) = \frac{k_N^p}{2}\left(\frac{p_t(n)}{p_{t-1}(n)} - \pi\right)^2 D_{N,t}$	
The Government and the Equilibrium	
$G_t = \left(\frac{g_t}{1+g_t}\right)Y_t = T_t$	G_t : Government purchases. We assume a public sector that consumes a fraction T_t of the output of each good, being $g_t = -\log(1-T_t)$
$E_t\{\mathcal{F}_{t,t+1}\prod_{t+1}\Omega_{t+1}\} = Q_t$	Y_t : Aggregate output
$(\sum_t - 1)(C_t - hC_{t-1}) =$	The present discounted real value of future financial wealth equals the current level of the real stock-price index
$\xi Q_t + (1-\xi)E_t\{\mathcal{F}_{t,t+1}\prod_{t+1}\sum_{t+1}(C_{t+1}hC_t)\}$	State equation for aggregate consumption
$Q_t = E_t\{\mathcal{F}_{t,t+1}\prod_{t+1}[Q_{t+1} + D_{t+1}]\}$	Standard pricing equation is micro-founded on the consumer's optimal behavior

Table 2. DSGE-VAR Model

Functions	Variables
The model VAR	
$y_t^V = c + B_1 y_{t-1}^V + \ldots + B_p y_t - p^V + u_t$	y_t^V: represent an nH × 1 vector corresponding to endogenous variables for t = 1..., T
	c: Group of terms
	p: The VAR lag length
	[B1,..., Bp]: Parameter matrices
	ut: The vector of forecast errors defined by the multivariate normal distribution $N(0; \sum u)$

(*continued*)

Table 2. (*continued*)

Functions	Variables
Vector of VAR variables	
$y_t^{v'}=100\times[\Delta log(Y_{T,t}),\Delta log(Y_{NT,t}),\Delta log(C_t),\Delta log(GDP_t),4\Delta log(P_t),R$ $\Delta log(TSP_t),\Delta log(\widehat{VaR}_{i,t,\tau})]$	YT,t: The production in the tourism sector YNT,t: The production in the non-tourism sector Ct: Per capita real consumption $GDPt$: Per capita real GDP Pt: Applies the GDP deflator Rt: The federal funds rate adjusted at the annual rate EXt: The trade-weighted nominal exchange rate in Spain TSP_t: The tourism stock prices index for Spanish companies $\widehat{VaR}_{i,t,\tau}$: Estimation of the systemic interdependence among Spanish tourism stocks
The DSGE-VAR estimation	
$Y^v=X^v\Phi+u_t$	Y_t^v be a $T\times nH$ matrix with each row consisting of $y_t^{v'}$ Xv be a $T\times k$ matrix with the *t-th* row containing in $x_t^{v'}\equiv\left[1,y_{t-1}^{v'},...,y_{t-p}^{v'}\right]$ where $k\equiv 1+p\times nH$ ϕ: The maximum-likelihood estimator is calculated according to DSGE parameters vector
DSGE parameters vector	
$\tilde{\Phi}(\theta)=\left(\lambda TT_{x^vx^v}(\theta)+X^{v'}X^v\right)^{-1}\left(\lambda TT_{x^vx^v}(\theta)+X^{v'}Y^v\right)$	θ: Vector consisting of the DSGE parameters EDh: The expectation operator conditional on the DSGE parameter vector θ

3   Empirical Results

The sample period in the valuation of the model has been from 1992Q1 to 2021Q3, during which data from the Spanish economy have been used. These data have been extracted from the Federal Reserve Economic Data (FRED) of the Federal Reserve Bank of St. Louis, Eurostat, and SABI (Iberian Balance Sheet Analysis System of Bureau Van Dijk). Once the posterior distribution has been estimated, it is also useful to perform an estimation of the so-called Marginal Data Density (MDD) for DSGE models $p(Y)=\int p(Y/\theta)p(\theta)d\theta$. . In this study, the posterior moments are estimated with the three models proposed (DSGE, VAR, and DSGE-VAR). We employ the Metropolis algorithm to simulate the posterior distribution to evaluate the accuracy of the models, running these algorithms 10,000 times and calculating the means and standard deviations of the posterior moment estimates in all runs. Tables 3 report the results of the estimates obtained by the different models with the prior distribution previously inserted and the posterior distribution from the estimation. To guarantee greater robustness in the estimates, three stages of the configuration of the DSGE model described above have been carried out. Table 3 shows the MDD estimates after the estimation of the models developed. These results demonstrate the greater stability offered by the DSGE-VAR model compared to the rest, especially in light of the deviations obtained for three settings. The results of the new Kalman filter with Prior Update improve the results of the Classical Kalman Filter, just as it improves the precision results shown in previous works [1, 5, 6], even if it is small and irregular samples like the one used in the present study.

Table 3. Log MDD Estimates (Base Model)

		Classical Kalman Filter		Bayesian Kalman Filter with Prior Update	
N	Model	MEAN (Log MDD)	STD (Log MDD)	MEAN (Log MDD)	STD (Log MDD)
Prior Distribution					
100	DSGE	−1728.812	0.81	−1357.738	0.49
	VAR	−1754.258	0.87	−1384.483	0.55
	DSGE-VAR	−1711.593	0.76	−1274.524	0.41
Posterior Distribution					
500	DSGE	−1625.851	0.74	−1236.104	0.39
	VAR	−1664.593	0.82	−1286.342	0.44
	DSGE-VAR	−1572.294	0.71	−1238.592	0.32

Note: N is the sample size; STD is the standard deviation

Table 4 provides the mean and standard deviation of the prior distributions of each parameter for Spain. The mean of the posterior distributions and the range of the 90% interval estimated by the Bayesian approach are presented. The estimation results of some of the structural parameters, such as β, δ, α_3, α_4, and h, work as the prior means according to the structure of the optimal equations used in some previous works [13]. The parameter α_1 increases from 0.41 to 0.58 but α_2 decreases from 0.50 to 0.11. By comparison with the non-tourism and public services sectors, the output of the tourism sector continues to be more labor-dependent, which is consistent with the realities of the tourism sector. On the other side, the coefficients of leisure (v_1), private land (v_2), and intertemporal substitution (σ) are estimated as 2.06, 1.97, and 1.98, which after dividing the unity by these results shows us the following elasticities 0.485, 0.508 and 0.505 respectively. In the three cases, the elasticities are less than 1, which is in line with previous works [13]. The substitute elasticity between tourism and non-tourism goods (θ_1) is 0.42.

Finally, to carry out a forecast evaluation, the three versions of the model (DSGE, VAR, DSGE-VAR) are estimated with the final configuration of the model used by [3] with out-of-sample data, with a horizon of one year. For this, the root-mean-square error (RMSE) is estimated to analyze the deviation obtained outside the sample by the BKPU filter. Table 5 shows the RMSE results obtained from the posterior distributions by the different models estimated by the Classical Kalman Filter and Bayesian Kalman Filter with Prior Update, respectively. These results also show greater precision of the new proposed Kalman filter with out-of-sample data, and also improve the precision shown in previous works that performed a forecast evaluation with out-of-sample data [4]. These simulations with out-of-sample data obtain robust and stable precision results, which would rule out possible parameter misspecification estimated by the new Kalman filter used (BKPU), a concern shown by previous works [3]. It also shows an improvement in precision results compared to other filters used by recent works such as the central

Table 4. Estimations results for Spain (Main components)

		Prior Distribution	Posterior Distribution	90% Interval	
				Low	High
Physical Capital Depreciation Rate	δ	Beta (0.03,0.00)	0.03	0.01	0.04
Output Elasticity of Physical Capital in the Tourism Sector	$\alpha 1$	Beta (0.41,0.10)	0.58	0.56	0.63
Output Elasticity of Human Capital in the Tourism Sector	$\alpha 2$	Beta (0.50,0.10)	0.11	0.07	0.16
Habit Persistent	h	Beta (0.81,0.01)	0.79	0.73	0.84
Elasticity of Leisure	v_1	Gamma (2.00,0.10)	2.06	2.03	2.11
Elasticity of Private Land	v_2	Gamma (2.00,0.10)	1.97	1.96	2.01
Elasticity of Intertemporal Substitution	σ	Gamma (2.00,0.10)	1.98	1.97	2.00
Substitute Elasticity between Tourism, Non-tourism Goods and Public Services	θ_1	Gamma (0.42,0.10)	0.42	0.36	0.47
Substitute Elasticity between FDI and Domestic Investment	θ_2	Gamma (1.45,0.10)	1.34	1.28	1.40
Substitute Elasticity between Tourism and Non-tourism Imports	θ_3	Gamma (0.40,0.10)	0.43	0.39	0.44
Price Elasticity of Tourism Exports (Absolute)	$\theta_{EX,T}$	Gamma (0.40,0.10)	0.45	0.42	0.47
Price Elasticity of Non-tourism Exports (Absolute)	$\theta E,$	Gamma (0.20,0.10)	0.28	0.24	0.33
Income Elasticity of Tourism Exports	ωT	Gamma (0.75,0.10)	1.08	1.05	1.15
Income Elasticity of Non-tourism Exports	ωNT	Gamma (0.30,0.10)	0.07	0.02	0.09

(continued)

Table 4. (*continued*)

		Prior Distribution	Posterior Distribution	90% Interval	
				Low	High
Autoregressive Coefficient of Return Rate	θtr	Beta (0.80,0.10)	0.84	0.80	0.90
Elasticity of Price in the Taylor Rule	θ_p	Gamma (1.70,0.10)	1.83	1.81	1.85
Elasticity of GDP in the Taylor Rule	θy	Gamma (0.15,0.05)	0.17	0.16	0.24
Elasticity of Tourism Exports in Human Capital Accumulation	a_T	Gamma (0.25,0.10)	0.51	0.47	0.56
Elasticity of Non-exports of the Tourism Sector in Human Capital Accumulation	b_T	Gamma (0.05,0.01)	0.06	0.01	0.11
Scale Effect of Human Capital Accumulated by the Tourism Sector	πT	Gamma (0.30,0.10)	0.24	0.23	0.26
Elasticity of Non-tourism Exports in Human Capital Accumulation	$a\,T$	Gamma (0.30,0.10)	0.47	0.46	0.51
Elasticity of Non-exports in the Non-tourism sector of Human Capital Accumulation	$B\,T$	Gamma (0.05,0.01)	0.05	0.01	0.07
Scale Effect of Human Capital Accumulated by the Non-tourism Sector	$\Pi\,T$	Gamma (0.30,0.10)	0.42	0.41	0.45
Depreciation Rate of Human Capital	δ_H	Gamma (0.05,0.01)	0.08	0.02	0.12
Spill-over Effect of Public Service on Tourism Productivity	$\varphi P,$	Gamma (0.10,0.01)	0.14	0.09	0.17

(*continued*)

Table 4. (*continued*)

		Prior Distribution	Posterior Distribution	90% Interval	
				Low	High
Spill-over Effect of Tourism Physical Capital on its Productivity	φ_T	Gamma (0.05,0.01)	0.03	0.01	0.04
Congestion Effect of Physical Capital on Tourism Productivity	$\varphi C,$	Gamma (0.06,0.01)	0.05	0.01	0.09
Spill-over Effect of Public Service on Non-tourism Productivity	$\varphi P,$	Gamma (0.10,0.01)	0.14	0.10	0.17
Spill-over Effect of Non-tourism Physical Capital on its Productivity	$\varphi\ T$	Gamma (0.05,0.01)	0.08	0.03	0.12

difference Kalman filter [1, 3] and Quadratic Kalman filter [6]. Finally, the average run time of the Classical Kalman Filter for this estimate with data outside the sample is 0.23 min, while the same estimate is made by the BKPU method in a time of 0.11 min.

Table 5. Prior and posterior distributions

Classical Kalman Filter				
	2018Q1	2018Q2	2018Q3	2018Q4
DSGE	0.78	0.82	0.84	0.88
VAR	0.85	0.88	0.92	0.96
DSGE-VAR	0.74	0.75	0.78	0.82
Bayesian Kalman Filter with Prior Update				
	2018Q1	2018Q2	2018Q3	2018Q4

(*continued*)

Table 5. (*continued*)

Classical Kalman Filter

	2018Q1	2018Q2	2018Q3	2018Q4
DSGE	0.43	0.44	0.47	0.48
VAR	0.52	0.52	0.53	0.55
DSGE-VAR	0.38	0.41	0.41	0.43

4   Conclusions

This research provides an additional simulation procedure for the estimation of DSGE and VAR models. It is demonstrated that, when properly adjusted to DSGE and VAR models, the BKPU technique is more robust than other commonly used algorithms, such as the Classical Kalman Filter. After a comparison of simulations with these two Kalman filters carried out successfully on three scenarios of a medium-scale Keynesian DSGE model, our results reveal high robustness of the BKPU algorithm for small samples with irregular data and possible cases of statistical misspecification, which has been a matter of concern shown by the literature in the DSGE and VAR model estimation. The results obtained in our research are valid both for the RMSE results as a criterion for out-of-sample data and for the marginal data density as a criterion to measure the fit of the in-sample models. Given the high accuracy shown by this new algorithm, this study also implies an improvement in the optimization of the calculation of macroeconomic forecasts, since it is not necessary to use any available resources or to carry out an extensive specification of the DSGE models. In addition, our research provides an important contribution to the literature on the tourism sector through a DSGE model, both in the estimation of the prior and posterior distribution. In this model, we analyse the effect of stock price volatility, systemic risk, and tourism productivity in the tourism economy. The estimation results reveal that a 10% increase in tourism productivity can improve the value-added of the tourism sector by 1.15% and increase GDP growth by about 0.74%. Given that Spain is an important tourism country, any increase in tourism development will increase GDP by a considerable proportion. Likewise, whereas an increase in tourism productivity leads to a rise in tourism prices, an increase in tourism consumption, and, in theory, a drop in tourism investment, the positive effect on other sectors produces different consequences. Furthermore, the estimation results also reveal that a 10% increase in systemic risk decreases the value-added of the tourism sector by 1.04%, in turn declining GDP growth by approximately 1.06%. Considering that one of the main sources of income in Spain is derived from the tourism sector, it is very important to consider the systemic risk, which is caused by the failure in payments by one or more members of the market system. This can lead to a generalized market collapse, particularly affecting companies in the tourism industry. Lastly, we observe a slight increase in non-tourist exports, but a small fall in consumption of these non-tourist products and a more persistent decline in investment in our Tourism productivity model.

Furthermore, the accuracy results show how the extended DSGE-VAR model is better than the previous DSGE model in the analysed country, both in the estimation of the prior and posterior distribution. These results show the higher stability provided by the DSGE-VAR model compared to the others, especially in comparison to the deviations obtained for three settings. The results of the new Kalman filter with prior updates improve the results of the classical Kalman filter, as well as improve the accuracy results shown in previous works, even when dealing with small and irregular samples like the one used in the present study. It would be an interesting idea as future research to compare this new Kalman application to another economic and finance in order to check the superiority shown in DSGE models.

In summary, this study offers a significant opportunity to contribute to the field of macroeconomic analysis, since the results obtained have important implications for public institutions and other interest groups. A policymaker is usually only really interested in a restricted number of available resources when making a macroeconomic forecast, based only on the variables of interest, an issue that our model has introduced by improving the optimization of macroeconomic forecast calculations. In addition, the BKPU technique can be extended to a wide variety of problems for which the Classical Kalman Filter has been previously applied. For instance, the study of the transmission of monetary policy shocks across economic areas and the tourism sector, the construction of measures of core inflation, and the natural rate of unemployment in country settings. Therefore, our study has relevant implications for monetary policy since the exchange rate, tourism stock prices and tourism productivity have a significant impact on the business cycle.

References

1. Aruoba, S.B., Cuba-Borda, P., Higa-Flores, K., Schorfheide, F., Villalvazo, S.: Piecewise-linear approximations and filtering for DSGE models with occasionally binding constraints. International Finance Discussion Papers 1272 (2020)
2. Martínez-Martín, J., Morris, R., Onorante, L., Piersanti, F.M.: Merging structural and reduced-form models for forecasting: opening the DSGE-VAR box. ECB Working Papers, No 2335 (2019)
3. Martín-Moreno, J.M., Pérez, R., Ruiz, J.: Exploring the sources of Spanish macroeconomic fluctuations: an estimation of a small open economy DSGE model. Int. Rev. Econ. Financ. **45**, 417–437 (2016)
4. Herbst, E., Schorfheide, F.: Tempered particle filtering. Journal of Econometrics **210**(1), 26–44 (2019)
5. Ivashchenko, S., Gupta, R.: Forecasting using a nonlinear DSGE model. J. Central Banking Theory and Practice **2**, 73–98 (2018)
6. Minford, P., Xu, Y., Zhou, P.: How good are out of sample forecasting tests on DSGE models? Italian Economic Journal **1**(3), 333–351 (2015)
7. Ca'Zorzi, M., Kolasa, M., Rubaszek, M.: Exchange rate forecasting with DSGE models. J. International Economics **107**, 127–146 (2017)
8. Yang, Y., Zhang, H., Chen, X.: Coronavirus pandemic and tourism: dynamic stochastic general equilibrium modeling of infectious disease outbreak. Annals of Tourism Research **83**, 102913 (2020)
9. Alaminos, D., León-Gómez, A., Sánchez-Serrano, J.R.: A DSGE-VAR analysis for tourism development and sustainable economic growth. Sustainability **12**, 3635 (2020)
10. Dehghannasiri, R., Esfahani, M.S., Qian, X., Dougherty, E.R.: Optimal Bayesian Kalman filtering with prior update. IEEE Trans. Signal Process. **66**(8), 1982–1996 (2018)

11. Cole, S.J., Milani, F.: The misspecification of expectations in New Keynesian models: a DSGE-VAR approach. Macroecon. Dyn. **23**(3), 974–1007 (2019)
12. Alaminos, D., Ramírez, A., Fernández-Gámez, M.A., Becerra-Vicario, R., Cisneros-Ruiz, A.J., Solano-Sánchez, M.A.: Estimating DSGE models using multilevel sequential monte carlo in approximate bayesian computation. J. Sci. Ind. Res. **79**(1), 21–25 (2020)
13. Liu, A., Song, H., Blake, A.: Modelling productivity shocks and economic growth using the Bayesian dynamic stochastic general equilibrium approach. Int. J. Contemp. Hosp. Manag. **30**(11), 3229–3249 (2018)

An Expert System to Detect and Classify CNS Disorders Based on Eye Test Data Using SVM and Nature-Inspired Algorithms

U. S. Samarasinghe and M. K. A. Ariyaratne[✉]

Department of Computer Science, Faculty of Applied Sciences, University of Sri
Jayewardenepura, Nugegoda, Sri Lanka
mkanuradha@sjp.ac.lk

Abstract. Changes in eye movements have a strong relationship with
the changes in the brain. Several medical studies have revealed that in
most CNS disorders, ocular manifestations are often associated with brain
symptoms. To date, computational intelligence has not been used to study
the relationship between eye movements and brain disorders. We propose
a support vector machine (SVM) based machine learning solution to identify, five disorders related to the central nervous system; Amyotrophic Lateral Sclerosis (ALS), Multiple Sclerosis (MS), and Alzheimer's Disease
(AD), Parkinson's Disease (PD), and Schizophrenia. Apart from the SVM,
the proposed solution handles two major problems which occur in the data
preprocessing stage; insufficiency of real eye test data and finding optimal
features set for a particular disorder. An algorithm is developed to generate
synthetic data and to find the optimal features set for a particular disorder,
a solution based on particle swarm optimization is proposed. We trained
the SVM models using the generated synthetic data and tested with the
real data. The proposed system based on SVMs with linear, polynomial,
and RBF kernels were able to identify the stages of the disorders, as diagnosed in medical studies. The SVM with the RBF kernel worked with an
accuracy of 97% in identifying the existence of a CNS disorder. In classifying the stages of ALS, the linear kernel worked with an accuracy of 77%
while the polynomial kernel worked with an accuracy of 100%, 90%, and
64% in classifying stages of MS, AD, and Schizophrenia. For PD, SVMs
with all kernels gave an accuracy of 96%. The results are encouraging, giving sufficient evidence that the proposed system works better. We further
illustrate the viability of our method by comparing the results with those
obtained in previous medical studies.

Keywords: SVM · Linear Kernel · RBF Kernel · Polynomial Kernel ·
PSO · GA · CNS disorders · Synthetic data · Feature selection

1 Introduction

The eye is the window of the brain. Having gone through this thought, many
studies in the field of medicine proved the connection between eye movements

L. Rutkowski et al. (Eds.): ICAISC 2022, LNAI 13589, pp. 182–194, 2023.
https://doi.org/10.1007/978-3-031-23480-4_15

and cognitive processes such as decision making, attention, and memory. Eye movements not only reflect such connections to the brain but are also useful in examining the brain and diagnosing its diseases. It has been recognized that the changes in eye movements have a connection to the changes in the brain as a result of either aging or diseases like neurodegeneration. Eye movements therefore can be considered an important indicator for the early detection of some brain disorders. The retina of the eye and optic nerve extending from the diencephalon are considered part of the Central Nerve System (CNS). Consequently, the eye is considered an extension of the brain. Studying various ocular changes in patients with CNS disorders emphasizes the strong connection between the brain and the eye. The studies reveal that, in many CNS disorders, ocular manifestations often precede symptoms in the brain [13].

In practice, many CNS disorders could be early diagnosed through ocular manifestations. London et al. studied various ocular changes through ophthalmological assessments in patients with CNS disorders. In their work, they worked on identifying Stroke, Multiple Sclerosis (MS), Parkinson's Disease (PD), and Alzheimer's Disease (AD) [13]. The findings include that some ocular manifestations are not specific to a particular brain disease. Further, many studies examined specific ocular manifestations for particular diseases. Mukarjee et al. worked on finding ocular changes in patients with Amyotrophic Lateral Sclerosis (ALS) [14]. ALS is a rapidly progressive neurodegenerative disease. Concerning the thickness of the retinal layers, they found that the retinal layers are thinner in the eyes of ALS patients in comparison with healthy ones. Thus, they were successful in finding the connection between the ocular changes and the disease progression of ALS. In another work, Bernardo Sanchez-Dalmau and others studied visual impairment in multiple sclerosis (MS) [18]. Visual impairment is a common feature of MS patients, and the research focused on monitoring the disease by ganglion cell volume and the inner plexiform layer (GCIPL). Elena-Salobrar-Garcia et al. has researched finding degeneration in the retina and optic nerve of patients with Alzheimer's Disease (AD) [17]. They worked on studying three groups of patients with AD. As a result, they were successful in finding the correlation between AD progression and cognitive decline of retina layer thickness. Kromer and the team worked on finding changes in the retinal vasculature in Parkinson's Disease (PD) patients and controls [10]. The study shows potential differences in vascular in the retina between PD patients and healthy controls, which may cause vision problems in PD patients. In recent work, Liu et al. studied reductions in retinal nerve layers (RNFL) in schizophrenia [12]. They revealed the biological etiology of cognitive impairment in patients with schizophrenia associated with a decrease in RNFL thickness.

By looking at all this medical research carried out recently, we observed the following pathways and improvements needed for the research, working on finding the relationships between the eye and CNS disorders.

- The amount of data seems to be not enough to arrive at the conclusions precisely.
- Computational intelligence has not been applied to find the relationships between eye and the CNS disorders.

Since eye test data are the main concern in finding the relationship, why not using the state of the art techniques such as machine learning?. This thought motivated us to find a new perspective on how to use computational intelligence to solve the matter. Thus, finding the relationship between eye test data and the CNS disorders for the early diagnosis has become the main goal of this study. Here we address the problem of detecting CNS disorders based on the eye test data using support vector machines. Further, the problems of having fewer data and the identification of the most prominent features for a particular disorder were addressed using two nature-inspired meta-heuristic algorithms; genetic algorithms (GA) and particle swarm optimization algorithm (PSO).

For the convenience of the reader, the paper is structured as follows. Section 2 provides an introduction to the algorithms used in the study and the importance of each algorithm to the study. Datasets used in the study are also discussed in Sect. 2. Section 3 introduces the proposed expert system and a complete description of the adopted method/algorithm of the solution. Section 4 presents the results obtained both from nature-inspired algorithms and the SVM. Conclusions are finalized in Sect. 5.

2   Materials and Methods

Here we will brief the used algorithms and each algorithm's contribution to the proposed work. Two nature-inspired algorithms; GA and PSO with one machine learning algorithm; SVM with linear, polynomial and RBF kernels are mainly used to implement the proposed expert system.

2.1   Genetic Algorithms (GA)

Genetic Algorithm (GA) is a meta-heuristic algorithm based on the Darwinian theory of evolution [7]. The genetic algorithm (GA) is a meta-heuristic algorithm based on the Darwinian theory of evolution. The main idea is to allow better parents to reproduce more. Chromosomes indicate the individuals in GA which are typically feasible solutions to the problem to be solved. A set of solutions will generate a population of one generation. A fitness function specified for the problem can be used to evaluate the goodness of the population members and crossover will take place probabilistically based on the fitness values. Crossed solutions will then get mutated again with a pre-defined mutation probability. Helping exploitation, crossover allows new individuals to have blocks of genes from their parents while mutation helps exploration by adding random tweaks to the solutions. With these simple steps, GA enables millions of researchers to be successful with its usage. It has been widely used in many research in different fields of studies [2–4,16].

Synthetic Data Generation Using GA. In our work, we used GA basically to generate data to handle the problem of having insufficient data to train the SVM. From the medical work published, we were able to find eye test data of around 22–386 people, which is not enough to train our machine learning algorithm. We used GA to produce synthetic data which imitates all the characteristics of the original data. The data set is high dimensional (dimension/features go from 22 to 68). Thus, we used the following process in generating the data.

- For a given data set, data points with missing data were separated.
- Using the min-max scaler function, all the features of a given data set were taken to be on the same scale.
- We consider one feature at a time to find the best range for the data in a particular feature.
  - By looking at the range of the feature we separate the range into (n/2) partitions. Here 'n' denotes the number of records we have regarding a particular disease. (an example: for ALS, we have 11 records, so initially we have (11/2) = 5 partitions for a particular feature). Apart from that, there are two special cases.
    1. If the number of records is not sufficient (n = 3), we will take the value as it is.
    2. If the number of records is large (>60), to reduce the computation effort, we took the $\sqrt{n}$.
  - From all the data points in the data set, we find the counts for each partition (variation of the value of that feature within the partitions) for that feature.
  - Using the counts, we evaluate (the fitness) goodness of the partition according to the following equation.

$$F = \frac{\text{\# of data points in the feature belonging to partition}}{\text{sum of distance between data points in the partition}} \quad (1)$$

  - In accordance with the fitness of the participations, we try to combine and make wider partitions by following the below steps.
    * The initial iteration will combine consecutive partitions (2 at a time) and evaluate the fitness.
    * If the combined partitions are good, we accept the combination, otherwise, we keep the original.
    * This process will continue the number of iterations and present the final partitions for the selected feature.
- From this, we find the best ranges to generate synthetic data for each feature in the data set.
- By finding the best partitions for all the features, we then use an algorithm based on GA, to find the optimal partition combinations for all the features in the data set. Optimal value combinations for the features are essential to regenerate test data.
- Existing patient data are used to evaluate the goodness of different partition combinations of different features.
- Once we get the optimal value combinations, we find the missing values for the separated real patient data in the pre-processing stage.

- We check the most suitable combination for a particular record with missing values.
- For the missing feature in that record, we take the average value of the other records in its most suitable combination.
- Now, these new records are also contributed to the data generation process.
- After that, we use the following process to regenerate training data.
  - Select the combination of partitions to generate data.
  - We generate 20 records per iteration.
  - To generate a record we follow the following process.
    1. for each feature we calculate, the lower bound (LB), the upper bound (UB), and the average (μ) value.
    2. randomly select (n/K) records, where n=number of records, K is a constant (we took K = 3)
    3. Find the closest record to the μ, randomly select one option from LB and UB.
    4. Generate a random number in between those two as the generated data for the particular feature.
    5. Repeat the process for all the features, to complete a record.
    6. Hundred (100) such iterations (consider as a single run) were performed and the best 20 records whose having a minimum average error with original mean values are added as new records to the data set.
    7. This is repeated until we generate a sufficient number of synthetic data for a particular partition combination.

2.2 Particle Swarm Optimization (PSO)

Natural optimized behavior of some species such as fish and bird-inspired the concept behind the PSO [8]. According to the concept, the population is consist of particles; solutions to the problem at hand, who wonder in the solution space based on the velocities. Velocity is calculated based on the particle's current solution, the previous best solution (known as **Pbest**), and the best solution of the group (known as **Gbest**). Eventually, the group reach the better areas in the solution space. The goodness of a solution is measured by the fitness function defined to the problem. The simple implementation behind its vast capability let it usable for a wide variety of optimization problems [6,11,15]. The standard functions in the canonical PSO are used in the study to update the velocity and the position of a particle.

Features Selection Using PSO Algorithm. For this work, PSO has been used to select the most prominent features for each CNS disorder. We consider detecting 05 CNS disorders in this study. The existing eye test data has a feature range that goes from 22 to 68. Since all the features are not necessary for detecting a particular CNS disorder, feature selection has been carried out. For some disorders data points/records are given stages-wise (For example AD). For the feature selection method, we used a scoring method. The process is as follows.

Table 1. Calculating scores of each feature

	Feature 1	Feature 2	Feature 3	Feature 4
Stage 1	Mean F1S1	Mean F2S1	Mean F3S1	Mean F4S1
Stage 2	Mean F1S2	Mean F2S2	Mean F3S2	Mean F4S2
Stage 3	Mean F1S3	Mean F2S3	Mean F3S3	Mean F4S3

Table 2. Implementing a fitness of a particle

	F1	F2	F3	F4
Score vector	0.92	0.61	0.68	1.21
·Particle value = 0.8	1	0	0	1

- Stage wise separate the patient records.
- Using the separated patient records, for each stage find the average value for each feature.
- Based on the average values obtained for each feature at each stage, we calculate a score value

$$\text{score of feature x} = \frac{\sum_{i=2}^{n}(\text{mean FxS1} - \text{mean FxS(i)})}{\# \text{ Stages} - 1} \tag{2}$$

where n = number of stages
- After finding the score vector of the features, we find the upper bound (UB) and the lower bound (LB) of the score vector.
- We generate n number of particles (n = 10), randomly in between LB and UB.
- For a given particle we calculate the fitness as follows.
  1. We generate a binary vector comparing the particle and the score vector.
  2. We define the following parameters γ and θ where.

$$\gamma = \frac{(\text{silhouette\_score of selected feature}/\text{Selected features})}{(\text{silhouette\_score of All feature}/\text{All features})}$$

θ = Spearman correlation of the selected features[1].

$$\text{Fitness of the particle} = \frac{\gamma}{\theta} \tag{3}$$

- At the end of the run, we select the features given by the **Gbest** particle as the prominent features set for the desired disease.

2.3   Support Vector Machines (SVM)

Support Vector Machines (SVMs) is a supervised learning algorithm that is useful in solving both classification and regression problems. It was developed at AT&T Bell Laboratories by Vladimir Vapnik and colleagues [5]. Given a set of training examples, each was given as belonging to one of several categories, an SVM training algorithm builds a model that assigns new examples to one category or the others. SVMs can efficiently perform a non-linear classification using what is called the kernel trick, implicitly mapping their inputs into high-dimensional feature spaces which we have used here.

In this study, 6 SVM models were trained with Linear, polynomial, and RBF (Radial Basis Function) kernels. Linear Kernel is used when the data is Linearly separable, that is, it can be separated using a single line. It is one of the most common kernels to be used. It is mostly used when there are a Large number of Features in a particular Data Set. A polynomial Kernel is a generalized representation of the linear kernel. RBF kernels are the most generalized form of kernelization and are one of the most widely used kernels due to their similarity to the Gaussian distribution. The RBF kernel function for two points X_1 and x_2 computes the similarity or how close they are to each other. In this research, the first SVM model is trained using all three kernels to identify whether a given eye test record reveals the presence of any type of CNS disease. Identification of such presence can be tested in other trained models, separately for each disorder to identify which disorder is present in the data.

2.4   Dataset Used for the Study

The dataset we used has data that can be used to identify five CNS disorders; Amyotrophic Lateral Sclerosis (ALS) [14], Multiple Sclerosis (MS) [18], Alzheimer's Disease (AD) [17], Parkinson's Disease (PD) [10], and Schizophrenia [12]. Table 3 shows the nature of the real data.

Table 3. Description of data

CNS disorder	# of patients (# of patients with missing values in it)	# of features	Stages of the disorder
ALS	23 (11)	49	Severe, moderate, mild
MS	119 (94)	49	Secondary Progressive (SP), Primary Progressive (PP), Relapsing Remitting (RR)
AD	42 (10)	22	Severe AD (sAD), Mild to Moderate AD (mAD), Amnestic mild cognitive impairment (aMCI)
PD	49 (40)	32	Based on duration of the disorder
Schizophrenia	386 (176)	49	Based on duration of the disorder

3   Proposed Expert System

In this section, we will describe the complete working of the proposed expert system. The pseudo-code given in `Algorithm 2` further elaborate the idea.

Algorithm 1 : Pseudo code of the Proposed Expert System

1: Begin;

2: Data Preprocessing: remove missing data, feature scaling

Require: : Synthetic data generation

Require: : Prominent features selection using PSO

Require: : Implementation of ML models to classify data

3: Train SVM's with RBF Kernal separately to diagnose the disorders using the generated synthetic data and the selected prominent features for each disorder.

4: Test the models.

5: End;

Basically, before implementing the expert system (six ML models we describe), we addressed the data generation and feature selection procedures. These two can be stated as the two main challenges of the study.

4 Experimentation and Results

The experiments performed in this study are detailed here. Section 4.1 is dedicated to the results of the process of synthetic data generation. Under this section, it is explained the results of the partitioning process, the creative combination of partitioning and synthetic data generation. Section 4.2 explained the results of feature selection using PSO. Finally, in Sect. 4.3, the accuracy of trained models is detailed.

All the work of this research has been carried out on an Intel Core i5 laptop, with 3.4 GHz and a RAM of 8 GB. Python has been used as the programming language. Five CNS disorders are used to get results.

Table 4. Number of partitions in each feature of all the disorders

Disorder name	Disorder stage	Number of all the patients	Number of patients with non missing values	Number of partitions
ALS	-	23	11	5
AD	mACI	14	11	5
	mAD	7	6	3
	sAD	9	3	3
	Control	12	12	6
MS	SP	3	2	2
	PP	2	1	1
	RR	114	22	11
Parkinson	-	49	9	4
Schizophrenia	Patient	222	138	11
	Control	164	68	8

Table 5. Initial partitioning of features in ALS

0.0–0.2	0.2–0.4	0.4–0.6	0.6–0.8	0.8–1.0

Table 6. Partitioning of 1st feature of ALS after employing GA based partition algorithm

0.0–0.2	0.2–0.4	0.4–0.6	0.6–0.8	0.8–1.0

Table 7. Partitioning of 2nd feature of ALS after employing GA based partition algorithm

0.0–0.19	0.19–0.39	0.39–0.79	0.79–0.99

4.1   Synthetic Data Generation

1. feature partitioning

   Feature scaling is performed to set each feature value between 0 and 1. Then for a particular feature, the features range (0–1 range) is divided into (number of patients/2) parts. But, if the number of patients is equal to 1 then each feature number of partitions equals the number of patients. Also, the number of patients is equal to 1 then, no need to partition.

   Ex:- ALS CNS disorder has records belonging to 23 patients. 11 patients out of 23 are non-null values. Thus, each feature is partitioned into 5 partitions. Table 5 shows the initial partitions taken for the ALS disorder. Table 6 and Table 7 shows the partitions obtained for the two features in the ALS data set after executing the algorithm for 100 iterations.

   After 100 iterations 1st feature of ALS is still partitioned to 5 while 2nd feature is reduced to 4 partitions.

   The Fig. 1 shows the distribution of data before and after the generation of synthetic data for all the five disorders. For the visualization purpose only, here we applied principal component analysis (PCA) to reduce the number of features.

4.2   Prominent Feature Selection

Here we present the number of features selected via PSO and the number of features given in medical research claiming to be the most prominent features for a particular CNS disorder (Table 8).

4.3   Identification of the CNS Disorders

We trained 6 SVMs using linear, polynomial, and RBF kernels to identify the CNS disorders. One SVM will detect a patient record to identify whether the

Table 8. Number of features selected by PSO and in medical research

Disorder	Number of selected features using PSO	Number of selected features in medical research
ALS	14	6 [14]
MS	21	25 [18]
AD	14	8 [10]
PD	14	11 [9]
Schizophrenia	49	45 [12]

record contains details to suspect a CNS disorder. The other five (05) SVMs are dedicated to identifying a specific disorder and the stage of the disorder. We have trained the SVMs using all three kernels to identify the most suitable one. Table 9 summarizes the testing accuracies of the six SVMs with linear, polynomial, and RBF kernels.

Table 9. Summary of the testing accuracy of the six SVMs with linear, polynomial and RBF kernels

Designed SVM for	Testing accuracies		
	Linear Kernal	RBF Kernal	Poly Kernal
	Parameters—C = 0.01	Parameters—C = 0.001 Gamma = 0.1	Parameters—C = 1°C = 3
Detecting an availability of CNS disorder	0.68	0.96	0.73
ALS	0.77	0.54	0.68
MS	0.96	0.89	1.00
PD	0.96	0.96	0.96
AD	0.86	0.26	0.9
schizophrenia	0.61	0.62	0.64

The results indicate that linear Polynomial and Linear kernels performed better than the RBF kernel in almost all cases. Since the primary purpose of the study is to divide data into two parts and that seems to be aligned well with the capability of the linear and polynomial kernels.

When it is about identifying an existence of a disorder, the RBF Kernal has performed well. For the five CNS disorders, we can identify the most suitable Kernels as follows (Table 9).

1. AD - Polynomial Kernel
2. ALS - Linear Kernel
3. MS - Polynomial Kernel
4. PD - Polynomial/Linear/RBF Kernels
5. Sysorphenia - Polynomial Kernel

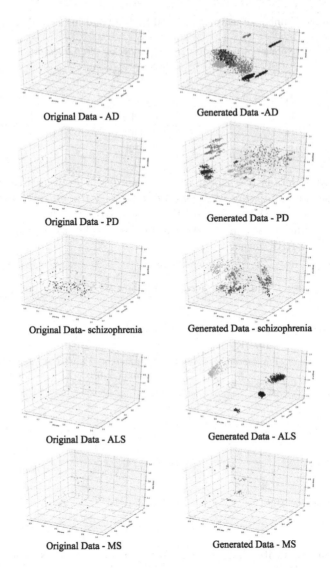

Fig. 1. Distribution of data before and after generation of synthetic data for the five disorders

5   Conclusions and Future Work

This paper presents one of the very first attempts of using computational intelligence to identify CNS disorders using eye test data. Much medical research reveals that there is a strong relationship between CNS disorders and the different problems in the human eye. To obtain a fair solution to find this relationship using Artificial Intelligence, we proposed a solution based on SVM. Implementation of a such solution lacked two major requirements, which we had to address before moving to the proposed solution. The first of such is the lack of suitable eye test data to train and test the SVM models. The second is to identify the most prominent features of a given CNS disorder to improve the accuracy of the SVM models.

We proposed GA based synthetic data generation model and a feature selection model based on PSO to overcome the problems. Figure 1 shows the progress of the synthetic data generation process. Table 8 shows how the selected features map with the features given in the medical research. There can be seen some slight differences, and further research may be helpful to find whether there is a connection between the newly selected features from our work and the CNS disorders.

SVMs with three kernel functions were used to identify the existence of a CNS disorder and if so, to identify each of the CNS disorders, AD, ALS, MS, PD, and Sysorphenia. All the kernels worked well with slight differences in the accuracies. As one of the first attempts of computational intelligence approach in detecting CNS disorders via test data, results are encouraging. We hope to improve the results in the future by adding more real data to the data set. Further, we hope that the synthetic data we generated and the features we have identified are useful for future researchers as well.

References

1. Spearman Rank Correlation Coefficient, pp. 502–505. Springer, New York (2008). https://doi.org/10.1007/978-0-387-32833-1_379
2. Brindle, A.: Genetic algorithms for function optimization, University of Alberta (1980)
3. Chakraborti, N.: Genetic algorithms in materials design and processing. Int. Mater. Rev. **49**(3–4), 246–260 (2004)
4. Chang, C.K., Christensen, M.J., Zhang, T.: Genetic algorithms for project management. Ann. Softw. Eng. **11**(1), 107–139 (2001)
5. Cortes, C., Vapnik, V.: Support-vector networks. Mach. Learn. **20**(3), 273–297 (1995)
6. Hajihassani, M., Armaghani, D.J., Kalatehjari, R.: Applications of particle swarm optimization in geotechnical engineering: a comprehensive review. Geotech. Geol. Eng. **36**(2), 705–722 (2018)
7. Holland, J.H.: Genetic algorithms. Sci. Am. **267**(1), 66–73 (1992)
8. Kennedy, J., Eberhart, R.: Particle swarm optimization. In: Proceedings of ICNN'95-International Conference on Neural Networks, vol. 4, pp. 1942–1948. IEEE (1995)
9. Kim, J.I., Kang, B.H.: Decreased retinal thickness in patients with Alzheimer's disease is correlated with disease severity. PLoS ONE **14**(11), e0224180 (2019)
10. Kromer, R., et al.: Evaluation of retinal vessel morphology in patients with Parkinson's disease using optical coherence tomography. PLoS ONE **11**(8), e0161136 (2016)
11. Kulkarni, R.V., Venayagamoorthy, G.K.: Particle swarm optimization in wireless-sensor networks: a brief survey. IEEE Trans. Syst. Man Cybernet. Part C (Appl. Rev.) **41**(2), 262–267 (2010)

12. Liu, Y., Huang, L., Tong, Y., Chen, J., Gao, D., Yang, F.: Association of retinal nerve fiber abnormalities with serum CNTF and cognitive functions in schizophrenia patients. Peer J. **8**, e9279 (2020)

13. London, A., Benhar, I., Schwartz, M.: The retina as a window to the brain-from eye research to CNS disorders. Nat. Rev. Neurol. **9**(1), 44–53 (2013)

14. Mukherjee, N., McBurney-Lin, S., Kuo, A., Bedlack, R., Tseng, H.: Retinal thinning in amyotrophic lateral sclerosis patients without ophthalmic disease. PLoS ONE **12**(9), e0185242 (2017)

15. Rana, S., Jasola, S., Kumar, R.: A review on particle swarm optimization algorithms and their applications to data clustering. Artif. Intell. Rev. **35**(3), 211–222 (2011)

16. Renner, G., Ekárt, A.: Genetic algorithms in computer aided design. Comput. Aided Des. **35**(8), 709–726 (2003)

17. Salobrar-García, E., et al.: Changes in visual function and retinal structure in the progression of Alzheimer's disease. PloS ONE **14**(8), e0220535 (2019)

18. Sanchez-Dalmau, B., et al.: Predictors of vision impairment in multiple sclerosis. PLoS ONE **13**(4), e0195856 (2018)

Assessment of Semi-supervised Approaches Applied to Convolutional Neural Networks

Cristiano N. de O. Bassani[1]([⊠])(iD), Prisicla T. M. Saito[2]([⊠])(iD), and Pedro H. Bugatti[1]([⊠])(iD)

[1] Federal University of Technology - Paraná (UTFPR), Cornélio Procópio PR, Alberto Carazzai Avenue, 1640, Curitiba 86300-000, Brazil
crisbassani94@gmail.com, pbugatti@utfpr.edu.br
[2] Federal University of São Carlos (UFSCar), Rodovia Washington Luís, km 235, São Carlos, SP 13565-905, Brazil
priscilasaito@ufscar.br

Abstract. Convolutional neural newtorks have been presenting great results regarding image classification in different contexts. As known in the literature, this kind of network demands a great volume of labeled samples to reach a good convergence during the training process. However, in real scenarios labeled samples are scarce because the labeling process is costly and time consuming. Moreover, it is highly susceptible to errors when accomplished by human specialists. This fact impairs in a great extent the applicability of convolutional neural networks in several contexts where there is a lack of labeled samples. Thus, in this work we focus on mitigate this issue. To do so, we apply the semi-supervised paradigm to the convolutional neural networks. We proposed the aggregation of two semi-supervised techniques from literature with this kind of deep learning networks and analyse their behavior. We performed an extensive assessment of this aggregation. We also considered the transfer learning approach in the process to verify its generalization under the semi-supervised paradigm. Our experiments, with three public datasets, testify that our proposed aggregation obtained better results, gains of up to 88% in accuracy, when compared with the supervised paradigm.

Keywords: Convolutional neural networks · Semi-supervised learning · Machine learning · Self-training · Co-training

1 Introduction

It is well known that there are several machine learning approaches in the literature, each one is better applicable than others in different problems and situations (i.e. no-free-lunch theorem). Each machine learning paradigm (e.g. supervised, unsupervised, semi-supervised, among others) has a range of applicability and usability [4,7].

To start studying the applicability of a given approach, it is necessary to restrict a specific paradigm and find the desired problem to be solved. In this

© The Author(s), under exclusive license to Springer Nature Switzerland AG 2023
L. Rutkowski et al. (Eds.): ICAISC 2022, LNAI 13589, pp. 195–205, 2023.
https://doi.org/10.1007/978-3-031-23480-4_16

work, we focus on the semi-supervised paradigm applied to the convolutional neural networks (CNNs). This kind of neural network has been presenting great results w.r.t. image classification process. However, as known in the literature, CNNs require great volume of labeled data to train a learning model. This fact impairs in a great extent their applicability in several contexts where there is a lack of labeled samples. Then, we performed an extensive assessment of two semi-supervised approaches applied to CNNs.

Hence, our work aims to carry out analyses between two main semi-supervised approaches on public image datasets. According to our experiments, we can note that pseudo-labeled samples can generate results similar to supervised learning paradigm (which requires 100% of labeled samples). We evaluated different types of metrics such as accuracy, precision, recall and f1-measure to obtain an wide and deep analyses of the CNNs aggregated with the semi-supervised approaches.

As an example of this, in [1] the authors used a variation of the optimum-path forest classifier to obtain better performance w.r.t. against other semi-supervised approaches from the literature. However, their approach has several aspects to be explored, such those mentioned in [4]. For instance, different from our comparative study, the authors' approach performed a pre-training of the CNNs with pseudolabels and then applied their learned weights to a second one. The point is that it can generate a greater footprint or even possible overfitting problems. Another difference is that, in our work, we use the baseline transfer learning approach through the ImageNet [2,10] in order to obtain clean and straightforward analyses.

Contributions: In summary, our contributions are threefold: (i) performed an assessment of self-training and co-training semi-supervised approaches regarding their behavior aggregated with CNNs; (ii) considered the transfer learning approach in the process to verify its generalization under the semi-supervised paradigm; (iii) performed analyses considering different metrics and using three public datasets.

2   Background

Related to the semi-supervised paradigm, our work considered the Self-training [1] and the Co-training [4] approaches. The reason to define such approaches is that they are widely-used in the literature. The Self-training approach is the most basic and well-known approaches w.r.t. semi-supervised paradigm. It makes use of the pseudo-labeling process to classify the unlabeled samples, then these unlabeled samples are incrementally added to the training model. The other approach that we evaluate was the so-called Co-training. It also uses the pseudo-labeling process. However, it uses multiple learning models training on the same samples to cross-validate the labeling process.

Regarding the convolutional neural networks, in this paper, we used then as an end-to-end process aggregated to the semi-supervised approaches. To do so, we considered two state-of-the-art architectures called Xception [8] and ResNet152_V2 [3]. The Xception architecture is used in several works. For instance, to perform image classification of human proteins from a hybrid model, among others. ResNet152_V2 was used, for example, to predict pneumonia disease from lungs x-ray images. In [9], CNNs was also used to classify cervical cells.

3 Proposed Approach

In this work, we associated the semi-supervised paradigm with end-to-end CNNs to classify images from 3 public datasets. To do so, we used Xception and ResNet152_V2 that are two different state-of-the-art CNN architectures. We set the same hyperparameters for both architectures (e.g. learning rates, optimizer, among others), image resizing, regularization and data normalization.

First, we considered the semi-supervised Self-training approach with the Xception architecture. Self-training selects a portion of the labeled samples to train the CNN. Then, a learning model is generated to classify the unlabeled samples of the training set. Each one of these samples, labeled with pseudo-labels, are then incrementally aggregated to the labeled set to retrain the CNN. Figures 1 and 2 show the pipeline of the proposed Self-training approaches. Figure 1 shows the pseudo-label generation process for unlabeled samples. Figure 2 illustrates the re-training considering the pseudo-labeled samples.

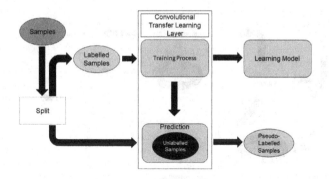

Fig. 1. Pseudo labeling in the self-training process using transfer learning

Regarding the Co-training approach, it makes use of two different CNNs (i.e. Xception and ResNet152_V2) to generate the pseudo-labels together with the mutual validation process of the predicted labels by both CNNs. We just used the samples labeled equally between both architectures. This second semi-supervised approach is different from the first one. In this case, the same samples are submitted to two CNNs (end-to-end) simultaneously. Then, it generates two pseudo-label sets, that are going through a mutual validation to assign a pseudo-label to a given unlabeled sample. The mutual validation block works as follows: if the pseudo-labels (outputs of both architectures) are equal, the sample is inserted into the set of pseudo-labeled samples. Otherwise it does not enter such a set. Once the pseudo-labeling is finished, the following networks, previously trained are reloaded and the new training set (labeled set plus the pseudo-label set) is submitted to train the CNN.

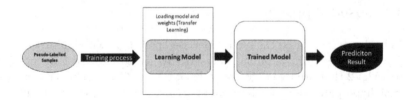

Fig. 2. Training phase of the self-training process

It is important to note that, initially, considering both approaches, all CNNs were initialized with weights (transfer learning) from ImageNet (pre-trained). In this case, Fig. 3 shows the pseudo-labeling process performed by the proposed Co-training approach. The retraining process is identical as the Self-training (see Fig. 2).

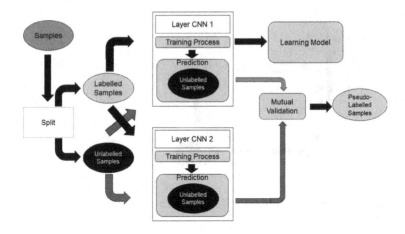

Fig. 3. Pseudo labeling in the co-training process using transfer learning

To evaluate both proposed approaches, we first divided each image dataset into a training and a test set, considering 80% and 20%, respectively. Then, the training set was splitted in labeled and unlabeled sets. To better analyse the proposed approaches we considered different splits w.r.t. the labeled set (10%, 20%, 30%, 40% and 50%). This same methodology was applied 10 times (stratified hold-out) to generated mutually exclusive sets and provide statistical analyses w.r.t. the evaluation process.

4   Experiments

4.1   Datasets' Descriptions

The experiments were performed considering the public datasets Papsmear [9], MAMMOSET [5] and WHOI-Plankton [6]. The Papsmear dataset comes from the Kaggle platform database, which is based on a set of cell images categorized into different subclasses of abnormal and normal cells.

The MAMMOSET dataset comprises regions of interest (ROIs) obtained from mamographic exams. It is composed of 4 classes related to calcification and mass findings considering benign or malign nodules. The third dataset, named WHOI-Plankton, presents 103 classes. It is composed of millions of images from different types of marine plankton. However, due to space limitations and computational cost, in our experiments, we chose a random subset of classes 14 classes.

In the Papsmear dataset, the images have dimensions ranging from 53×129 pixels to 252×28 pixels, as well as balanced sample amounts. The images are in the RGB color space with 8-bit depth. On the other hand, the MAMMOSET dataset has images (ROIs) in gray scale, with equal dimensions of 1000×1000 pixels. Finally, the WHOI-Plankton dataset presents grayscale images with varying dimensions. Figures 4, 5 and 6 show examples of images from each class considering the Papsmear, MAMMOSET and WHOI-Plankton datasets, respectively.

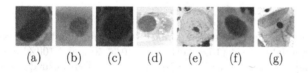

Fig. 4. Examples of each class from the Papsmear dataset. (a) carcinoma. (b) light. (c) moderate. (d) normal columnar. (e) normal intermediate. (f) severe dysplastic. (g) normal superficial

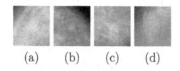

Fig. 5. Examples of each class from the MAMMOSET dataset. (a) malignant mass. (b) benign mass. (c) malignant calcification. (d) benign calcification

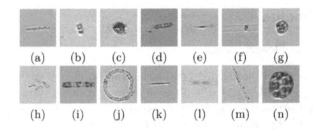

Fig. 6. Examples of each class from the WHOI-Plankton dataset. (a) Cerataulina. (b) Chaetoceros. (c) Ciliate-mix. (d) Corethron. (e) Cyliandrotheca. (f) Dactyliosolen. (g) Dino30. (h) Dinobryon. (i) Guinardia delicatula. (j) Guinardia striata. (k) Leptocylindrus. (l) Pseudonitzschia. (m) Rhizosolenia. (n) Thalassiosira.

4.2 Results

In order to analyze our proposed approaches (Self-training and Co-training aggregated with CNNs) we compared them against the supervised paradigm, considering the three datasets aforementioned.

Table 1 shows the results obtained considering the Papsmear dataset. According to the experiments, our approaches presented better results (all metrics), considering 50% of labeled samples when compared with the supervised one (100% of labeled samples). For instance, the Co-training CNN approach obtained a gain of up to 15% of accuracy.

Table 1. Results obtained by each approach (Co-training and Self-training) for each percentage of the labeled set (10%, 20%, 30%, 40%, 50% and 100%) considering each metric (accuracy, precision, recall and f1-score) for the Papsmear dataset

Labeled	Approach	Accuracy	Precision	Recall	F1-Score
10%	Co-Training	37.71%±0.255	42.03%±0.258	39.74%±0.267	38.69%±0.253
	Self-Training	33.43%±0.317	41.51%±0.365	37.29%±0.351	33.11%±0.297
20%	Co-Training	38.86%±0.262	40.91%±0.253	40.91%±0.276	39.09%±0.248
	Self-Training	33.43%±0.305	37.57%±0.319	35.26%±0.321	32.43%±0.282
30%	Co-Training	38.14%±0.280	38.71%±0.262	40.20%±0.294	38.34%±0.270
	Self-Training	35.86%±0.314	38.89%±0.315	37.77%±0.329	34.91%±0.294
40%	Co-Training	40.14%±0.273	41.54%±0.262	42.26%±0.287	40.00%±0.253
	Self-Training	37.14%±0.350	40.49%±0.364	40.17%±0.379	35.83%±0.341
50%	Co-Training	**45.00%±0.276**	**44.83%±0.261**	**47.37%±0.291**	**44.83%±0.273**
	Self-Training	39.86%±0.345	44.43%±0.332	42.00%±0.363	38.46%±0.306
100%	Supervised	38.86%±0.290	42.03%±0.311	44.00%±0.322	40.66%±0.303

Table 2. Results obtained by each approach (Co-training and Self-training) for each percentage of the labeled set (10%, 20%, 30%, 40%, 50% and 100%) considering each metric (accuracy, precision, recall and f1-score) for the Mammoset dataset

Labeled	Approach	Accuracy	Precision	Recall	F1-score
10%	Co-Training	31.56% ± 0.115	33.70% ± 0.113	33.25% ± 0.121	32.55% ± 0.104
	Self-Training	30.03% ± 0.330	28.85% ± 0.286	31.55% ± 0.348	24.60% ± 0.267
20%	Co-Training	34.75% ± 0.118	36.50% ± 0.106	36.50% ± 0.125	35.88% ± 0.108
	Self-Training	26.52% ± 0.328	14.45% ± 0.240	27.95% ± 0.346	17.80% ± 0.252
30%	Co-Training	35.83% ± 0.121	39.10% ± 0.132	37.95% ± 0.128	37.60% ± 0.114
	Self-Training	35.08% ± 0.305	36.05% ± 0.281	36.90% ± 0.321	31.10% ± 0.239
40%	Co-Training	**37.55% ± 0.128**	**39.60% ± 0.113**	**39.40% ± 0.135**	**38.65% ± 0.110**
	Self-Training	36.64% ±0.352	33.10% ± 0.167	38.50% ± 0.371	31.45% ± 0.216
50%	Co-Training	36.59% ± 0.166	39.15% ± 0.137	38.50% ± 0.175	37.30% ± 0.134
	Self-Training	37.00% ± 0.319	38.70% ± 0.215	38.85% ± 0.336	32.60% ± 0.200
100%	Supervised	30.49% ± 0.365	21.80% ± 0.251	33.75% ± 0.405	23.95% ± 0.265

Table 3. Results obtained by each approach (Co-training and Self-training) for each percentage of the labeled set (10%, 20%, 30%, 40%, 50% and 100%) considering each metric (accuracy, precision, recall and f1-score) for the WHOI-Plankton dataset

Labeled	Approach	Accuracy	Precision	Recall	F1-score
10%	Co-Training	82.43% ± 0.124	87.70% ± 0.109	86.73% ± 0.131	86.49% ± 0.100
	Self-Training	78.93% ± 0.172	85.84% ± 0.137	83.09% ± 0.181	82.79% ± 0.131
20%	Co-Training	84.00%±0.115	88.67%±0.086	88.43%±0.122	88.09%±0.091
	Self-Training	85.71%±0.113	90.87%±0.085	90.17%±0.119	89.94%±0.084
30%	Co-Training	85.29%±0.106	90.39%±0.076	89.70%±0.113	89.49%±0.073
	Self-Training	87.79%±0.095	93.29%±0.082	92.36%±0.100	92.21%±0.069
40%	Co-Training	86.29%±0.106	91.31%±0.067	90.76%±0.112	90.50%±0.068
	Self-Training	**88.71%±0.095**	**94.17%±0.073**	**93.36%±0.100**	**93.13%±0.065**
50%	Co-Training	87.00%±0.102	91.91%±0.068	91.53%±0.108	91.29%±0.072
	Self-Training	88.71%±0.110	94.01%±0.065	93.39%±0.116	92.97%±0.076
100%	Supervised	84.86%±0.082	94.70%±0.060	94.17%±0.090	94.01%±0.060

This same behavior was observed when analyzing the Mammoset and WHOI-Plankton datasets (Tables 2 and 3). For example, considering the results obtained from the Mammoset dataset (Table 2), the Co-training CNN approach obtained a gain of up to 19% of accuracy with just 40% of labeled samples, when compared with the supervised one. Regarding the precision and F1 score, the gains were even higher (45% and 38%, respectively). The Self-training CNN also presented better results than the supervised paradigm.

Table 4. Results obtained by each approach (Supervised, Co-training, and Self-training) for each class considering each metric (accuracy, precision, recall and f1-score) for the Papsmear dataset

Class	Metrics	Supervised	Co-training 50%	Self-training 50%
0	Accuracy	13.00% ± 0.178	5.00% ± 0.035	14.00% ± 0.204
	Precision	10.20% ± 0.143	19.00% ± 0.122	12.00% ± 0.098
	Recall	14.40% ± 0.198	5.20% ± 0.039	14.80% ± 0.217
	F1-score	12.00% ± 0.166	8.20% ± 0.057	12.20% ± 0.135
1	Accuracy	16.00% ± 0.219	45.00% ± 0.145	34.00% ± 0.245
	Precision	23.80% ± 0.223	35.40% ± 0.048	48.20% ± 0.158
	Recall	31.00% ± 0.298	47.40% ± 0.154	35.80% ± 0.258
	F1-score	26.40% ± 0.241	40.00% ± 0.073	36.40% ± 0.139
2	Accuracy	25.00% ± 0.145	20.00% ± 0.117	19.00% ± 0.129
	Precision	26.60% ± 0.080	26.40% ± 0.178	48.40% ± 0.307
	Recall	26.80% ± 0.174	21.00% ± 0.124	20.00% ± 0.136
	F1-score	23.00% ± 0.052	23.00% ± 0.139	23.60% ± 0.114
3	Accuracy	37.00% ± 0.256	47.00% ± 0.044	14.00% ± 0.171
	Precision	22.00% ± 0.020	39.60% ± 0.062	8.20% ± 0.096
	Recall	30.20% ± 0.268	49.60% ± 0.050	14.80% ± 0.181
	F1-score	21.20% ± 0.104	44.00% ± 0.051	10.20% ± 0.120
4	Accuracy	70.00% ± 0.252	83.00% ± 0.067	92.00% ± 0.044
	Precision	73.80% ± 0.178	75.00% ± 0.067	67.00% ± 0.178
	Recall	80.00% ± 0.294	87.40% ± 0.072	96.80% ± 0.048
	F1-score	76.00% ± 0.241	80.60% ± 0.065	77.80% ± 0.150
5	Accuracy	74.00% ± 0.147	77.00% ± 0.057	70.00% ± 0.183
	Precision	92.40% ± 0.107	88.20% ± 0.068	95.20% ± 0.074
	Recall	79.00% ± 0.146	81.00% ± 0.057	73.80% ± 0.191
	F1-score	84.00% ± 0.099	84.00% ± 0.019	81.00% ± 0.098
6	Accuracy	37.00% ± 0.130	38.00% ± 0.103	36.00% ± 0.386
	Precision	45.40% ± 0.108	30.20% ± 0.053	32.00% ± 0.247
	Recall	46.60% ± 0.201	40.00% ± 0.110	38.00% ± 0.407
	F1-score	42.00% ± 0.102	34.00% ± 0.067	28.00% ± 0.230

Analyzing the results from the WHOI-Plankton dataset, the Self-training CNN reached the best results (88.71% of accuracy) with 40% of labeled samples, while the supervised one achieved an accuracy of 84.86%.

Table 5. Results obtained by each approach (supervised, co-training, and self-training) for each class considering each metric (accuracy, precision, recall and f1-score) for the Mammoset dataset

Class	Metrics	Supervised	Co-training 40%	Self-training 50%
0	Accuracy	14.20%±0.317	29.40%±0.068	62.60%±0.377
	Precision	12.80%±0.286	44.80%±0.019	46.40%±0.162
	Recall	15.80%±0.353	30.80%±0.072	66.00%±0.398
	F1-score	14.20%±0.318	36.40%±0.050	43.60%±0.249
1	Accuracy	15.74%±0.161	26.38%±0.055	16.60%±0.166
	Precision	24.20%±0.230	21.60%±0.038	22.00%±0.182
	Recall	17.20%±0.177	27.60%±0.056	17.20%±0.174
	F1-score	20.00%±0.198	23.80%±0.037	19.20%±0.176
2	Accuracy	46.60%±0.442	40.60%±0.064	35.60%±0.347
	Precision	24.80%±0.269	44.00%±0.054	31.60%±0.315
	Recall	51.60%±0.490	42.60%±0.064	37.40%±0.365
	F1-score	31.20%±0.296	43.20%±0.050	34.00%±0.335
3	Accuracy	45.40%±0.437	53.80%±0.091	33.20%±0.302
	Precision	25.40%±0.284	48.00%±0.040	54.80%±0.389
	Recall	50.40%±0.486	56.60%±0.099	34.80%±0.317
	F1-score	30.40%±0.288	51.20%±0.025	33.60%±0.307

In order to better testify the results obtained by our approaches, we show in Tables 4, 5 and 6, respectively, all the metrics for each class from the datasets Papsmear, Mammoset and WHOI-Plankton, considering the best cases.

For instance, Table 4 shows the results for each class considering the supervised approach, the Co-training CNN with 50% of labeled samples and the Self-training also with 50% of labeled samples (i.e. best cases from Table 1). Analyzing the results, we can see that the Co-training CNN presented better results for classes 1 and 3 to 6 in comparison with the supervised one. For instance, for class 1 the accuracy gain was 2.8 times better.

Considering Table 5, the Co-training CNN with 40% of labeled samples presents better results for classes 0, 1 and 3. It is worth to mention that the Self-training CNN (50% of labeled samples) achieved for class 0 an accuracy 4.4 times better than the supervised one. Finally, when analyzing Table 6, our approaches presented the best results for almost every class. These results also lead us to a possible future work to generate an ensemble from both approaches.

Table 6. Results obtained by each approach (Supervised, Co-training, and Self-training) for each class considering different metrics for the WHOI-Plankton dataset

Class	Metrics	Supervised	Co-training 50%	Self-training 40%
0	Accuracy	61.00% ± 0.108	60.00% ± 0.061	64.00% ± 0.102
	Precision	100.00% ± 0.000	90.40% ± 0.059	100.00% ± 0.000
	Recall	67.80% ± 0.117	63.20% ± 0.065	67.40% ± 0.107
	F1-score	80.20% ± 0.089	74.00% ± 0.035	79.80% ± 0.076
1	Accuracy	89.00% ± 0.022	78.00% ± 0.075	93.00% ± 0.044
	Precision	100.00% ± 0.000	90.20% ± 0.092	96.20% ± 0.060
	Recall	98.80% ± 0.027	81.80% ± 0.080	97.80% ± 0.049
	F1-score	99.40% ± 0.013	85.60% ± 0.072	96.80% ± 0.032
2	Accuracy	84.00% ± 0.022	89.00% ± 0.022	89.00% ± 0.022
	Precision	96.40% ± 0.033	94.00% ± 0.064	93.20% ± 0.053
	Recall	93.00% ± 0.022	93.80% ± 0.027	93.80% ± 0.026
	F1-score	94.60% ± 0.025	93.60% ± 0.038	93.20% ± 0.021
3	Accuracy	84.00% ± 0.022	93.00% ± 0.027	81.00% ± 0.082
	Precision	95.40% ± 0.047	94.00% ± 0.042	93.40% ± 0.049
	Recall	93.00% ± 0.022	98.00% ± 0.027	85.20% ± 0.084
	F1-score	94.20% ± 0.018	95.80% ± 0.011	88.80% ± 0.048
4	Accuracy	90.00% ± 0.000	95.00% ± 0.000	95.00% ± 0.000
	Precision	100.00% ± 0.000	94.00% ± 0.042	100.00% ± 0.000
	Recall	100.00% ± 0.000	100.00% ± 0.000	100.00% ± 0.000
	F1-score	100.00% ± 0.000	96.80% ± 0.020	100.00% ± 0.000
5	Accuracy	90.00% ± 0.000	95.00% ± 0.000	95.00% ± 0.000
	Precision	98.00% ± 0.027	97.00% ± 0.045	100.00% ± 0.000
	Recall	100.00% ± 0.000	100.00% ± 0.000	100.00% ± 0.000
	F1-score	98.80% ± 0.016	98.40% ± 0.023	100.00% ± 0.000
6	Accuracy	90.00% ± 0.000	93.00% ± 0.027	95.00% ± 0.000
	Precision	97.20% ± 0.063	94.20% ± 0.062	98.00% ± 0.027
	Recall	100.00% ± 0.000	98.00% ± 0.027	100.00% ± 0.000
	F1-score	98.40% ± 0.036	96.00% ± 0.042	98.80% ± 0.016
7	Accuracy	88.00% ± 0.027	88.00% ± 0.057	92.00% ± 0.044
	Precision	82.40% ± 0.051	85.60% ± 0.047	77.60% ± 0.058
	Recall	97.60% ± 0.033	92.60% ± 0.062	96.80% ± 0.048
	F1-score	89.40% ± 0.030	88.80% ± 0.034	86.00% ± 0.041
8	Accuracy	90.00% ± 0.000	95.00% ± 0.000	95.00% ± 0.000
	Precision	94.00% ± 0.042	98.00% ± 0.027	89.80% ± 0.065
	Recall	100.00% ± 0.000	100.00% ± 0.000	100.00% ± 0.000
	F1-score	96.80% ± 0.020	98.80% ± 0.016	94.40% ± 0.037
9	Accuracy	85.00% ± 0.035	84.00% ± 0.074	91.00 ± 0.054
	Precision	89.00% ± 0.070	88.60% ± 0.097	96.00% ± 0.041
	Recall	94.20% ± 0.039	88.20% ± 0.078	95.60% ± 0.060
	F1-score	91.40% ± 0.038	88.00% ± 0.044	95.40% ± 0.015
10	Accuracy	90.00% ± 0.000	95.00% ± 0.000	91.00% ± 0.065
	Precision	98.00% ± 0.027	89.80% ± 0.064	87.20% ± 0.096
	Recall	100.00% ± 0.000	100.00% ± 0.000	95.80% ± 0.069
	F1-score	98.80% ± 0.016	94.60% ± 0.036	90.40% ± 0.040
11	Accuracy	83.00% ± 0.027	86.00% ± 0.022	84.00% ± 0.108
	Precision	93.20% ± 0.045	92.20% ± 0.029	97.80% ± 0.030
	Recall	92.00% ± 0.027	90.20% ± 0.027	88.40% ± 0.116
	F1-score	92.60% ± 0.013	91.40% ± 0.025	92.40% ± 0.066
12	Accuracy	85.00% ± 0.035	85.00% ± 0.050	89.00% ± 0.065
	Precision	89.40% ± 0.032	85.00% ± 0.097	94.60% ± 0.052
	Recall	94.20% ± 0.039	89.40% ± 0.055	93.60% ± 0.070
	F1-score	91.80% ± 0.022	86.80% ± 0.050	94.00% ± 0.053
13	Accuracy	79.00% ± 0.054	82.00% ± 0.097	88.00% ± 0.027
	Precision	92.80% ± 0.027	93.80% ± 0.065	94.60% ± 0.005
	Recall	87.80% ± 0.059	86.20% ± 0.101	92.60% ± 0.032
	F1-score	89.80% ± 0.045	89.40% ± 0.047	93.80% ± 0.016

5   Conclusions

In this paper, we proposed the aggregation of two semi-supervised techniques with convolutional neural networks and analyse their behavior. Moreover, we performed an extensive assessment of this aggregation.

Our obtained results testify the great advantage of such aggregation considering the Self-training CNN and the Co-training CNN. Besides, we show that the pseudo-labeling process joined with the transfer learning technique can reach better results when compared with the supervised paradigm in different image contexts. Our approaches achieved an accuracy 4.4 times better when analyzing one class of the datasets, and with just half of labeled samples (i.e. 50%).

Thus, pseudo-labeling of unlabeled samples can not only help the training process in a great extent, but also can provide higher efficacy (i.e. better accuracy, among other metrics) and efficiency (fewer labeled samples). In future works, we intend to propose the ensemble of our both approaches.

References

1. Chen, J., Feng, J., Sun, X., Liu, Y.: Co-training semi-supervised deep learning for sentiment classification of MOOC forum posts. Symmetry **12**(1), **8** (2020)
2. Deng, J., Dong, W., Socher, R., Li, L.J., Li, K., Fei-Fei, L.: ImageNet: a large-scale hierarchical image database. In: 2009 IEEE Conference on Computer Vision and Pattern Recognition, pp. 248–255 (2009)
3. Elshennawy, N.M., Ibrahim, D.M.: Deep-pneumonia framework using deep learning models based on chest x-ray images. Diagnostics **10**(9), 649 (2020)
4. van Engelen, J.E., Hoos, H.H.: A survey on semi-supervised learning. Mach. Learn. **109**(2), 373–440 (2019). https://doi.org/10.1007/s10994-019-05855-6
5. Oliveira, P., de Carvalho Scabora, L., Cazzolato, M., Bedo, M., Traina, A., Jr, C.: MAMMOSET: an enhanced dataset of mammograms. In: Dataset Showcase Workshop - DSW at the Brazilian Symposium on Databases, pp. 1–11 (October 2017)
6. Orenstein, E., Beijbom, O., Peacock, E., Sosik, H.: WHOI-plankton- a large scale fine grained visual recognition benchmark dataset for plankton classification. Tech Report, pp. 1–2 (2015)
7. Rosenberg, C., Hebert, M., Schneiderman, H.: Semi-supervised self-training of object detection models. In: 2005 Seventh IEEE Workshops on Applications of Computer Vision (WACV/MOTION'05) - Volume 1. vol. 1, pp. 29–36 (2005)
8. T.R., Shwetha, Thomas, S.A., Kamath, V., Niranjana, K.B.: Hybrid xception model for human protein atlas image classification. In: 2019 IEEE 16th India Council International Conference (INDICON), pp. 1–4 (2019)
9. Zhang, L., Lu, L., Nogues, I., Summers, R.M., Liu, S., Yao, J.: Deeppap: deep convolutional networks for cervical cell classification. IEEE J. Biomed. Health Inform. **21**, 1633–1643 (2017)
10. Zhuang, F., et al.: A comprehensive survey on transfer learning. CoRR abs/1911.02685 (2019)

Bi-Space Search: Optimizing the Hybridization of Search Spaces in Solving the One Dimensional Bin Packing Problem

Derrick Beckedahl[(✉)] and Nelishia Pillay

University of Pretoria, Pretoria, South Africa
u18319557@tuks.co.za, npillay@cs.up.ac.za

Abstract. Search methodologies essentially explore a solution space to solve optimization problems. As the field has developed the effectiveness of exploring other spaces has been established. For example genetic programming explores the program space. Similarly, hyper-heuristics explore the heuristic space. In previous work the advantage of switching search between different spaces rather than working in a single space has been illustrated. This paper extends this work by presenting the bi-space search which optimizes when the switch between spaces should take place. The bi-space search employs iterated local search to optimize when to switch between the solution and heuristic spaces in solving discrete optimization problems. The performance of the bi-space search is compared to searching the solution space only and a hyper-heuristic searching the heuristic space. Both the solution and heuristic space searches employ iterated local search to explore the solution and heuristic space respectively. All three searches are evaluated on the one dimensional bin packing problem. The bi-space search was found to outperform the solution space search and hyper-heuristic.

Keywords: Bi-space search · Hyper-heuristic · Solution space search · One dimensional bin packing

1 Introduction

At their inception search methodologies essentially focused on exploring the solution space to solve optimization problems. However, more recently exploration of other spaces have proven to be more effective than exploring the solution space for certain problem domains. For example, genetic programming [7] searches the program space, the program found is usually then executed to produce a solution to the problem. Hyper-heuristics [8] explore the heuristic space which maps to the solution space to solve the problem at hand. Exploring an alternative space for a particular problem domain may overcome the challenges of working in the solution space directly, e.g. a rugged fitness landscape.

© The Author(s), under exclusive license to Springer Nature Switzerland AG 2023
L. Rutkowski et al. (Eds.): ICAISC 2022, LNAI 13589, pp. 206–217, 2023.
https://doi.org/10.1007/978-3-031-23480-4_17

In previous work the potential of switching between spaces in solving a problem rather than exploring the solution or heuristic space was illustrated [2]. The study compared three approaches, namely, the sequential approach which explored each space sequentially, the interleaving approach which switched between exploring the heuristic and solution space and the concurrent approach working in both the heuristic and solution space at the same time by exploring the heuristic space to identify construction heuristics to use to create a solution and a move operator to explore the solution space. A genetic algorithm was used to determine the construction heuristics and move operator to use. The research presented in this paper investigates the bi-space search for optimizing at which point to switch the search between the heuristic and solution space. The bi-space search employs an iterated local search to perform the optimization. The performance of the bi-space search (BSS) is compared to solution space search (SSS) and heuristic space search (HSS) which explore the solution and heuristic spaces separately.

Hence the main contribution of this research is optimizing when to switch between the heuristic and solution spaces, i.e. the hybridization of spaces, in solving the one dimensional bin packing problem.

All three searches have been applied to solving the one dimensional bin packing problem (1BPP), however the bi-space search proposed could be applied to any discrete optimization problem. The 1BPP has been selected to allow for a performance comparison with previous work [2]. The study has revealed that the BSS outperformed both SSS and HSS applied separately in solving the 1BPP.

The following section presents the 1BPP and previous approaches applied to solve the problem. Section 3 describes the bi-space search, including the solution space and heuristic space searches. The experimental setup used to evaluate the performance of the three searches is explained in Sect. 4. Section 5 compares the performance of the three searches in solving the 1BPP. Finally, Sect. 6 summarizes the findings of the study and describes future extensions of this work.

2   1BPP and Related Work

The one dimensional bin packing problem (1BPP) involves placing a set of n items, each with associated weights w_i, into a minimum number of bins of capacity C. All n items must be placed in a bin and no bin can exceed its capacity. It is a thoroughly researched domain, with numerous variants directly applicable to industry [2,3].

The majority of the literature for solving the 1BPP employs heuristic techniques [10]. In [3] a *consistent neighbourhood search* approach (CNS_BP), which tries to iteratively reduce the number of bins from an initial feasible solution, is proposed for solving the 1BPP. A variation of the grouping genetic algorithm, referred to as the *Grouping Genetic Algorithm with Controlled Gene Transmission* (GGA-CGT), is proposed by [9] in which new genetic operators are presented which allow for intelligent selection of the genes that are transmitted. Numerous approaches aimed specifically at solving the 1BPP exist in the literature and the reader is referred to [10] for a comprehensive overview.

[2] presents three different approaches for searching across spaces each of which are demonstrated using the heuristic and solution spaces to solve the 1BPP. As previously stated, the approaches are referred to as the sequential approach (SSA) which sequentially searches the heuristic space followed by the solution space, the interleaving approach (ISA) which consecutively alternates between searching the heuristic and solution spaces, and the concurrent approach (CSA) which attempts to search the two spaces simultaneously by using a genetic algorithm to optimize the sequence of construction heuristics and a move operator (for exploring the solution space).

The research presented in this paper differs from that of [2] by using a single point search as opposed to a multi-point search (genetic algorithm). Further, the research here explores the solution space by means of an iterated local search rather than a move operator. Additionally, the heuristic space contains both constructive and perturbative heuristics compared to the use of only constructive heuristics in previous work.

3   Bi-space Search Algorithms

Search techniques are traditionally restricted to a single search space (e.g. heuristic or solution space). Each search space has its own advantages and disadvantages. The aim behind bi-space (and more generally n-space) search is to try to maximize the advantages and minimize the disadvantages of each of the spaces by optimizing the points at which each space is searched. This section describes the bi-space search to optimize when to switch between the solution and heuristic spaces as well as the searches for exploring the solution and heuristic spaces separately.

In this research, a simple iterated local search (ILS) procedure is used for the bi-space search as well as to search the heuristic and solution spaces. The pseudocode for the iterated local search and local search procedures are provided in Algorithms 1 and 2 respectively. The manner in which the *perturb* operator (referred to in Line 2) is applied is dependent upon the representation being used, and therefore differs for each of the three searches. The subsections which follow describe, for each search space, the representation of a candidate solution in the ILS, and the implementation of the *perturb* operator.

Algorithm 1. Iterated Local Search (ILS) Pseudocode

1: Create initial candidate solution S
2: **for** $i \leftarrow 1, N$ **do**   ▷ N = number of iterations
3:      $S' \leftarrow LocalSearch(S)$ ▷ perform local search on S
4:      $S \leftarrow S'$ ▷ update S for the next iteration
5: **end for**
6: **return** S

Algorithm 2. Local Search Pseudocode

---

    input: candidate solution S ▷ e.g. heuristic combination

1: **for** $i \leftarrow 1, M$ **do**   ▷ M = number of perturbations
2:     $S' \leftarrow perturb(S)$ ▷ perform a perturbation on S
3:     **if** S' is fitter than S **then**
4:         $S \leftarrow S'$ ▷ update S if there is an improvement
5:     **end if**
6: **end for**
7: **return** S

3.1 Solution Space Search (SSS)

Individuals in the solution space search are candidate solutions for the problem at hand. In the case of the bin packing problem this would be the set of bins together with their respective packing configurations, as well as any items that still need to be packed. The *perturb* operator was implemented as follows. If the solution is a partial one (i.e. there are still items unpacked), it is randomly decided whether to build or perturb (rearrange the currently packed items in) the solution. If the decision is to build, then the best fit decreasing heuristic is used to place an unassigned item.

Otherwise perturbation occurs as follows. The least-filled bin is deleted and its items are added to the set labeled *free*. Iterating over the remaining bins in the (partial) solution, one then attempts to first swap any two items in *free* with any two items in the bin. After all bins have been attempted, the process is repeated attempting to swap two packed items with a single *free* item. A third and final pass is performed swapping one *free* item with one packed item. A swap is only accepted if it reduces the residual capacity of a bin. After all swaps have been attempted, the remaining items in the set *free* are repacked using the First Fit Descending construction heuristic. Pseudocode for the solution space perturbation is provided in Algorithm 3.

3.2 Heuristic Space Search (HSS)

Heuristic space candidate solutions consist of a string of (constructive or perturbative) low-level heuristics (e.g. *"FNHED"*). The perturbation operator replaces a random substring in the heuristic string with a new randomly generated sequence of low-level heuristics. The lengths of both the new substring and that being replaced are randomly determined, subject to respective maximum values specified as parameters. If the perturbed combination of heuristics exceeds a parameter-specified limit it is truncated to the appropriate length.

Evaluation of a given heuristic sequence is performed as follows. Starting with a specified solution space candidate solution (e.g. a packing configuration of items and bins in the case of 1BPP), or an empty candidate solution (all items

---

Algorithm 3. Solution Space *Perturb* Operator

1: *free* ← items from least filled bin
2: **for** $i \leftarrow 1, n$ **do**   ▷ $n =$ number of remaining bins
3:    Swap two items packed in the ith bin with two items in *free* if the residual capacity of the bin is reduced after the swap
4: **end for**
5: **for** $i \leftarrow 1, n$ **do**   ▷ $n =$ number of remaining bins
6:    Swap two items packed in the ith bin with one item in *free* if the residual capacity of the bin is reduced after the swap
7: **end for**
8: **for** $i \leftarrow 1, n$ **do**   ▷ $n =$ number of remaining bins
9:    Swap one item packed in the ith bin with one item in *free* if the residual capacity of the bin is reduced after the swap
10: **end for**
11: **while** *free* contains items **do**
12:    remove the first item from *free* and pack it using the First Fit Descending construction heuristic
13: **end while**

---

unpacked) if none is specified, each heuristic in the sequence is consecutively applied until the end of the sequence is reached. The entire sequence is applied in order to ensure that perturbative heuristics are still considered after obtaining a complete solution.

For example, given the heuristic sequence $C_1 C_2 P_3 P_4$ where C_i and P_i are arbitrary constructive and perturbative heuristics respectively. If during evaluation a complete solution is obtained after applying C_2 and evaluation of the sequence is then terminated, the perturbative heuristics P_3 and P_4 would be neglected.

3.3   Bi-space Search (BSS)

A candidate solution in the bi-space search consists of a sequence of components, where each component specifies a search space (i.e., solution or heuristic) as well as iteration and perturbation limits for ILS and local search, respectively, in said space (e.g., [2-H-2] corresponds to 2 iterations of ILS in the heuristic space).

The perturbation operator for bi-space search replaces a random subsequence of components with a new randomly generated sequence, with the lengths of both the replaced and replacement sub-sequences being randomly determined, subject to respective maximum values taken as parameters. The search in the heuristic and solution spaces, as specified by the different components, is as described previously in the relevant subsections (Sects. 3.1 and 3.2, respectively).

When evaluating a given component sequence, the empty solution (i.e., no items packed) is used as the starting point for searching the space specified by the first component (e.g. heuristic space). The solution found at the end of the specified search (from the first component) is used as the starting point for the search specified by the next component in the sequence. This process is repeated until all components in the sequence have been evaluated.

For example, evaluation of the component sequence *[2-H-2]; [3-S-1]; [2-S-5]* occurs as follows. Two iterations of ILS, with 2 perturbations per iteration, are performed in the heuristic space. During the search process each heuristic combination is applied to the empty candidate solution for 1BPP (no packed items). The best heuristic combination found at the end of the search is then used to construct a 1BPP candidate solution. This solution is then used as the starting point for three iterations of ILS, with one perturbation per iteration, in the solution space. The result of this solution space search is subsequently passed as the starting point for evaluating the third component, namely a further search in the solution space consisting of two ILS iterations with five perturbations each.

4   Experimental Setup

A total of five different benchmark datasets for the one-dimensional bin packing problem (1BPP) were used. The breakdown of the datasets are as follows: the Falkenauer [5] set, comprised of the so called uniform and triplet classes, with 80 instances per class; the Scholl et al. [12] instances, separated into easy (720 instances), medium (480 instances) and hard (10 instances) categories; the two classes of 100 instances each proposed by Schwerin and Wäscher [13]; the 17 instances of Wäscher and Gau [14]; and the set of 28 difficult instances (referred to in the literature as *hard28*) used in [11]. All instances were downloaded from *BPPLIB*[1].

The fitness of candidate solutions was determined using the function proposed by Falkenauer [5], which is to minimize:

$$f_{BPP} = 1 - \frac{\sum_{i=1}^{N}(F_i/C)^2}{N} \tag{1}$$

where N is the number of bins, F_i the sum of the item sizes in the ith bin and C the capacity of each bin. In order to distinguish between partial and complete solutions, a value of one was added to the fitness value for each item still to be packed.

4.1   Heuristic Space

The set of low-level heuristics used in the heuristic space was as follows [1,6,8]:

- **First-Fit Decreasing**: The remaining items are sorted in descending order and the largest item is placed into the first feasible bin. If no such bin exists the item is placed into a new bin.
- **Best-Fit Decreasing**: The remaining items are sorted in descending order and the largest item is placed into the bin with the minimum free space after the item is packed. If no feasible bin exists the item is placed in a new bin.

---

[1] http://or.dei.unibo.it/library/bpplib.

- **Next-Fit Decreasing**: The remaining items are sorted in descending order and the largest item is placed in the current bin if feasible, else in a new bin.
- **Worst-Fit Decreasing**: Similar to the *Best-Fit* heuristic, with the condition being to maximize the free space (rather than minimize).
- **Minimum Bin Slack (MBS)**: The set of remaining items is searched for a subset of items which completely fills a single bin. If no such subset exists, then the subset with the minimum free space (slack) is used.
- **Relaxed MBS (R-MBS)**: The *MBS* heuristic with an allowed slack (as opposed to a full bin).
- **Time-Bounded R-MBS (TBR-MBS)**: A time-limited variant of the *R-MBS* heuristic.
- **Reallocate Fullest**: The fullest bin is deleted and its items repacked using the *Best-Fit* heuristic.
- **Reallocate Emptiest**: The emptiest bin is deleted and its items repacked using the *Best-Fit* heuristic.
- **Split Fullest**: Place each item from the fullest bin into its own new bin.
- **Split Emptiest**: Place each item from the emptiest bin into its own new bin.
- **Move Largest Fullest**: Use *Best-Fit* to repack the largest item from the fullest bin (the item cannot be placed in its original bin).
- **Move Smallest Fullest**: Use *Best-Fit* to repack the smallest item from the fullest bin (the item cannot be placed in its original bin).
- **Move Largest Emptiest**: Use *Best-Fit* to repack the largest item from the emptiest bin (the item cannot be placed in its original bin).
- **Move Smallest Emptiest**: Use *Best-Fit* to repack the smallest item from the emptiest bin (the item cannot be placed in its original bin).

In order to reduce runtimes, a pre-processing simple reduction procedure [3] was applied before each of the approaches whereby pairs of items, whose weights sum to the capacity of a bin, were packed. Additionally, the search process was stopped if a solution reached a simple lower bound value for the number of bins, calculated as $LB = \left\lceil \frac{\sum_{i=1}^{n} w_i}{C} \right\rceil$, where w_i is the weight of the ith item and C is the bin capacity.

5   Results and Analysis

This section presents and compares the results obtained for each of the three approaches. Table 1 shows, for each approach, the number of instances where the optimum solution was, the number of instances with a near-optimum (one bin more) solution and the number of instances that were more than one bin from the optimum. The results are grouped per dataset, with the best performing approach in bold. The average runtimes per problem instance are presented in Table 2.

From Tables 1 and 2 it can be seen that searching across the heuristic and solutions spaces (i.e., bi-space search) produces better quality solutions when compared to searching a single space. With the exception of two instances, the bi-space search consistently produced the best, or tied for best, solutions. However,

Table 1. The number of problem instances, for each of the data set categories, which were solved to optimality or near-optimality (one bin from the optimum), for bi-space (BSS), heuristic space (HSS) and solution space (SSS) searches. The best performing algorithm is shown in bold.

Prob. Set	BSS				HSS				SSS			
	Opt	Opt.-1	Sum	Rem	Opt	Opt.-1	Sum	Rem	Opt	Opt.-1	Sum	Rem
Falkenauer_T (80)	0	9	9	71	0	0	0	80	0	1	1	79
Falkenauer_U (80)	35	37	72	8	10	23	33	47	20	33	53	27
Scholl_1 (720)	671	44	715	5	589	80	669	51	618	80	698	22
Scholl_2 (480)	408	41	449	31	252	117	369	111	367	61	428	52
Scholl_3 (10)	5	5	10	0	0	0	0	10	0	0	0	10
Schwerin_1 (100)	100	0	100	0	1	99	100	0	100	0	100	0
Schwerin_2 (100)	100	0	100	0	9	89	98	2	97	3	100	0
Hard28 (28)	5	23	28	0	5	23	28	0	5	23	28	0
Wäscher (17)	12	5	17	0	2	15	17	0	9	8	17	0
Total (1615)	1336	164	1500	115	868	446	1314	301	1246	215	1431	184

Table 2. The average runtime (in seconds) per problem instance, for each of the data set categories, for the bi-space (BSS), heuristic space (HSS) and solution space (SSS) searches.

Prob. Set	BSS	HSS	SSS
Falkenauer_T (80)	22.28	0.40	0.10
Falkenauer_U (80)	81.87	0.46	0.24
Scholl_1 (720)	26.58	0.18	0.07
Scholl_2 (480)	5.44	0.16	0.03
Scholl_3 (10)	14.34	0.68	0.15
Schwerin_1 (100)	0.22	0.10	0.01
Schwerin_2 (100)	1.86	0.11	0.02
Hard28 (28)	16.34	0.26	0.16
Wäscher (17)	5.43	0.34	0.12
Total (1615)	19.18	0.20	0.06

as is shown by Table 2, the increased solution quality is obtained at the expense of considerably longer runtimes. This is evident from Table 3, shows that although the bi-space is generally the best performing of the three approaches, it also averages the longest runtimes.

The non-parametric Friedman test was used (as proposed by [4]) to determine the statistical significance of the differences in both the solution quality (distance to optimum) and runtime. Using the average rank values presented in Table 3 the F-statistics evaluate to $F_{soln} = 242.93$ and $F_{time} = 2727.96$ for the solution quality and runtime performance metrics respectively. Under the F-distribution,

Table 3. The average rank, using both solution quality and runtime as a performance metric, for the bi-space (BSS), heuristic space (HSS) and solution space (SSS) searches. Averages were calculated across all problem instances.

Rank Aspect	BSS	HSS	SSS
Solution	1.702	2.402	1.895
Runtime	2.718	2.133	1.149

with an α-level of 5%, the critical value is $F_{0.05} = 3.00$, hence the null hypothesis of algorithm equivalence can be rejected in both cases.

For the Nemenyi post-hoc test the critical difference evaluates to $CD = 0.082$ and therefore all three approaches are significantly different from one another when using either the runtime or the solution quality as a performance metric. Therefore the significant improvement in performance of the BSS is contrasted with a significant increase in runtime. With this in mind one needs to consider the need for quality in relation to any time constraints.

5.1   Comparison with Previous Bi-Space Search Approaches

Table 4 compares the performance of the three approaches with those taken from the literature [2], namely the *sequential, interleaving* and *concurrent* search approaches as well as a genetic algorithm based selection constructive hyper-heuristic (SSA, ISA, CSA and GAHH respectively). The average runtimes per problem instance are reported in Table 5. From Tables 4 and 5 it can be seen that the performance of bi-space search is comparable to that of the concurrent search approach, with bi-space search having considerably shorter runtimes. Applying the non-parametric Friedman test, using the ranking averages in Table 6, shows there is a significant difference in the algorithms. Using bi-space search as the control in the Holm's post-hoc test (with an α-level of 5%) shows that, for the case of solution quality as a performance metric, there is statistically no significant difference between the bi-space search and the CSA approach.

Table 4. The number of problem instances, for each of the data set categories, which were solved to optimality or near-optimality (one bin from the optimum), for each of the methods tested. The results in bold indicate the best performing algorithm.

Approach	Scholl_1 (720)			Scholl_2 (480)			Scholl_3 (10)			Total (1210)		
	Opt	Opt. + N. Opt.	Rem	Opt	Opt. + N. Opt.	Rem	Opt	Opt. + N. Opt.	Rem	Opt	Opt. + N. Opt.	Rem
GAHH	590	672	48	242	363	117	0	0	10	832	1035	175
SSA	626	689	31	281	375	105	0	0	10	907	1064	146
ISA	632	711	9	371	438	42	2	8	2	1005	1157	53
CSA	673	**717**	3	**428**	**468**	12	**6**	**10**	0	**1107**	**1195**	15
BSS	671	715	5	408	449	31	5	10	0	1084	1174	36
HSS	589	669	51	252	369	111	0	0	10	841	1038	172
SSS	618	698	22	367	428	52	0	6	4	985	1132	78

However, when using runtime as the performance metric for ranking, the difference between the two approaches becomes statistically significant.

Table 5. The average runtime (in seconds) per problem instance, for each of the methods tested.

Approach	Scholl_1	Scholl_2	Scholl_3	Total
GAHH	13.5	6.0	12.7	10.5
SSA	21.6	30.6	34.4	25.3
ISA	3185.5	2507.3	997.4	2898.4
CSA	817.5	1589.8	1761.1	1131.6
BSS	26.6	5.4	14.3	18.1
HSS	0.18	0.16	0.68	0.18
SSS	0.07	0.03	0.15	0.05

Table 6. The average rank, using both solution quality and runtime as a performance metric, for the sequential, interleaving and concurrent search approaches, a genetic algorithm based selection constructive hyper-heuristic (SSA, ISA, CSA and GAHH respectively) and bi-space (BSS), heuristic space (HSS) and solution space (SSS) searches.

Rank Aspect	GAHH	SSA	ISA	CSA	BSS	HSS	SSS
Solution	4.64	4.35	3.71	3.47	3.49	4.53	3.82
Runtime	3.50	4.61	6.32	6.57	3.73	2.10	1.18

Given that the CSA approach is not competitive with the state of the art [2] and that there is no significant difference in performance between bi-space search and the CSA, it can be concluded that the bi-space search is not competitive with the state of the art. This is to be expected as the aim of this research was not to solve the 1BPP, but rather to demonstrate the effectiveness in searching across more than one space as opposed to only searching a single space. For this reason only simple, general search techniques were employed, in contrast to the more sophisticated techniques tailored specifically toward solving the 1BPP that are used in the state of the art. In spite of using only a simple technique and less optimization, the bi-space search was still able to perform equivalently and with significantly reduced runtimes when compared to the CSA approach which employed a comparatively more sophisticated genetic algorithm.

6 Conclusions and Future Work

A bi-space search (BSS) using an iterated local search to optimise the switching between the heuristic and solution spaces was proposed. This was compared with heuristic space search (HSS) and solution space search (SSS), as well as with previous techniques proposed for search across spaces [2]. The one dimensional bin packing problem (1BPP) was used to compare the performance of the different approaches.

Experimental results showed that BSS performed significantly (with a 95% confidence) better than either the HSS or SSS. However this was contrasted with equally significant longer runtimes. While the runtimes for the BSS were not excessively large in and of themselves, further investigation needs to be conducted into how the runtimes scale both with problem complexity as well as with the search technique employed in the respective search spaces.

When comparing the BSS with previous work [2], experimental results showed no significant difference in performance between the BSS and the best-performing concurrent search approach (CSA), however a significant decrease in runtimes was evident in the BSS when compared with each of the sequential, interleaving and concurrent approaches (SSA, ISA and CSA respectively).

The fact that BSS, which uses a simple iterated local searches, was able to compete with CSA, which uses a comparatively sophisticated genetic algorithm, indicates that better optimization for deciding on switching between search spaces could allow for lower optimization (and therefore computational overhead) in the individual spaces. Future work will further investigate this hypothesis, as well as additional techniques for switching between spaces.

Acknowledgements. This work was funded as part of the Multichoice Research Chair in Machine Learning at the University of Pretoria, South Africa. The authors acknowledge the Centre for High Performance Computing (CHPC), South Africa, for providing computational resources toward this research.

References

1. Bai, R., Blazewicz, J., Burke, E.K., Kendall, G., McCollum, B.: A simulated annealing hyper-heuristic methodology for flexible decision support. 4OR **10**(1), 43–66 (2012)
2. Beckedahl, D., Pillay, N.: A study of bi-space search for solving the one-dimensional bin packing problem. In: Rutkowski, L., Scherer, R., Korytkowski, M., Pedrycz, W., Tadeusiewicz, R., Zurada, J.M. (eds.) ICAISC 2020. LNCS (LNAI), vol. 12416, pp. 277–289. Springer, Cham (2020). https://doi.org/10.1007/978-3-030-61534-5_25
3. Buljubašić, M., Vasquez, M.: Consistent neighborhood search for one-dimensional bin packing and two-dimensional vector packing. Comput. Oper. Res. **76**, 12–21 (2016)
4. Demšar, J.: Statistical comparisons of classifiers over multiple data sets. J. Mach. Learn. Res. **7**, 1–30 (2006)
5. Falkenauer, E.: A hybrid grouping genetic algorithm for bin packing. J. Heurist. **2**(1), 5–30 (1996)
6. Fleszar, K., Hindi, K.S.: New heuristics for one-dimensional bin-packing. Comput. Operat. Res. **29**(7), 821–839 (2002)
7. Koza, J.R.: Genetic Programming On the Programming of Computers by Means of Natural Selection. MIT (1992)

8. Pillay, N., Qu, R.: Hyper-Heuristics: Theory and Applications. Natural Computing Series, Springer International Publishing (2018). https://doi.org/10.1007/978-3-319-96514-7

9. Quiroz-Castellanos, M., Cruz-Reyes, L., Torres-Jimenez, J., Gómez S., C., Huacuja, H.J.F., Alvim, A.C.: A grouping genetic algorithm with controlled gene transmission for the bin packing problem. Comput. Operat. Res. **55**, 52–64 (2015)

10. Scheithauer, G.: one-dimensional bin packing. In: Introduction to Cutting and Packing Optimization. ISORMS, vol. 263, pp. 47–72. Springer, Cham (2018). https://doi.org/10.1007/978-3-319-64403-5_3

11. Schoenfield, J.: Fast, exact solution of open bin packing problems without linear programming. Tech. rep, US Army Space and Missile Defense Command, Huntsville, Alabama, USA (2002)

12. Scholl, A., Klein, R., Jürgens, C.: Bison: A fast hybrid procedure for exactly solving the one-dimensional bin packing problem. Comput. Operat. Res. **24**(7), 627–645 (1997)

13. Schwerin, P., Wäscher, G.: The bin-packing problem: A problem generator and some numerical experiments with ffd packing and mtp. Int. Trans. Oper. Res. **4**(5–6), 377–389 (1997)

14. Wäscher, G., Gau, T.: Heuristics for the integer one-dimensional cutting stock problem: A computational study. Operat. Res. Spektrum **18**(3), 131–144 (1996)

Assessing the Sentiment of Book Characteristics Using Machine Learning NLP Models

Paweł Drozda[1]([⊠])[iD] and Krzysztof Sopyła[2]

[1] Faculty of Mathematics and Computer Science, University of Warmia and Mazury in Olsztyn, Olsztyn, Poland
pdrozda@matman.uwm.edu.pl
[2] Literacka Ltd, Olsztyn, Poland

Abstract. This paper presents a new approach for processing the entire books with Natural Language Processing algorithms. In particular, we proposed methods to evaluate books in terms of assessing the intensity of the book's soft features, such as fantastic, touching, suspenseful, etc. Using Bag of Words and TF/IDF, we embedded books and conducted classification experiments to determine the most appropriate parameters for classifying the intensity of features. The obtained results showed, that in the considered problem the Random Forests algorithm fitted the best, achieving accuracy of 95% and F1 measure of 89%. The evaluation also included the selection of the best converter and data aggregation method.

Keywords: Text representation · Word embeddings · Long text classification · Bag of words

1 Introduction

One of the most famous leisure activities in the world is reading books. Due to the unflagging popularity of various types of books, there are a huge number of new titles every month, which makes it much more difficult for readers to choose the right books. This results in the need to create recommendation and book exploration systems that allow you to select from the maze of books, those appropriate for a given reader at a given moment, especially for entities renting and selling books.

The idea of recommendation systems is well known for many years and is applied in various domains such as: movie recommendations [24,29], e-commerce [12,15,16], news recommendations [33,34], career path recommendations [8,11,21], tourism [1] and many others. In addition, many methods have been developed, used in various situations, allowing for recommendations, among which the following should be mentioned: collaborative filtering [10,16], reinforcement learning [6,7,28], sentiment analysis [1,26], deep learning [9,10,26] or clustering [1].

The most recommendation systems for books, which are based on reviews or a shopping cart, meet the main problem of all recommendation systems,

L. Rutkowski et al. (Eds.): ICAISC 2022, LNAI 13589, pp. 218–231, 2023.
https://doi.org/10.1007/978-3-031-23480-4_18

which is a cold start. This means that if there appears a new item, there is no information about it until no one buys it or gives any reviews, there is no way to rate it in the recommendation system. Moreover, solutions which are based on collaborative filtering methods, taking into account the purchase history, prepare incorrect recommendations very often. This is mainly related to the contents of the basket of previous activities, where book purchases can be made for many family members, the items refer to various purposes, for example, a travel guide and a crime story, or additional accessories appear in the cart in addition to book items. Finally, in the case of books, the additional difficulty is the length of the text, which causes problems with the use of many machine processing algorithms that require a lot of time and hardware resources.

To overcome the aforementioned issues in the field of book recommendation, this paper proposes a new approach to the problem under consideration.

There were implemented methods which check the intensity of different characteristics of books, based on their content. Having defined a different level of selected features (such as realistic, fantastic, terrifying, touching), the overall description of a particular book is derived, which can be an input for the content-based recommendation system. Even for new books, the user can find an appropriate one indicating the features of books that interest him at a given time (for example, if he is sad, he would like to read something that will be optimistic and funny). Moreover, the methods implemented in this work allow identifying books with a similar intensity of features to those previously read by the user. This allows readers to easily find a book with a similar moodiness.

In addition, it is a very big challenge to analyze the entire content of the book and to determine the severity of individual features, due to the fact that books are relatively long to label and to be analyzed using machine learning algorithms. Therefore, the well known fast word embeddings (such as fasttext, bag of words, or word2vec) should be considered for training of the state of the art algorithms. In the case of more resource-consuming methods, such as deep neural networks, appropriate adaptation should be applied in order to make possible the training of the model based on long content of books.

In this paper, we first proposed Bag of Words with TF/IDF embeddings and the use of leading classification algorithms: SVM, Random Forests and Ridge Classifier. With these solutions, the models were trained and tuned, allowing us to determine the level of intensity of particular features of the book. The preparation of such models made it possible to overcome the problem of a cool start, which in our case means that when new books appear in the available collection, there is no need for a preliminary, manual evaluation of the book, since the prepared models will allow determining the marking of individual parameters. To the best of the author's knowledge, this is the only attempt to define the characterization of books in NLP research. The conducted experiments allowed to achieve an accuracy of 95%, while the F1 measure, which was much more important in the research, reached the level of 89%. The achieved results prove an appropriate selection of models and machine learning algorithms for extracting the intensity of the soft features of books and indicate the correct direction of overcoming the problem of cold start in recommendation systems.

The rest of the paper is organized as follows: Sect. 2 reviews current techniques for sentiment analysis and natural language processing. In Sect. 3, input datasets and proposed methods are introduced. The details of the experimental sessions validating introduced solutions are presented in Sect. 4. Section 5 concludes the paper.

2   Related Work

The goal of natural language processing is to try to understand human expression by machines and to create the possibility of preparing conclusions from the analyzed text. There is a huge number of possible applications of natural language processing in various fields. The vast majority of solutions where natural language processing is necessary require word or whole document embeddings in the first phase. This is due to the fact that machine learning algorithms require vectors composed of a sequence of numbers as an input set, which is ensured by the embedding process. The main and most frequently used methods of text encoding include: Bag of Words, Word2Vec, GloVe, FastText and Transformers Neural Networks.

The first application that has developed a lot and gained importance recently is the implementation of NLP solutions in chatbots. The main purpose of a such implementation is to replace the interlocutor on the company side with a bot that should imitate the natural conversation. The papers [18,31] describe the chatbot solution in medical domain. Authors of paper [31] introduced the study of possibility to introduce the NLP tools in categorization of dialog data of patients with inflammatory bowel diseases for the chatbot development. As the main method of text annotation the standard bag of words was used. Different approach was presented in paper [18] where the main effort was directed to prepare a solution for appropriate medical specialty recommendation for patients. Authors proposed pipeline consisting in deep learning long short-term memory (LSTM) models for natural language processing. As a final result the smartphone AI chatbot was produced. Another noteworthy example of research in the field of chatbots is paper [17], where the NLP was implemented for collage chatbot software, which main goal was to answer the typical questions to students.

Another very important direction in the development of natural language processing methods is Question Answering domain, which main goal is similar to chatbot systems, where both are focused on preparing reliable answer for user questions. Authors of [27] introduce system, which prepare the abstractive and extractive summarizations of scientific publications for COVID-19 questions. As main methods for summarizations, the BART and ALBERT solutions were proposed. Another use of deep learning models can be found in [32], where LSTM Recurrent Neural Networks for seq2seq model were introduced.

The next large area where NLP is implemented is text classification. J. Howard and S. Ruder in [14] implemented an effective transfer learning method derived from computer vision, which can be applied to different NLP tasks. The proposed method bases on deep learning techniques and consists of pretraining the Language Model on Wikitext-103 and Language Model fine-tuning for specific task. On the other hand, the introduction of bidirectional long-term memory and convolutional neural networks was made in [19] for less general use, which is news text classification. A similar direction of research was proposed by the authors of paper [20], where also convolutional neural networks in the classification of the text were introduced, but this time it concerned law texts.

The ubiquitous social media are also an important direction in the development and application of methods based on natural language processing. Implementation of real-time traffic reporting systems [30] can be an example of NLP implementations in social media. In addition, the authors of the field also directed research to consider security issues. The paper [22] proposes a model that allows for the identification of environmental crimes, while the research in [5] studies were focused on detecting denial-of-service attacks.

In addition to the succinctly described fields of application, NLP is also used in sentiment analysis for texts, which should be considered the closest research directions to the one presented in this study. The authors of [4] conducted research which main aim was to determine whether it is possible to transfer the conclusions drawn when assessing the sentiment for short texts to long texts as well. For example, is it possible to predict sentiment for long reviews on the basis of short labeled with sentiment. A different approach was taken in [3], where the dataset of tweets was analyzed in context of sentiment defining. The authors used basic classification algorithms such as Naive Bayes, K-NN, Decision Tree, Random Forest and Random Tree to train the model on both unbalanced and balanced datasets. It should be noted, however, that the research was conducted on the basis of very short texts, which made experiments much easier compared to the processing of long texts. An interesting approach to sentiment analysis was proposed in [25]. The paper introduces a combination of convolutional neural networks (CNN) and bi-directional long short-term memory (BiLSTM) models with Doc2vec embeddings, which are suitable for long text opinion analysis. The experimental session contained a comparison of the proposed solution to the convolutional neural networks, long short-term memory and bi-directional long short-term memory models with Word2vec / Doc2vec embeddings. The comparison was based on medium-length article dataset from French newspapers. Other examples of the use of natural language processing algorithms in the analysis of the sentiment of short texts can be found in [2,13].

The last direction of research development, similar to the main objectives of this study, is multi-labeling in the context of books [23]. The authors proposed the usage of the BERT architecture to classify books in German based on metadata such as the authors, book title and blurb content. The purpose of the classification was to assign appropriate genre labels to individual books.

It should be noted that all above-mentioned solutions refer directly to the short texts, where the processing time of individual elements of the input dataset during embedding process is very short. In contrast, this study takes into account during the experiments whole books in predicting intensity of individual features, where the transformation of the input text into a numerical vector was a very difficult and time-consuming task and to the best of the authors' knowledge it is a first attempt to process such long texts.

3   Datasets

This chapter describes the methods of collecting and preparing the input dataset. To create a ground truth dataset of input data, over 30 000 books in Polish were collected from the repositories of the project partners. Then, using the crowd-sourcing system, approximately 73 000 assessments of individual features were obtained. Among the features that have been assessed are, for instance: how strong a book is amusing, historical, informative, fantastic, terrifying, erotic, soulful, etc. A form for the evaluation of Harry Potter books in the web crowd-sourcing system is presented in Fig. 1.

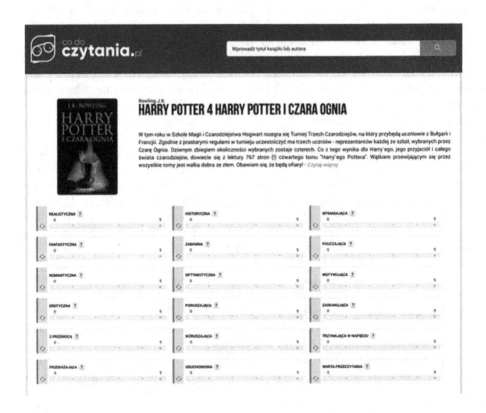

Fig. 1. Evaluation of Harry Potter book

The evaluators included both ordinary users reading books and a group of experts consisting of librarians, experts from the Literacka company and an expert from the publishing houses. To obtain the best possible training dataset, the values assigned by experts in the input set were weighted twice. In addition, an expert in the field of literary studies was also hired to annotate the test dataset.

$$y(x) = \begin{cases} 0 & \text{if } x \text{ is lower than } 2.00, \\ 1 & \text{if } x \text{ is between } 2.00 \text{ and } 4.00, \\ 2 & \text{otherwise} \end{cases} \tag{1}$$

Moreover, due to the fact that the number of ratings of individual features for individual books differed to a large extent, as shown in Table 1, a converter was used to compensate for the differences in the number of ratings.

Table 1. Number of ratings for features

Title	Amusing	Informative	Fantastic	...	Terrifying	Soulful	Touching
book1	132	134	133	...	130	129	123
book2	22	19	13	...	25	20	23
book3	154	160	143	...	150	160	154
book4	211	214	209	...	200	219	223
book5	13	14	10	...	13	9	12
book6	452	434	455	...	440	409	423
book7	7	4	3	...	9	14	9
book8	142	138	123	...	136	149	133
book9	132	134	135	...	130	129	134

From among the prepared books with scored features, depending on the experiments and the selection of the characteristics to be tested, between 4,000 and 8,000 books were taken into account for input dataset creation. Then for each book, a grading vector consisting of score averages or median grades was derived.

The experiments presented in the next section used classification algorithms in the field of machine learning. For this reason, for classification purposes, it was necessary to properly transform the dataset. The preprocessing of data consisted in assigning one of three labels to each rating which allows for three-class classification. In each case, the inputs are mapped to the following classes: no feature, medium intensity, high intensity. During the experiments, three different mappings were used, determining different class boundaries. They were presented by means of Formulas (1), (2) and (3) and called Regular Converter, Symmetric Converter and Data Adjusted Converter, respectively.

Finally, after the described transformations, an initial dataset was prepared, which was used in the experimental session. The example of input dataset preparing process is shown in Fig. 2.

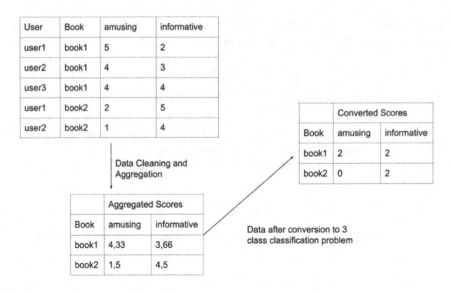

Fig. 2. Process of input dataset preparation

$$y(x) = \begin{cases} 0 & \text{if } x \text{ is lower than } 1.50, \\ 1 & \text{if } x \text{ is between } 1.50 \text{ and } 3.50, \\ 2 & \text{otherwise} \end{cases} \tag{2}$$

$$y(x) = \begin{cases} 0 & \text{if } x \text{ is lower than } 1.66, \\ 1 & \text{if } x \text{ is between } 1.66 \text{ and } 3.66, \\ 2 & \text{otherwise} \end{cases} \tag{3}$$

It should be noted that for each experiment, the test dataset was the same and consisted of 105 labelled books by an expert in literary studies. Such preparation and determination of the test dataset was mainly dictated by the fact that, the partial purpose of the experiments was to include books in the training dataset that had real scores for all the features considered, which could only be ensured by evaluating them by experts. It allowed for the preparation of a comprehensive picture of the assessment of the sentiment of soft features.

4   Experiment Results

The main goal of the experimental session was to evaluate the previously described algorithms for assessing the sentiment of soft books' features using a carefully prepared dataset composed of the contents of the entire book.

The experiments were divided into four separate tasks, where the impact of different elements for the classification effectiveness were taken into account.

Simple and fast Bag of Words models were chosen as embedding method to perform classification of intensity of individual soft books' features into three classes (lack of feature, medium intensity, high intensity).

To obtain a book feature value, we have tested two data aggregation approaches based on user assessments (mean and median). Moreover, three different mappings were taken into account, which were used to determine the boundaries of individual intervals for classes (converters described in the previous section). Finally, after the initial analysis of the state of the art classification algorithms, where the effectiveness of the models was tested with the use of F1 and accuracy measures, it turned out that RandomForestClassifier, LinearSVC, RidgeClassifier were the best for the problem under consideration.

As part of the experimental session, we examined the impact of the above-mentioned elements on the quality of classification of individual soft characteristics. We conducted 324 experiments with different configurations of the converter, aggregation measure and classifier.

4.1   Impact of Converter

First, we examined the influence of the converter used on the obtained results. For each of the available features, the maximum values of the F1 measure, achieved during the classification, were defined. The results for selected features are presented in Table 2.

Table 2. F1 classification measure for different converters

Feature	Data Adjusted	Regular	Symmetric
Erotic	0.777	0.780	0.756
Fantastic	0.876	0.844	0.822
Historical	0.676	0.676	0.706
Motivating	0.693	0.647	0.672
Optimistic	0.734	0.734	0.738
Stirring	0.503	0.503	0.490
Informative	0.604	0.604	0.600
Terrifying	0.594	0.576	0.643
Realistic	0.619	0.591	0.619
Romantic	0.871	0.871	0.849
Suspenseful	0.623	0.598	0.607
Spiritual	0.888	0.888	0.843
Worth reading	0.673	0.678	0.615
Demanding	0.773	0.707	0.746
Touching	0.717	0.717	0.714
With violence	0.645	0.629	0.671
Funny	0.797	0.797	0.775
Surprising	0.539	0.539	0.580

Despite the fact that none of the converters achieved the best results in all the experiments, it should be stated that among the used functions for determining the limits of output classes, Converter Data Adjusted turned out to be the best globally. Among the examined features, it achieved the highest values for 12 out of 18.

4.2   Impact of Classification Algorithm

In the next step, the focus was directed on the global determination of which in the group of classification algorithms is the most effective in determining the intensity of soft features. As in the case of the converter analysis, also for the classifiers, the maximum values of the F1 measure were determined for all the considered features, obtained as a result of the experiments. The most significant results are presented in Table 3.

Table 3. F1 classification measure for different algorithms

Feature	Linear SVC	Random forest	Ridge
Erotic	0.777	0.780	0.756
Fantastic	0.876	0.876	0.826
Historical	0.706	0.706	0.671
Motivating	0.651	0.693	0.672
Optimistic	0.734	0.738	0.697
Stirring	0.470	0.503	0.471
Informative	0.604	0.604	0.594
Terrifying	0.643	0.643	0.624
Realistic	0.606	0.619	0.606
Romantic	0.871	0.871	0.864
Suspenseful	0.623	0.623	0.619
Spiritual	0.864	0.888	0.825
Worth reading	0.615	0.678	0.615
Demanding	0.773	0.763	0.756
Touching	0.714	0.717	0.702
With violence	0.671	0.671	0.665
Funny	0.724	0.797	0.724
Surprising	0.580	0.565	0.554

As can be seen, the Random Forest classifier should be considered the best among the considered solutions. It achieved the highest values of the F1 measure for 16 out of 18 features. Moreover, it should be noted that in many cases the linear SVM classifier achieves exactly the same results as the Random Forest, and even in two cases F1 is higher. Among the considered solutions, the Rigde Classifier stands out significantly, as it did not reach the highest value in any of the cases, reporting results 1 to 6% points lower.

4.3   Analysis of Soft Features

Interesting regularities can be observed in the analysis of the learning of individual soft traits. 18 features were taken into account in the study, where the results illustrated by F1 and accuracy are presented in Table 4.

Table 4. Classification F1 and accuracy measures for different features

Feature	Max F1	Max accuracy
Erotic	0.780	0.830
Fantastic	0.876	0.943
Historical	0.706	0.719
Motivating	0.693	0.733
Optimistic	0.738	0.830
Stirring	0.503	0.734
Informative	0.604	0.713
Terrifying	0.643	0.711
Realistic	0.619	0.963
Romantic	0.871	0.856
Suspenseful	0.623	0.676
Spiritual	0.888	0.837
Worth reading	0.678	0.874
Demanding	0.773	0.844
Touching	0.717	0.745
With violence	0.671	0.790
Funny	0.797	0.768
Surprising	0.580	0.733

It can be seen that features such as fantastic, erotic, romantic, spiritual, demanding, funny achieve the best quality of the classification model. On the other extreme, there are such features as: surprising, informative, realistic or stirring, where the achieved results oscillate around 50–60%, which is a low threshold in the classification. This may be due to the fact that better classified descriptors seem to be much easier to define and not very subjective. On the other hand, those with low performance are very subjective, so it is often difficult to determine the intensity of such a feature, which may result in low learnability.

4.4   Impact of Aggregation Function

The last element considered in this phase of the experiments was the influence of the use of the mean and the median as a function aggregating the scores of individual books on the quality of the classification. The obtained results are summarized in Table 5. For individual soft features, the F1 measure was indicated using the median and the mean, with the distinction of individual converters.

Table 5. F1 measure classification for different aggregations

Feature	Mean Agg	Median Agg
Erotic	0.780	0.759
Fantastic	0.876	0.814
Historical	0.618	0.706
Motivating	0.693	0.638
Optimistic	0.688	0.738
Stirring	0.472	0.503
informative	0.595	0.604
terrifying	0.643	0.624
realistic	0.619	0.591
romantic	0.863	0.871
suspenseful	0.623	0.607
spiritual	0.821	0.888
worth	0.678	0.615
demanding	0.773	0.723
touching	0.688	0.717
with	0.645	0.671
funny	0.775	0.797
surprising	0.458	0.580

After analyzing the results, it should be stated that the use of the median does not significantly affect the choice of converter, due to the fact that the values of the F1 measure for individual features and various converters are mostly very similar. This is not the case when the mean is used as a measure of data aggregation. Then, the results achieved for individual converters differ significantly. Additionally, none of the measures gained global advantage because for many features it turned out that it was better to use the median for a similar number of features, the mean performs better.

5   Conclusions and Future Work

In this study, we proposed a new approach to processing and analyzing entire books. We evaluated different classification algorithms, different data converters and the methods of data aggregation for determining the intensity of soft features for each book based on its content. The conducted experiments showed promising results, allowing the classification of book features at the level of accuracy over 90%.

In further studies, we intend to check the possibility of using deep learning algorithms in book analysis as well as the possibility of using the developed solutions in real world recommendation systems.

Acknowledgments. This work is part of the project No POIR.01.01.01-00-1118/17 "Automatic Reviewer - Advanced Book Recommendation System" funded by the National Centre for Research and Development.

References

1. Abbasi-Moud, Z., Vahdat-Nejad, H., Sadri, J.: Tourism recommendation system based on semantic clustering and sentiment analysis. Expert Syst. Appl. **167**, 114324 (2021)

2. Almjawel, A., Bayoumi, S., Alshehri, D., Alzahrani, S., Alotaibi, M.: Sentiment analysis and visualization of amazon books' reviews. In: 2019 2nd International Conference on Computer Applications & Information Security (ICCAIS), pp. 1–6. IEEE (2019)

3. Alshamsi, A., Bayari, R., Salloum, S., et al.: Sentiment analysis in english texts. Adv. Sci. Technol. Eng. Syst. J. **5**(6), 1683–1689 (2020)

4. Amplayo, R.K., Lim, S., Hwang, S.w.: Text length adaptation in sentiment classification. In: Asian Conference on Machine Learning, pp. 646–661. PMLR (2019)

5. Chambers, N., Fry, B., McMasters, J.: Detecting denial-of-service attacks from social media text: Applying nlp to computer security. In: Proceedings of the 2018 Conference of the North American Chapter of the Association for Computational Linguistics: Human Language Technologies, Volume 1 (Long Papers), pp. 1626–1635 (2018)

6. Chen, S.Y., Yu, Y., Da, Q., Tan, J., Huang, H.K., Tang, H.H.: Stabilizing reinforcement learning in dynamic environment with application to online recommendation. In: Proceedings of the 24th ACM SIGKDD International Conference on Knowledge Discovery & Data Mining, pp. 1187–1196 (2018)

7. Chen, X., Li, S., Li, H., Jiang, S., Qi, Y., Song, L.: Generative adversarial user model for reinforcement learning based recommendation system. In: International Conference on Machine Learning, pp. 1052–1061. PMLR (2019)

8. Dave, V.S., Zhang, B., Al Hasan, M., AlJadda, K., Korayem, M.: A combined representation learning approach for better job and skill recommendation. In: Proceedings of the 27th ACM International Conference on Information and Knowledge Management, pp. 1997–2005 (2018)

9. Da'u, A., Salim, N.: Recommendation system based on deep learning methods: a systematic review and new directions. Artif. Intell. Rev. **53**(4), 2709–2748 (2020)

10. Fu, M., Qu, H., Yi, Z., Lu, L., Liu, Y.: A novel deep learning-based collaborative filtering model for recommendation system. IEEE Trans. Cybern. **49**(3), 1084–1096 (2018)

11. Gugnani, A., Misra, H.: Implicit skills extraction using document embedding and its use in job recommendation. In: Proceedings of the AAAI Conference on Artificial Intelligence, vol. 34, pp. 13286–13293 (2020)

12. Guo, Y., Yin, C., Li, M., Ren, X., Liu, P.: Mobile e-commerce recommendation system based on multi-source information fusion for sustainable e-business. Sustainability **10**(1), 147 (2018)

13. Hassan, A., Mahmood, A.: Deep learning approach for sentiment analysis of short texts. In: 2017 3rd International Conference On Control, Automation And Robotics (ICCAR), pp. 705–710. IEEE (2017)

14. Howard, J., Ruder, S.: Universal language model fine-tuning for text classification. arXiv preprint arXiv:1801.06146 (2018)

15. Hwangbo, H., Kim, Y.S., Cha, K.J.: Recommendation system development for fashion retail e-commerce. Electron. Commer. Res. Appl. **28**, 94–101 (2018)
16. Jiang, L., Cheng, Y., Yang, L., Li, J., Yan, H., Wang, X.: A trust-based collaborative filtering algorithm for e-commerce recommendation system. J. Ambient. Intell. Humaniz. Comput. **10**(8), 3023–3034 (2019)
17. Lalwani, T., Bhalotia, S., Pal, A., Rathod, V., Bisen, S.: Implementation of a chatbot system using ai and nlp. International Journal of Innovative Research in Computer Science & Technology (IJIRCST), vol. 6(3) (2018)
18. Lee, H., Kang, J., Yeo, J., et al.: Medical specialty recommendations by an artificial intelligence chatbot on a smartphone: development and deployment. J. Med. Internet Res. **23**(5), e27460 (2021)
19. Li, C., Zhan, G., Li, Z.: News text classification based on improved bi-lstm-cnn. In: 2018 9th International Conference On Information Technology In Medicine And Education (ITME), pp. 890–893. IEEE (2018)
20. Li, P., Zhao, F., Li, Y., Zhu, Z.: Law text classification using semi-supervised convolutional neural networks. In: 2018 Chinese Control and Decision Conference (CCDC), pp. 309–313. IEEE (2018)
21. Ma, X., Ye, L.: Career goal-based e-learning recommendation using enhanced collaborative filtering and prefixspan. Int. J. Mobile Blended Learn. (IJMBL) **10**(3), 23–37 (2018)
22. Manna, R., Pascucci, A., Zarino, W.P., Simoniello, V., Monti, J.: Monitoring social media to identify environmental crimes through nlp. a preliminary study. In: CLiC-it (2020)
23. Ostendorff, M., Bourgonje, P., Berger, M., Moreno-Schneider, J., Rehm, G., Gipp, B.: Enriching bert with knowledge graph embeddings for document classification. arXiv preprint arXiv:1909.08402 (2019)
24. Reddy, S.R.S., Nalluri, S., Kunisetti, S., Ashok, S., Venkatesh, B.: Content-based movie recommendation system using genre correlation. In: Satapathy, S.C., Bhateja, V., Das, S. (eds.) Smart Intelligent Computing and Applications. SIST, vol. 105, pp. 391–397. Springer, Singapore (2019). https://doi.org/10.1007/978-981-13-1927-3_42
25. Rhanoui, M., Mikram, M., Yousfi, S., Barzali, S.: A cnn-bilstm model for document-level sentiment analysis. Mach. Learn. Knowl. Extract. **1**(3), 832–847 (2019)
26. Rosa, R.L., Schwartz, G.M., Ruggiero, W.V., Rodríguez, D.Z.: A knowledge-based recommendation system that includes sentiment analysis and deep learning. IEEE Trans. Industr. Inf. **15**(4), 2124–2135 (2018)
27. Su, D., Xu, Y., Yu, T., Siddique, F.B., Barezi, E.J., Fung, P.: Caire-covid: a question answering and query-focused multi-document summarization system for covid-19 scholarly information management. arXiv preprint arXiv:2005.03975 (2020)
28. Tang, X., Chen, Y., Li, X., Liu, J., Ying, Z.: A reinforcement learning approach to personalized learning recommendation systems. Br. J. Math. Stat. Psychol. **72**(1), 108–135 (2019)
29. Walek, B., Fojtik, V.: A hybrid recommender system for recommending relevant movies using an expert system. Expert Syst. Appl. **158**, 113452 (2020)
30. Wan, X., Lucic, M.C., Ghazzai, H., Massoud, Y.: Empowering real-time traffic reporting systems with nlp-processed social media data. IEEE Open J. Intell. Trans. Syst. **1**, 159–175 (2020)
31. Zand, A., et al.: An exploration into the use of a chatbot for patients with inflammatory bowel diseases: retrospective cohort study. J. Med. Internet Res. **22**(5), e15589 (2020)

32. Zhang, X., Chen, M.H., Qin, Y.: Nlp-qa framework based on lstm-rnn. In: 2018 2nd International Conference on Data Science and Business Analytics (ICDSBA), pp. 307–311. IEEE (2018)
33. Zhu, Z., Li, D., Liang, J., Liu, G., Yu, H.: A dynamic personalized news recommendation system based on bap user profiling method. IEEE Access **6**, 41068–41078 (2018)
34. Zihayat, M., Ayanso, A., Zhao, X., Davoudi, H., An, A.: A utility-based news recommendation system. Decis. Support Syst. **117**, 14–27 (2019)

Autoencoder Neural Network for Detecting Non-human Web Traffic

Marcin Gabryel[✉], Dawid Lada, and Milan Kocić

Spark Digitup, Plac Wolnica 13 Lok. 10, 31-060 Kraków, Poland
`marcin.gabryel@sparkdigitup.com`

Abstract. In this paper, a neural network model is presented to identify fraudulent visits to a website, which are significantly different from visits of human users. Such unusual visits are most often made by automated software, i.e. bots. Bots are used to perform advertising scams or to do scraping, i.e., automatic scanning of website content frequently not in line with the intentions of website authors. The model proposed in this paper works on data extracted directly from a web browser when a user or a bot visits a website. This data is acquired by way of using JavaScript. When bots appear on the website, collected parameter values are significantly different from the values collected during usual visits made by human website users. However, just knowing what values these parameters have is simply not enough to identify bots as they are being constantly modified and new values that have not yet been accounted for appear. Thus, it is not possible to know all the data generated by bots. Therefore, this paper proposes a neural network with an autoencoder structure that makes it possible to detect deviations in parameter values that depart from the learned data from usual users. This enables detection of anomalies, i.e., data generated by bots. The effectiveness of the presented model is demonstrated on authentic data extracted from several online stores.

Keywords: Autoencoder · Neural networks · Deep learning · Web traffic · Ad fraud

1 Introduction

Many websites are not sufficiently secured against fraudulent activity. One of the main types of online crime and fraud in the last few years has become so-called traffic fraud, i.e. page view or click fraud, which consists in increasing revenue from online advertising by automatically generating artificial page views, clicks or filling out online forms. This procedure generates real financial revenue or benefits resulting from losses made by competitive companies. Globally, ad fraud brings multi-billion losses to the advertising industry [6, 7].

The advertising system involves three main parties to this business:

1. Publishers, who provide resources for advertising traffic. They provide network services to users, i.e., they generate advertising traffic when a user visits their sites. When a user visits a publisher's site, it creates an opportunity to display one or more ads to that user.

L. Rutkowski et al. (Eds.): ICAISC 2022, LNAI 13589, pp. 232–242, 2023.
https://doi.org/10.1007/978-3-031-23480-4_19

2. Advertisers who buy web traffic to deliver their ads to an audience,
3. Affiliates that are an intermediary between publishers and advertisers. A publisher can sell advertising space to interested advertisers and this is usually done through an intermediary in the form of an affiliate partner.

Different types of ad fraudulent activity may be committed by:

1. Publishers who are billed per result. Billing can be done in several forms: CPC - cost-per-click, CPM - cost-per-mille, CPL cost-per-lead or CPS - cost-per-sale. The most common online frauds in online advertising are related to effectiveness advertising, which means that they are settled for the volume of clicks on an advertisement. Apart from "clicking the ads" the effect of dishonest publishers actions may be also fraudulent leads. These are generated by filling in forms with incorrect or false data. In extreme cases, there are also fictitious sales resulting in unjustified commission payments.
2. Intermediaries who offer advertisers to support their marketing efforts, e.g. affiliate partners, website owners, and sites with price checking services. Fraudulent activities include, among others, artificial generating of worthless clicks/leads, repeatedly clicking ads, entering false data into forms, automatic completion of applications without the actual knowledge of the person concerned, domain spoofing, creating websites that generate additional clicks, etc. Most of such procedures are generated by specially prepared software - bots.
3. Competition that competes with other companies in an unfair way. This is mainly done by fraudulent clicking on their competitor's ads in order to ensure better positioning of their own ads and lower cost per click. These actions consume the advertising budget and do not bring expected sales results, as a result, they limit the scale of the advertising campaign.

Another dishonest behavior in this field involves carrying out automated website scans, i.e. scraping a website, for example, in order to gain information about competitors' prices. Knowing the prices of other sellers makes it possible for you to keep adjusting the prices of your own products in order to attract more customers. There is also other content that may be taken from or scraped off other websites without the consent of their authors, such as product descriptions or pictures.

Many online stores do not have the ability to automatically monitor in real time the data informing them about users' behavior on their website and interpret it accordingly. In particular, this applies to automatic detection and notification of system administrators of the fact that unusual data deviating from the other existing data has appeared and displaying the nature of an anomaly. Such anomalies can be generated by automated software, i.e., bots which appear on the website pretending to show human behavior. Detection of anomalies requires implementation of additional software, which allows tracking user behavior on the website. This is possible through JavaScript and tracking events that occur during user interaction with elements of the website [9]. Collecting this data allows you to gather information about the typical behavior of a website user. The behavior of automatic programs downloading the page - bots - is completely different from human behavior. In most cases, during a bot's visit, the collected data is incomplete due to the fact that the software does not need to render the page layout very precisely or interact with particular elements of the page. There will also be unusual values among the collected data or unusual combinations of parameter values, which do not appear during a human visit to the website. As mentioned, the most frequent occurrence of these anomalies is related to the appearance of a new program - bot, however, a new version of the browser or an update of the operating system cannot be excluded in this case, either.

Detecting unusual data without the support of appropriate detection algorithms is extremely labor intensive. It requires manual search among hundreds of thousands of records. The number of unusual data increases as the number of users visiting the website increases. Manual data search is imperfect as a lot of information may be overlooked. Unusual data is mostly detected long after that fact already occurred or it is not detected at all. There are also off-the-shelf protections available that can, to some extent, distinguish the presence of a bot from a human presence (e.g., [1] or [10]). However, these mechanisms belong to third parties, their algorithms are unknown, and they often require a direct response from the user or just take time to perform an analysis. In contrast to these solutions, the model proposed in this paper allows a bot to quickly identify visits to a website by detecting abnormal parameter values (anomalies) taken directly from the browser during browsing. In addition, anomalies generated by other factors (invalidly embedded advertisements) can be detected. This will allow the online store administrator to quickly react to incidents occurring on the website. It also gives the possibility to detect errors related to quickly appearing successive versions of browsers and their backward compatibility. Identification of a visit as unusual, not coming from a human, protects against ad fraud and can be the basis for questioning the settlement with an affiliate network or publisher of online advertising. It will allow you to save money on online advertising costs.

This paper will present an artificial neural network model for automatic detection of anomalies in Internet traffic in real time. The paper consists of several sections. Section Two introduces similar solutions of using autoencoders for anomaly detection. In the next two Sections, the neural network model of the autoencoder structure is described and the proposed solution is presented. Section Five presents the results of experimental work performed on authentic data from online stores. The last Section summarizes the work carried out on the model so far and the plans for development.

2   Similar Solutions

According to numerous research papers (e.g., [2]), deep neural networks, and in particular deep autoencoder structures, can be successfully used to detect anomalies in data. The authors in publication [4] noted the problem of identifying network attacks associated with a large number of vulnerabilities in computer systems. For this purpose, they proposed a deep learning system for hacks detection. Their special concern was that the deep autoencoder should be able to avoid over-fitting and local optima. In [5], on the other hand, the problem of network anomaly detection was solved using a variational autoencoder. These methods differ from previously used statistical models and supervised machine learning techniques. Hacking detection capabilities of computer systems were also studied by the authors of [12]. They proposed the use of stacked ensembles consisting of neural network (SNN) and autoencoder (AE) models enriched with a novel hyperparameter optimization approach. Anomaly detection can also be useful in such fields in geochemistry. The authors of [8] train an autocoder network to encode and reconstruct populations of geochemical samples with unknown complex multivariate probability distributions. During the training, small probability samples contribute little to the autoencoder network. These samples can be recognized by the trained model as anomalous samples due to their comparatively higher reconstructed errors. In work [3], an autoencoder was used for a slightly different purpose - it was able to efficiently learn already noisy data and was able to remove outliers. This makes it possible to take advantage of the remarkable generalization ability of this neural network structure. The authors of [11], on the other hand, studied the method of detecting anomaly patterns in the data stream, among others, based on autoencoder and compared with other methods (Isolation Forest and Local outlier factor).

3 Autoencoder Structure

The main objective of the work described in this paper is to develop a model of the deep structure of the autoencoder, which with proper learning will allow anomaly detection, i.e., unusual values of parameters given to the input of the network. A correctly learned autoencoder is characterized by the fact that it can reproduce at its output almost without error what appears at its input. Its characteristic feature is that one of the middle layers contains a very small number of neurons and its task is to compress the data given to the input. This is done by the first of the autoencoder fragments - the encoder, which includes the first layers including the above-mentioned smallest layer. The compressed data is then passed to the decoder, whose task is to decode the data from the smallest layer and reconstruct it at the output. This network design means that data are not directly passed from the input to the output and that the neural network has generalization capabilities. The autoencoder, after learning, recognizes the input data by returning identical output values while generating a small reconstruction error, which is the difference between the expected (input) values and the values obtained during data reconstruction. Data with which the autoencoder has not been learnt cause generation of a large value of the reconstruction error. Checking the value of the received error gives the possibility to make classification of data into data known and data unknown to the autoencoder.

The structure of the autoencoder can be described as follows. The training set given is $X = \{x_1, x_2, \ldots, x_N\} \in R^m$ where x_i is m-dimensional feature vector, and N - the number of samples. The encoder maps input vector x_i to hidden representation $h_i \in R^n$ by using function f_θ according to the formula:

$$h_i = f_\theta(x_i) = s(Wx_i + b) \tag{1}$$

where $W \in R^{m \times n}$ is the set of weights, n is the number of units in hidden layer h_i, $b \in R^n$ is a bias vector, θ is set $\{W, b\}$, and $s(\cdot)$ is the assumed sigmoid function defined by the formula:

$$s(t) = \frac{1}{1 + exp^{-t}} \tag{2}$$

The decoder maps back values h_i obtained in the hidden layer to output vector $y_i \in R^m$ according to the formula:

$$y_i = g_\theta(x_i) = s(\acute{W}h_i + \acute{b})$$
(3)

where $\acute{W} \in R^{n \times m}$ are weights, $\acute{b} \in R^m$ is a bias vector and $\acute{\theta} = \{\acute{W}, \acute{b}\}$.

The autoencoder training involves minimizing the difference between input x_i and output y_i. To this end, a loss function expressed by the following formula is calculated:

$$L(x_i, y_i) = \|x_i - y_i\|^2 = \|x_i - s(\acute{W}s(Wx_i + b) + \acute{b})\|^2$$
(4)

The goal of the learning process is therefore to find optimal values for parameters θ and $\acute{\theta}$ allowing to minimize the error between the input and output for the entire training set:

$$\theta, \acute{\theta} = \underset{\theta, \acute{\theta}}{\mathrm{argmin}}\, L(x, y)$$
(5)

The autoencoder discussed above is considered for continuous data x. In this paper, the extracted data from websites are categorical in nature.

4   Proposed Solution

By monitoring the values of parameters retrieved from the browser while viewing a page, it allows to check whether we are dealing with a human or a bot mimicking human behaviour. The usual use of the browser by a human allows, during one page view, to retrieve a whole set of predefined parameter values. The majority of data have similar parameter values and no empty values. However, often some attempts to access a web page are unusual. It is done by various types of software, which thus creates the so-called non-human traffic. During such entries to the website most often monitored parameter values are significantly different from those collected so far and often there are numerous gaps in the values of collected parameters. Such data can come, for example, from software that tries to speciously simulate human behavior, is imperfect, or simply comes from a new version of the browser that no longer supports some functionality. The main goal of the solution proposed in this paper is to detect such anomalies, which is normally done manually by analysts tediously reviewing the data flowing from a monitored site.

Working on a model capable of classifying anomalies relies heavily on selecting of an appropriate autoencoder structure. In the case at hand, data generated during typical web page accesses by usual users are used for learning. When data values differing from the learned values are given to the neural network input, a large reconstruction error value will be generated. Selecting an appropriate threshold value for this error will allow the samples to be treated as correct (the reconstruction error value is below the threshold) or as abnormal (then the reconstruction error value will be above the set threshold).

The data used for the experiments described in this paper were obtained from several large online stores. During the initial analysis, it was shown that the vast majority of the data was typical, human-generated data. This data was treated as correct and was used to learn the autoencoder. The other part was the data that was treated as abnormal. These are mainly incomplete data that do not have all the parameter values. Such special cases occur when the collected data is obtained only at the end of the process of monitoring the website while the values cannot be retrieved at the start of the run of JavaScript. Another large group of atypical data are also records with empty values at the same time with long time spent on the page. Data marked as abnormal are used in the learning process in the validation and testing phases to check the performance of the autoencoder.

For the experiments we selected data coming from internet traffic connected with online advertising services. During the analysis of these data two separate groups of inputs were identified: those created at the moment of visiting the page after clicking on an advertisement (the process was called clickscan, for CPC and CPM ads accounting models) and those created at the moment of filling in the data in the contact form (the process was called leadscan, for CPL and CPS ads accounting models). Because of the differences detected, each of these groups was examined separately and for each of them a separate model was prepared to detect anomalies. In this paper we present a model dedicated to handling the first type of web traffic – clickscan - generated at the moment of navigating to an online store's website after clicking on an online ad.

The parameters monitored during a web page view and selected to search for anomalies for the clickscan case are: type of client (denoted as "client_id"), type of device ("is_mobile", "touch_enabled", "virtual_machine"), source of page entry ("source_id"), time spent on the page ("first_req_st2_close_seconds"), information whether the page was launched in a frame ("in_frame"), information about falsified data ("os_t_platformOK", "os_t_touchOK", "screen_tampering", "language_tampering", "br_t_evalOk", "br_t_subOk"), mouse movements ("counter", "has_ws_mouse", "has_ws_scroll", "mm_unique_points"), screen information ("scr_num"), ISP and connection data ("proxy_flag", "ip_type", "isp_is_suspicious", "is_known_crawler_bot"), number of pages visited in this session ("history_len"), sales information ("had_sale"), browser information ("browser_type_id", "notif_permission", "chrome_prop_not_set", "browser_type_name").

During the experiments many different autoencoder structures were tested. The most optimal model turned out to be built from five layers in the configuration 64–32–3–32-64. A schematic diagram of the structure of this model is shown in Fig. 1. In addition to the five mentioned layers, for proper operation the autoencoder has several additional modules which are used to process the input data and to obtain the corresponding values at the output. Since in most cases the input data are categorical, a transformation from index values to zero-one form is performed. In the figure the layer doing this is labeled one-hot-encoder. Each category is converted into a sequence of low bits (0) and single high bit (1).. The "1" is placed where the category information is encoded. At the output of the autoencoder, in order to obtain identical values to those given at the input, a dense layer dedicated to each parameter separately is given. This enables us to generate the same sequence of zeros and ones given on the input. This makes it easier to count the reconstruction error, which is the error showing the difference between the input and the output.

The input data is pre-generated by one-hot-encoder. This module is designed to convert the integer values of 28 parameters into zero-one (0–1) form. For the selected experimental data, a vector of 2476 numbers is given on the input. These are predominantly zeros, and only in 28 places appears "1". The task of the prepared model is to reproduce the identical vector in the output, so the last layer consists of 28 layers of neurons, each with a softmax activation function. This design makes it possible to generate one "1" for each input parameter and thus to be able to accurately represent the input values.

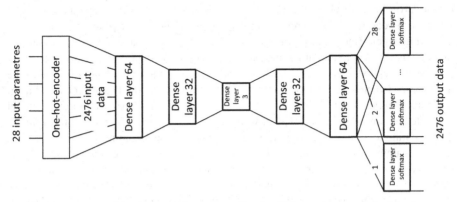

Fig. 1. Diagram of the design of the autoencoder for clickscan data anomaly detection.

5 Experimental Work

For the experiments, data was downloaded from nine online stores. About 98,000 records were selected for the experiments. The data was divided into three sets: the training set (about 40,000 samples), the validation set (38,000 samples) and the test set (about 20,000 samples). Each set has valid data (from a website visited by a human) and data identified as anomalies (2,500 samples in each case). The learning process was performed on a learning sequence, but only on the set of the data marked as correct. The autoencoder was learnt in such a way that during the prediction it generated data on the output that were expected to be similar to the input data. The reconstruction error was calculated as the difference between the input and output values. Assuming a certain threshold value, it can be assumed that the samples with an error below this value are treated as correct values, and those whose reconstruction error is above the threshold are treated as abnormal. Experimentally, the error limit was set at 0.02342. This value was an optimum chosen to minimize the number of misclassified samples. Figure 2 shows the error values for a few dozen random samples from the training, validation and test data. The error values were obtained during the prediction made by the learned autoencoder structure discussed in Sect. 4. The figure also indicates the value of the error threshold. The best results, obtained from all the experiments performed, are shown in Table 1. Non-standard data were used to calculate accuracy, precision, recall, and F1 score. Validation data were used to check how the model performs in distinguishing between correct and incorrect data during the learning. The test data were used to check the generalization capabilities of the autoencoder. The table additionally reports the error that was obtained at the end of the learning process (MSE error). The prediction process consisted of several steps: feeding the data onto the autoencoder input, their reconstruction at the output, calculation of the reconstruction error and evaluation of the value of this error in comparison to the assumed threshold.

Figures 3, 4 and 5 show the confusion matrices obtained for the training, validation and test sequences respectively. Data labeled as correct and abnormal are subjected to a classification that assigns them a predicted class. The values in these matrices made it possible to calculate accuracy, precision, recall and F1.

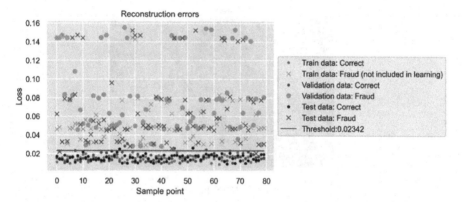

Fig. 2. Graph showing reconstruction error values for training, validation, and test samples.

Table 1. Results obtained by the best autoencoder for clickscan data.

Data type	MSE error	Accuracy	Precision	Recall	F1 score
Learning	0.0174	0.9971	0.9964	0.9994	0.9979
Validation	0.0495	0.9989	0.9988	0.9998	0.9993
Test		0.9966	0.9963	0.9993	0.9978

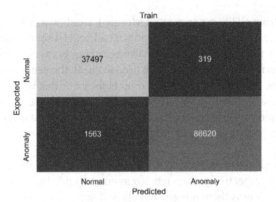

Fig. 3. Confusion matrix for the learning sequence.

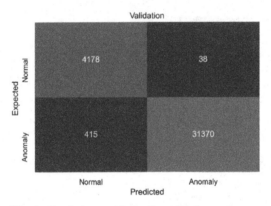

Fig. 4. Confusion matrix for the validation sequence.

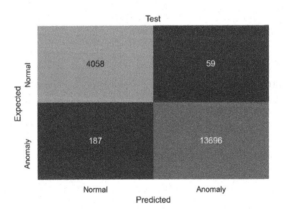

Fig. 5. Confusion matrix for the test sequence.

6 Summary

This paper proposes a model for detecting non-human/fake traffic on a website. An artificial neural network structure in the form of autoencoder was used to this end. Experimental studies performed on authentic data collected from online stores have proven the effectiveness of this method for detecting fraudulent traffic. When this solution is applied in practice, it could help advertisers to reduce financial losses related to fraud or to counteract unfair competition.

The discussed problem is planned to be tested and compared with a classifier based on a feedforward neural network with one output. However, it requires collecting a larger number of non-standard data. The algorithm can be implemented using also other mechanisms of computational intelligence: fuzzy systems, feedforward neural networks [13, 14], using parallel computing mechanisms and others [15, 16].

Acknowledgments. The presented results are obtained within the realization of the project "Traffic Watchdog 2.0 – verification and protection system against fraud activities in the on-line marketing (ad frauds) supported by artificial intelligence and virtual finger-print technology" financed by the National Centre for Research and Development; grant number POIR.01.01.01–00-0241/19–01.

References

1. Recaptcha https://www.google.com/recaptcha/about/
2. Goodfellow, I., Bengio, Y., Courville, A.: Deep Learning. MIT press (2016)
3. Zhou, C., Paffenroth, R.C.: Anomaly detection with robust deep autoencoders. In: Proceedings of the 23rd ACM SIGKDD International Conference on Knowledge Discovery and Data Mining (2017)
4. Farahnakian, F., Heikkonen, J.: A deep auto-encoder based approach for intrusion detection system. In: 2018 20th International Conference on Advanced Communication Technology (ICACT). IEEE (2018)
5. Nguyen, Q.P.: GEE: A Gradient-based Explainable Variational Autoencoder for Network Anomaly Detection (2019)
6. 2019. https://www.emarketer.com/content/digital-ad-fraud-2019
7. Barker, S.: Future Digital Advertising, Artificial Intelligence & Advertising Fraud 2019–2023, Juniper Research (2019)
8. Xiong, Y., Zuo, R.: Recognition of geochemical anomalies using a deep autoencoder network. Comput. Geosci. **86**, 75–82 (2016)
9. Gabryel, M., Grzanek, K., Hayashi, Y.: Browser fingerprint coding methods increasing the effectiveness of user identification in the web traffic. J. Artificial Intelligence and Soft Computing Res. **10** (2020)
10. Gabryel, M., et al.: Decision making support system for managing advertisers by ad fraud detection. J. Artificial Intelligence and Soft Computing Res. **11** (2021)
11. Kim, T., Park, C.H.: Anomaly pattern detection in streaming data based on the transformation to multiple binary-valued data streams. J. Artificial Intelligence and Soft Computing Res. **12**(1), 19–27 (2022)
12. Brunner, C., Kő, A., Fodor, S.: An autoencoder-enhanced stacking neural network model for increasing the performance of intrusion detection. J. Artificial Intelligence and Soft Computing Res. **12**(2), 149–163 (2022)

13. Bilski, J., Kowalczyk, B., Marjański, A., Gandor, M., Żurada, J.: A novel fast feedforward neural networks training algorithm. J. Artificial Intelligence and Soft Computing Res. **11**(4), 287–306 (2021). https://doi.org/10.2478/jaiscr-2021-0017

14. Bilski, J., Rutkowski, L., Smoląg, J., Tao, D.: A novel method for speed training acceleration of recurrent neural networks. Information Sciences **553**, 266–279 (2021). https://doi.org/10.1016/j.ins.2020.10.025

15. Grycuk, R., Scherer, R.: Novel fast binary hash for content-based solar image retrieval. In: 2020 International Joint Conference on Neural Networks (IJCNN). IEEE (2020)

16. Grycuk, R., Scherer, R.: Solar image hashing by intermediate descriptor and autoencoder. In: 2021 International Joint Conference on Neural Networks (IJCNN). IEEE (2021)

Edge Detection-Based Full-Disc Solar Image Hashing

Rafał Grycuk⬤, Patryk Najgebauer⬤, and Rafał Scherer(✉)⬤

Department of Intelligent Computer Systems, Częstochowa University of Technology,
Al. Armii Krajowej 36, 42-200 Częstochowa, Poland
{rafal.grycuk,patryk.najgebauer,rafal.scherer}@pcz.pl

Abstract. We propose a content-based solar image descriptor for fast retrieving similar images. The method is divided into three main stages: active region detection by using edge detection, representation learning and hash generation. The first step uses morphological operations for active region detection and afterwards Canny edge detection. In the learning step we use an unsupervised convolutional autoencoder in order to obtain the solar image hash. This process reduces hash length more than twelve times compared to the active region image matrix. The process of reducing the hash length is significant in reference to solar image retrieval process, in which we focus on calculating the distances between hashes. The performed experiments proved the efficiency of the proposed approach. The presented method has various potential, not only solar, applications. Moreover, the problem of searching of and retrieving solar flares has a significant impact on many aspects of life on Earth and beyond.

Keywords: Solar activity analysis · Solar image description · CBIR · Edge detection

1 Introduction

Solar activity, driven by the sunspot cycle and by transient aperiodic processes, creates space weather and has a profound impact on space- and ground-based technologies as well as the Earth's atmosphere. Moreover, our Planet's climate fluctuations on scales of centuries and longer depend partially on the Sun's behavior. Understanding and predicting the sunspot cycle remains one of the challenges in science with significant ramifications for space science and the understanding of magnetohydrodynamic phenomena on Earth and the Solar System. Computer-based Solar activity research is relatively new as it was possible with the advent of observatories and increasing computer computational power. In 2010 NASA launched the Solar Dynamics Observatory (SDO) as a part of the Living with a Star program. The goal was to provide data to research the connected Sun-Earth system and the impact of the Sun to life on Earth. One of the appliances that increased possibilities was the NASA Solar Dynamic Observatory (SDO) started several years ago. One of its subsystems is the Atmospheric

© The Author(s), under exclusive license to Springer Nature Switzerland AG 2023
L. Rutkowski et al. (Eds.): ICAISC 2022, LNAI 13589, pp. 243–251, 2023.
https://doi.org/10.1007/978-3-031-23480-4_20

Imaging Assembly (AIA), capturing every several seconds high-resolution images in eight different wavelengths. In the paper, we use a 4K resolution dataset prepared for image retrieval purposes by Kucuk et al. [10]. The dataset contains hundreds of thousands of full-disk images of the Sun with temporal and spatial features of the event records in the dataset. Usually, to classify or predict Sun state from its image, the methods used hand-crafted features. For example, Banda at al. in [1] identified ten different image parameters, that when extracted from Solar full-disk images, are the best representation of the Solar state. These parameters are also available in the dataset. In [3], the Lucene retrieval engine is adapted to retrieve solar images relying on descriptive solar features developed in [2]. These features were also used by Boubrahimi et al. in [4] and Ma et al. [11] to predict solar event trajectories. In [9] the authors use fuzzy rules to retrieve images. Our method speeds up solar image retrieval and classification in large databases of files. We use only image content to compute its representation, what is independent of human-based annotation or hand-crafted image features. We use a convolutional autoencoder to obtain a concise image hash, what is a form of approximate information retrieval [12] .

The rest of the paper is organized as follows. In Sect. 2 our method for generating solar image description and retrieval is described. Section 3 presents the results of experiments on real solar images, and Sect. 4 concludes the paper.

2   Edge Detection-Based Full-Disc Solar Image Hashing

In this section we describe a novel method for solar image description. The proposed algorithm can be used for image retrieval tasks. The dataset used for training are obtained from the Solar Dynamics Observatory. This dataset was created based on the images extracted from SDO and then published in the form of Web API by [10]. Although there are many resolutions available, we decided to use 2048×2048. The presented method can be divided into three main steps: active region detection based on edge detection, training and hash generation by using convolutional autoencoder, and solar image retrieval. The first stage can be divided into the following steps, namely, applying filters, morphological operations and thresholding for detecting active regions (AR), and, the most important, edge detection. The second stage takes edge detected solar images and use a convolutional autoencoder for training process. Here we apply solar image to the autoencoder as the training data. In the next step, only the encoder part of autoencoder is used for hash generation. As a result the encoded mathematical representation of the solar image is obtained. All the previous steps allow to significantly reduce the number of data for further processing, e.g. in the solar image retrieval process. In the last step we use our hashes to retrieve similar images. They are also used as a query in the image retrieval process. The dataset we use is unlabelled, therefore the process of defining the similarity between images is not trivial. Thus, we proposed a method for defining the similarity of the unlabelled solar images. This approach is used for evaluating the efficiency of the proposed method.

2.1   Active Region Detection Based on Edge Detection

After careful analysis of solar image we can distinguish Active Regions (AR, see. Fig. 1A). They are strongly related to solar flares and monitoring solar activity. AR's have various shapes, and due to the Sun's rotation the position of AR's change on consecutive images. The main task of active region detection process is to detect the AR's on the given solar image. This stage of the presented method determines the positions and shapes of active regions. The process of active region detection is composed of the several steps. In the first the solar image is converted to the greyscale; therefore since now the pixel intensities scale will be between [0..255]. In the next step we apply Gaussian blur filter in order to remove insignificant, small regions of the image. Therefore, since now on only important image features are analysed. The previous preprocessing steps are necessary in order to perform the thresholding stage. In the thresholding stage every pixel intensity is compared with the provided threshold th value. Based on that comparison we determine if the given pixel is a part of the active region area. Every pixel with value greater or equal than th is treated as an active region. In the next step we apply the morphological operations, namely, erosion and dilation. The erosion operator allows removing small objects, thus that only substantive objects remain. The dilation operator makes objects more visible, and removes the small holes in objects therefore, this operation allows emphasizing only the important areas of the active regions. The more descriptive information about morphological operations can be found in [7, 13]. After all the previous steps we apply edge detection procedure in order to determine edges of the active regions. We used the Canny edge detector (see [6]) for this purpose. As a result we obtain a set of objects with edges detected. The process of active region detection with edge detection was presented in the form of pseudocode, see Algorthim 1. The location and shape of active regions are very important in reference to the Coronal Mass Ejections (CME) and thus, on the solar flare prediction. The output image of the active region detection with edge detection is presented in Fig. 1.

INPUT: $SolarImage$
OUTPUT: $ActiveRegionDetectedImg$
$GrayScaleImg := ConvertToGrayScale(SolarImage)$
$BlurredImg := Blur(GrayScaleImg)$
$ThreshImg := Threshold(BlurredImg)$
$ErodedImg := Erode(ThreshImg)$
$ActiveRegionDetectedImg = Dilate(ErodedImg)$
$EdgeDetectActiveRegionImg = EdgeDetection(ActiveRegionDetectedImg)$
Algorithm 1: Algorithm for active region detection with edge detection.

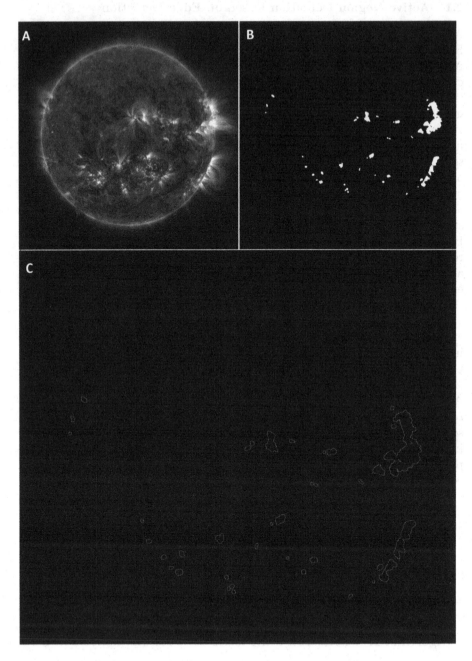

Fig. 1. Active region detection with edge detection, left up (A) image is the input solar image, right up (B) image is active region detection. The bottom image (C) is the output image with edge detected shapes of active regions.

2.2   Training and Hash Generation

In this section we describe learning process along with the encoding stage. The convolutional autoencoder was used for learning active region images. After learning stage we use only encoding layers for hash generation process. The main reason behind using an autoencoder is that the autoencoder is an unsupervised convolutional neural network and therefore no labelled data for training is required. In Table 1 we present an architecture of the convolutional autoencoder (network model). During the detailed analysis of Table 1 we determine that, we used an autoencoder with two convolutional layers with max-pooling, where *kernel_size* parameter is equal 2. *Kernel_size* parameter for convolutional layers is set to 3. After two convolutional layers with pooling, we have the latent space (bottleneck layer) which is the encoded layer. This stage allows us to get a mathematical representation of the image in the form of vector. By applying the hash generation stage, we reduce data volume from $1 \times 2048 \times 2048$ to $8 \times 170 \times 170$). After encoding layers, we have convolutional decoding layers. The decoding layers are used only for training. For a hash generation, we only use encoding layers. During the experiments, we used Python along with Pytorch tensor library. We applied the binary cross-entropy loss function. We empirically proved that 40 epochs are sufficient to obtain the required level of generalization and not cause the network over-fitting. After the learning process, every active region image is fed to the encoded layers of the autoencoder. As a result, encoded solar image hash is obtained. The hash length is 231200. Such hash can be used for content-based solar image retrieval applications (see Sect. 2.3).

Table 1. Tabular representation of the model.

Layer (type)	Output Shape	Filters (in, out)	Kernel size	Params no
$Input2d(InputLayer)$	[1, 2048, 2048]			
$Conv2d\_1(Conv2D)$	[16, 683, 683]	1,16	3	160
$ReLU\_1$	[16, 683, 683]			
$Max\_pooling2d\_1(MaxPool2D)$	[16, 341, 341]		2	
$Conv2d\_2(Conv2D)$	[8, 171, 171]	16,8	3	1160
$ReLU\_2$	[8, 171, 171]			
$Encoded(MaxPool2D)$	[8, 170, 170]		2	
$ConvTranspose2d\_1(ConvTranspose2D)$	[16,341,341]	8,16	3	1168
$ReLU\_4$	[16,341,341]			
$ConvTranspose2d\_2(ConvTranspose2D)$	[8, 1025, 1025]	8, 1	2	3208
$ReLU\_5$	[8, 1025, 1025]			
$ConvTranspose2d\_3(ConvTranspose2D)$	[1, 2048, 2048]	61, 1	2	33
$Decoded(Tanh)$	[1, 2048, 2048]			

2.3 Image Retrieval

In the last step, we perform the image retrieval process. The main assumption in our system is that every solar image has a hash assigned in our image database. The query is an image to which we compare every image stored in the database. More precisely, we compare the query image hash with other image hashes stored in the database. The comparison process is performed by some distance measure. The retrieval process is the following: Generate a hash for the query image in order to calculate the distance between the given image in the database and query image hash. Prior to this step we need to obtain hashes for all images in our dataset. In most cases it is a dataset with time interval images (e.g. from 01/01/2014 to 31/12/2014). In the next step we calculate the distance between the query image hash and every hash previously calculated in our dataset. As a result, we obtain a list of distances d, every distance calculated by the cosine distance measure presented below [8]:

$$\cos(Q_j, I_j) = \sum_{j=0}^{n} \frac{(Q_j \bullet I_j)}{\|Q_j\| \, \|I_j\|},$$

where $\bullet$ is dot product. The performed experiments proved that the chosen distance measure is the most suitable for the proposed hash. We also supported our decision by analysing similar methods [8]. As the next step we perform sorting by distance in the ascending order by the distance d. In the last step we take n images closest to the query. The obtained images are returned to the user as the retrieved images.

INPUT: *Hashes, QueryImage, n*
OUTPUT: *RetrivedImages*
foreach *hash* ∈ *Hashes* **do**
  | *QueryImageHash = CalculateHash(QueryImage)*
  | *D[i] = Cos(QueryImageHash, hash)*
end
SortedDistances = SortAscending(D)
RetrivedHashes = TakeFirst(n)
RetrievedImages = GetCorrespondingImages(RetrievedHashes)
Algorithm 2: Image retrieval steps.

---

3 Experimental Results

In this section we present the experimental results along with our approach for method evaluation. The dataset that we obtained from WebApi is composed of unlabelled data; thus, we had to develop an approach for evaluating the efficiency of our method. Therefore, we propose our approach to solve this issue. Based on that we used Solar rotation movement in order to determine similar images. The consecutive images within a small-time window (by default window is 6 min.) should contain similar data, based on that we can assume that those images are similar. The consecutive solar images should be slightly shifted. Firstly we need to order our images by creation date obtained from SDO. The API that we use allow fetching images with a 6-minute window, base on that we can assume image the similarity. The consecutive images with in the given window will be defined as similar. The main issue that we solve was adjusting the difference time window. After the series of experiments we determine that a 48-hour window is the most suitable. Let us analyse the following example. The image taken at 2012-02-15, 00:00:00 as a query image, we can now assume that 24 h before and 24 h after, the images are similar. This simple assumption allows us to determine the similar images (SI), and afterwards, perform experiments and evaluate efficiency. Every experiment consists of the following steps:

1. Execute image query and obtain the retrieved images.
2. For every retrieved image compare its timestamp with the query image timestamp.
3. If the timestamp is the 48-hour window, the image is similar to the query.

The process of determining the similar images (SI) and (NSI) allow defining the following performance measures: *precision, recall, F − measure* [5],[14]. These measures require defining the following sets:

- *SI* - set of similar images,
- *RI* - set of retrieved images for query,
- *PRI(TP)* - set of positive retrieved images (true positive),
- *FPRI(FP)* - false positive retrieved images (false positive),
- *PNRI(FN)* - positive not retrieved images,
- *FNRI(TN)* - false not retrieved images (TN).

Based on the previously defined set we adapted the state-of-the-art formulas to our needs

$$precision = \frac{|PRI|}{|PRI + FPRI|},$$ (1)

$$recall = \frac{|PRI|}{|PRI + PNRI|}.$$ (2)

$$F_1 = 2 * \frac{precision * recall}{precision + recall}.$$ (3)

The simulation results are presented in Table 2. As can be seen, values of F_1 proves effectiveness of our method. It should be noted that high value of the *precision* measure. The most of the solar images with close distance to the query were successfully retrieved. The solar images with farther distance are classified as positive not retrieved images (PNRI). This phenomenon is caused by the Sun's rotation motion, which causes that more active regions are missing or shifted. Many of the active regions change its position and shape and due to

Table 2. Experiment results for the proposed algorithm, performed on AIA images obtained from [10]. Due to lack of space, we present only a part of all queries.

Timestamp	RI	SI	PRI (TP)	FPRI (FP)	PNRI (FN)	Precision	Recall	F_1
2014-01-01 00:00:00	202	241	170	32	71	0.84	0.71	0.77
2014-01-08 03:00:00	390	481	337	53	144	0.86	0.7	0.77
2014-01-12 07:06:00	367	481	343	24	138	0.93	0.71	0.81
								...
2014-04-09 07:36:00	333	481	322	11	159	0.97	0.67	0.79
2014-04-12 10:42:00	374	481	338	36	143	0.9	0.7	0.79
2014-04-15 03:42:00	347	481	337	10	144	0.97	0.7	0.81
								...
2014-08-23 04:42:00	376	481	337	39	144	0.9	0.7	0.79
2014-08-27 16:48:00	385	481	344	41	137	0.89	0.72	0.8
2014-09-01 12:48:00	386	481	342	44	139	0.89	0.71	0.79
Avg						**0.910**	**0.70**	**0.79**

rotation movement even gone missing. Such situation can occur even during the 48-hour window. Such behaviour could have impact on obtained hash. Therefore, the distance to the query will be increased. The described case, was observed during the experiments. Based on that we assume that, lower values of *Recall* are caused by this phenomena. The entire simulation environment was developed in Python scripting language and PyTorch. The training stage took ≈ 9.5 hours for 83819 images. The hash creation process took 1.5 h. The average retrieval time is 1100 ms.

4   Conclusions

We proposed a novel approach for full-disc solar hashing designed mainly for content-based solar image retrieval. The method is divided on two three main stages: active region detection by using edge detection, learning and hash generation. The first step uses well known morphological operations for active regions' detection process and afterwards apply canny edge detection. In the learning step we use, an unsupervised convolutional autoencoder in order to obtain the solar image hash. This process reduces hash length more than 12 times over compared to active region image matrix. The process of reducing the hash length is significant in reference to solar image retrieval process, in which we focus on calculating the distances between hashes. The performed experiments presented in Table 2 proved the efficiency of the proposed approach. The presented method has various potential applications. Moreover, the problem of searching and retrieving solar flares, has significant impact on many aspects of life on Earth and beyond.

References

1. Banda, J.M., Angryk, R.A.: Selection of image parameters as the first step towards creating a cbir system for the solar dynamics observatory. In: 2010 International Conference on Digital Image Computing: Techniques and Applications, pp. 528–534. IEEE (2010)
2. Banda, J.M., Angryk, R.A.: Large-scale region-based multimedia retrieval for solar images. In: Rutkowski, L., Korytkowski, M., Scherer, R., Tadeusiewicz, R., Zadeh, L.A., Zurada, J.M. (eds.) ICAISC 2014. LNCS (LNAI), vol. 8467, pp. 649–661. Springer, Cham (2014). https://doi.org/10.1007/978-3-319-07173-2_55
3. Banda, J.M., Angryk, R.A.: Scalable solar image retrieval with lucene. In: 2014 IEEE International Conference on Big Data (Big Data), pp. 11–17. IEEE (2014)
4. Boubrahimi, S.F., Aydin, B., Schuh, M.A., Kempton, D., Angryk, R.A., Ma, R.: Spatiotemporal interpolation methods for solar event trajectories. Astrophys. J. Suppl. Ser. **236**(1), 23 (2018)
5. Buckland, M., Gey, F.: The relationship between recall and precision. J. Am. Soc. Inf. Sci. **45**(1), 12 (1994)
6. Canny, J.: A computational approach to edge detection. IEEE Trans. Pattern Anal. Mach. Intell. (6), 679–698 (1986)
7. Dougherty, E.R.: An introduction to morphological image processing. In: SPIE, 1992 (1992)
8. Kavitha, K., Rao, B.T.: Evaluation of distance measures for feature based image registration using alexnet. arXiv preprint arXiv:1907.12921 (2019)
9. Korytkowski, M., Senkerik, R., Scherer, M.M., Angryk, R.A., Kordos, M., Siwocha, A.: Efficient image retrieval by fuzzy rules from boosting and metaheuristic. J. Artif. Intell. Soft Comput. Res. **10**(1), 57–69 (2020)
10. Kucuk, A., Banda, J.M., Angryk, R.A.: A large-scale solar dynamics observatory image dataset for computer vision applications. Sci. Data **4**, 170096 (2017)
11. Ma, R., Boubrahimi, S.F., Hamdi, S.M., Angryk, R.A.: Solar flare prediction using multivariate time series decision trees. In: 2017 IEEE International Conference on Big Data (Big Data), pp. 2569–2578. IEEE (2017)
12. Salakhutdinov, R., Hinton, G.: Semantic hashing. Int. J. Approximate Reasoning **50**(7), 969–978 (2009). Special Section on Graphical Models and Information Retrieval
13. Serra, J.: Image analysis and mathematical morphology. Academic Press, Inc. (1983)
14. Ting, K.M.: Precision and recall. In: Encyclopedia of machine learning. Springer (2011). https://doi.org/10.1007/978-1-4899-7993-3_5050-2

Employee Turnover Prediction
From Email Communication Analysis

Marcin Korytkowski[1] ![ORCID], Jakub Nowak[1] ![ORCID], Rafał Scherer[1]([⊠]) ![ORCID], Anita Zbieg[2],
Błażej Żak[2], Gabriela Relikowska[2], and Paweł Mader[2]

[1] Czestochowa University of Technology, al. Armii Krajowej 36, Czestochowa, Poland
{marcin.korytkowski,jakub.nowak,rafal.scherer}@pcz.pl
[2] Network Perspective Ltd., Wroclaw, Poland
https://www.networkperspective.io/

Abstract. One of the biggest problems faced by companies is the sudden departure of employees from the company. Such events may even result in a serious paralysis of the functioning of enterprises in the event of resignation from work by people holding significant positions. Therefore, an extremely important issue is to develop techniques that will allow detecting the planned resignation of a given employee well in advance. Gaining knowledge about the factors influencing this type of events may allow for taking actions aimed at counteracting them. This work proposes a proprietary method based on the use of artificial neural networks to predict employees leaving work and to indicate which of the possible analyzed reasons are the most significant. Ultimately, the proposed system achieved an efficiency of 74 %.

1 Introduction

High employee turnover is currently a problem that affects most companies, and refers to tangible and intangible expenses, such as replacement costs, loss of productivity and tacit knowledge drain. Observed during the last decade, the change in the labour market from the employer's market to the employee's market has made employees more and more willing to change jobs. 47 million Americans voluntarily quit their jobs in 2021, and the voluntary turnover rate has raised more than 100% over the last 10 years [2]. It is expected that the problem will only grow under the name of the Great Resignation [6]. The problem of employees quitting their jobs is not only rising, but some companies face it more painfully than others. Why? Among many other factors (e.g., industry or culture), the studies show that employee turnover is contagious and can spread among employees working together [19]. Employees tend to imitate the turnover-related attitudes and behaviours of their co-workers, which means that the exit of an employee may stimulate additional incidents of turnover within the workplace. The turnover among co-workers not only greatly increases a person's turnover probability, but also the

The work was supported by The National Centre for Research and Development (NCBR), the project no POIR.01.01.01-00-1288/19; 2020-22.

L. Rutkowski et al. (Eds.): ICAISC 2022, LNAI 13589, pp. 252–263, 2023.
https://doi.org/10.1007/978-3-031-23480-4_21

peer effects may have a greater impact than own effects [23]. These findings are consistent with works on social diffusion [15,26], and network studies describing the spread of social phenomena, alike happiness [11], loneliness [3], obesity [4], or smoking [5]. The risk of quitting a job is associated not only with the exits of co-workers, but among other factors, it is also linked to workload, as it refers to burnout. In the last years triggered by the pandemic, employee workload has seriously raised – employees tend to experience up to 45% increase in workday span, 48% growth in chats, and 55% increase in number of meetings and calls per week [21]. This is probably not without impact on employee turnover, as the chances for burnout may increase two times for employees with the high workload, [13], and those who experience burnout may have nearly two times higher odds of leaving [14]. Both phenomena – employee turnover contagion, and employee workload – can be studied by using the data describing employee work-related interactions and communication patterns. Studies show, that using this kind of data can be successful in predicting employee turnover [8,9,18,25], but they have some limitations. Most studies in this field were conducted on data gathered with questionnaire surveys [8,9,18], which by nature are subjective and may contain biases. What might be even more important, the used data describes work-related interactions and communication patterns before the pandemic [8,9,18,25] when the work interactions were more stable and predictable in time, while current workplace experiences are changing very fast because of switching work models triggered by the pandemic (e.g., remote, hybrid, office), and other related factors (e.g., higher workforce mobility). In the current work, for the prediction of employee turnover we use longitudinal data describing employees work-related e-mail interactions from 28 months that cover the experiences of several thousand employees during the pandemic.

The rest of the paper is organized as follows. In Sect. 2 available data and their coding are described. Section 3 presents the results of experiments on a novel method of data selection based on a glial controller. Section 4 presents the application and experiments of recurrent neural networks for emploee turnover prediction and Sect. 5 concludes the paper.

2 Data and Model

To solve the problem described in Sect. 1, we implemented a system which architecture is shown in Fig. 1. The key element is the module responsible for collecting and analyzing data from various sources, including Google collaboration tools (Google Workspace).

The applicable legal provisions have forced the anonymization of the collected data. To this end, we used a hash function generated by the hash-based message authentication code algorithm (HMAC) [1]. The data connector continuously collects the following values:

1. ID of the message recipient (if the account is in the company's domain),
2. ID of the message sender (if the account operates in the company's domain),
3. Date the message was sent / received.

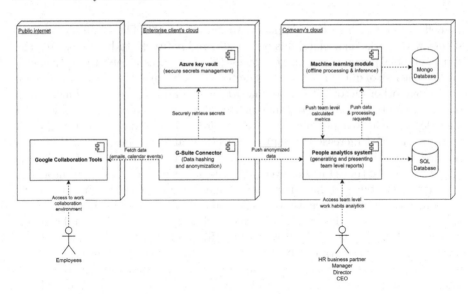

Fig. 1. The idea of the system.

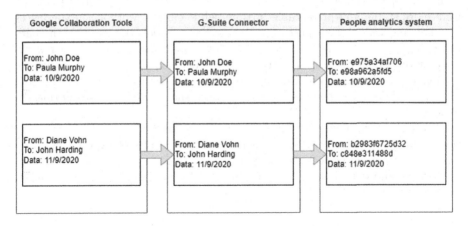

Fig. 2. Data flow through system components.

The Machine Learning module defined in Fig. 1 was built on the basis of a computing unit consisting of an I7 770K processor, 48GB RAM and four Nvidia 1080TI GPU cards. The Python environment in conjunction with the Pytorch 1.10 library was used for the calculations. Training of the machine learning models, described later in this article, took place offline. However, the final solution created is able to classify the input data in near real time.

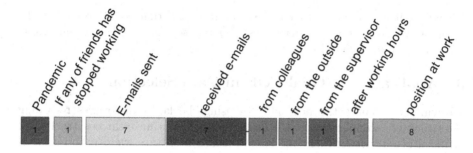

Fig. 3. Input data format.

2.1 Data Coding for Neural Networks

We used three neural network models: dense (MLP) networks, LSTM recurrent networks and networks based on GRU units. Figure 3 shows the features of the input vector. The output of particular types of networks were the values constituting the labels of one of the two classes (the employee will leave the job or the employee will stay in the team). Ultimately, the input vector consisted of 28 integer values. On the basis of previously collected information: ID of the recipient of the email message, ID of the sender of that message and the dates of its sending or receipt, the information in individual training vectors was supplemented with subsequent numbers. Including the first input stores a bit indicating a pandemic period or time without COVID virus, the next input, also of a bit-bit character, refers to information whether any of the persons with whom the analyzed employee exchanged messages has left work in the recent period. The next seven numbers contained the following information: how many messages were sent on each day of the week (broken down into monthly periods), i.e., how many messages someone sent on Mondays, Tuesdays, Wednesdays, etc. The next seven integers are the number of received messages with a division into days of the week (analogous to the previous seven inputs). Features 17 and 18 contained information, respectively, how many emails came from local traffic within other company associates and messages outside of the company. The 19th value included information on how many emails the employee received from the immediate supervisor. Input number 20 determines how many messages the current employee received outside working hours, i.e., in the hours (5:00 p.m. - 6:00 a.m.). The last, one-hot encoded value indicates the position held by the employee. Eight different positions were distinguished in the research.

The data described in this way was provided after appropriate transformation into three different deep learning models. The first of them was a dense network extended by the original concept of glial networks. For this case, the input data did not require any reorganization. In order to supply the above-mentioned features to the inputs of LSTM and GRU recurrent networks, they have been divided into predefined time intervals, i.e., the interval lasting one month. The last six months of work have been extracted for each user. Unfortunately, in some cases the database was incomplete and contained a smaller number of months.

Therefore, it was assumed that the minimum period that could be used in the training process is from 3 to 6 months. Our dataset contained data concerning sent emails from 28 months.

3   Application of Glial Cells in Data Selection

We will try to check the possibility of using the latest discoveries regarding the structure and operation of the human brain to optimize the construction of artificial neural networks. In the latest medical literature, we can find a lot of information about the impact of the so-called glial cells for building connections between neurons in the brain and for the flow of information. It should also be noted that in the last two years a lot of research has been undertaken in the field of the use of glial cells in the treatment of oncology, cardiology and dentistry. The pioneers in this area are Polish doctors from Wrocław, who, based on their knowledge of glial cells, reconstructed the severed spinal cord [24]. How important is the influence of these cells on the thought process can be found in the book "The other brain" by R. Douglas Fields [10]. Based on this information, we made an attempt to create a new neural network architecture, the purpose of which will be to supervise the process of building and training neural networks. This approach is analogous to the existing medical knowledge regarding the way the human brain works. It clearly shows that glial cells protect neuronal cells from damage and are responsible for the construction of connections between them. Therefore, new structures of neural networks have been developed, which in the learning process, by regulating the weight values, will turn on or off the connection data between neurons, and thus eliminate unnecessary neurons or groups of neurons. At this stage of knowledge, glial networks can be combined with dense networks and convolutional networks. Therefore, to determine the necessary parameters, a glial network connected to a deep network consisting of dense layers was used.

In this section, the concept of using glial cells will be presented to identify the features that have the greatest impact on the classification of employees, in accordance with the criterion defined above. For this purpose, we use the convolutional network structure from [7]. The diagram of the original network is presented in Fig. 5. The presented division into blocks is intended to indicate the places where glial cells were added (Fig. 5). The proposed concept assumes that the convolutional glial network consists of two parts (two convolutional networks): the so-called "Glia controller" (Fig. 5 symbol Glial) and CNN structures, each receiving the same input in the process of operation. The connection of the glia controller with the CNN is done by multiplying the output signal from the convolution block by the output from the controller dedicated to the block.

Training such a defined structure required the development of a dedicated learning method. On the basis of numerous simulations, it has been found that in order to obtain optimal results, the process should be the alternate training the CNN block and the "glial controller". In the presented model, two activation functions were used: sigmoidal (values in the range ¡0.1¿) for the glial outputs and ReLU in other cases.

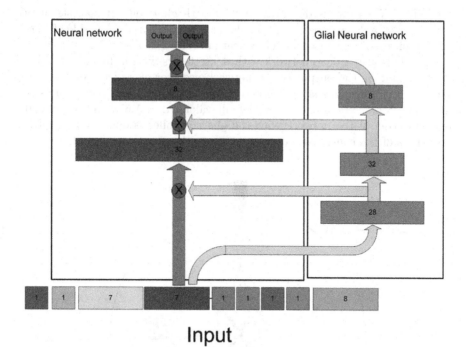

Fig. 4. Glial network to determine the significance of the network parameters.

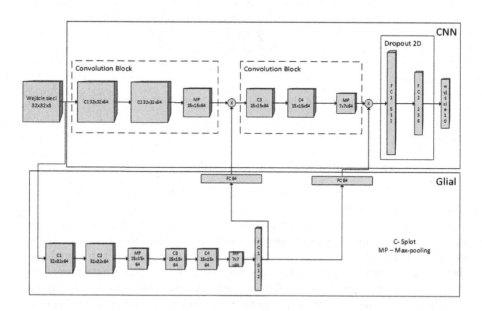

Fig. 5. Diagram of the propagation of the glial convolutional network signal.

Figure 4 shows the used neural network with glial cells. It consists of two deep nets with a dense layer. The first classic is marked on the left side of Fig. 4. It is a typical solution used for classification. The glia controller is drawn on the right. It is also a deep network with dense layers with a dimension at each layer the same as the output of the network used for classification. Moreover, each layer has a dropout layer [22] and uses SELU [16]. This function uses two additional parameters and we their default values: $\alpha = 1.673$ and $\lambda = 1.0507$. The connection of the network with the glia controller is done by multiplying the outputs of the controller with the inputs to the neural network, Fig. 6.

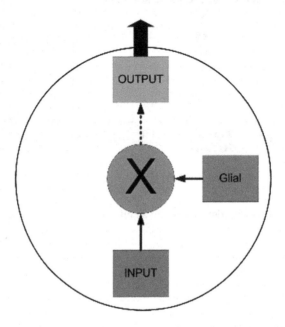

Fig. 6. Connection of the neural network with the glia controller.

3.1   Network Training with Glia Controller

The presented structure requires a non-standard method of network training. The training of this solution takes place alternately, which means that during one epoch only the glial part or the neural network is trained for classification. In the first stage, all weights are selected at random. Then the weight are frozen in the glial controller. They take part in the propagation of the signal, but their weights are not modified at this stage. Then the weights are unfreezed and the weights of the neural network are frozen. At this stage, we train only the glia controller, analogically to the previous situation, the weights of the neural network take part in the propagation of the signal but are not modified. The dropout layer also plays an important role in the learning process. For the first 20 epochs, this value is set to 0.3, which means that 30% of randomly selected signals from neurons are impermeable to the network. The number of epochs and the DropOut coefficient were selected empirically by repeating experiments. After exceeding the given number of epochs, DropOut is turned off and it is completely replaced by the glia driver. The network with DropOut turned off is trained for the next 20 epochs and then the structure training is finished. Throughout the learning process, the learning coefficient was 0.001 and momentum 0.9.

3.2   Selection of Input Parameters Using the Glial Network

When examining new, never-described data, one can encounter the problem of selecting the appropriate training data for the neural network. Researchers have repeatedly encountered the problem of irrelevance of data in machine learning [17] and there are many algorithms based on statistical methods for selecting training data. In the presented solution, we want the decision on the significance of the data to be made by the neural network itself, and more precisely by the glial controller. To check which features are less important for the proper operation of neural networks, we pass the signal through the neural network and check the output values of the glial cells in the first layer. For each of the 28 values it is possible to check when the feature was active or not. Figure 7 shows the average output values of all samples from the test dataset. On the basis of the presented values, we can determine the inputs which are considered less important in the classification process by the glial controller. Based on the results, it can be concluded that the 4th day of the messages sent and the three types of jobs were not significant in each case. The result of the classification of the entire network, depending on the experiment, ranged from 64% to 66% of correctly classified samples on the test dataset. In the next experiment we compare the use of all data and those selected by the glial controller.

4   Classification with the Use of Recursive Networks

The data on which the research was carried out has been aggregated at the level of individual months. In the case of data with a time distribution, it is worth using the approach that allows you to provide data for a neural network with a division into the time in which they are recorded. Thanks to the recurrent network, it will be possible to apply such a solution. In this section we use the LSTM-Long Short Term Memory cells [20] and GRU Gated Recurrent Unit [12,27]. A variable number of calls approach was used depending on the amount of data available for each user. In the collected data and also in the future in practical use, the HR service will be able to assess after three months of work whether there is a probability that the employee quits. Four approaches to the input data were applied; in each case, the same recurrent cell with the same number of hidden neurons was used:

1. The first one uses all the data that was used to train the network with the glial controller,

input	training 1	training 2	training 3	training 1	training 2	training 3
1	0,998816	0,999033	0,999342	1	1	1
2	0,999393	0,027302	0,999682	1	0	1
3	0,571981	0,982403	0,57816	1	1	1
4	0,708754	0,998963	0,724498	1	1	1
5	0,054335	0,196468	0,042862	0	0	0
6	0,996432	0,998777	0,997814	1	1	1
7	0,999266	0,998938	0,999609	1	1	1
8	0,997048	0,854333	0,998221	1	1	1
9	0,999471	0,999393	0,999726	1	1	1
10	0,996491	0,212694	0,997852	1	1	1
11	0,9993	0,99957	0,999628	1	1	1
12	0,995443	0,887671	0,997146	1	1	1
13	0,112142	0,99931	0,095386	0	1	0
14	0,999139	0,999715	0,999534	1	1	1
15	0,998926	0,982794	0,999408	1	1	1
16	0,999141	0,090876	0,999536	1	0	1
17	0,999644	0,998452	0,999822	1	1	1
18	0,099733	0,999139	0,083785	0	1	0
19	0,999388	0,998926	0,999679	1	1	1
20	0,998435	0,999141	0,999108	1	1	1
21	0,989673	0,982794	0,99304	1	1	1
22	0,991325	0,854333	0,994244	1	1	1
23	0,012718	0,212142	0,094978	0	1	0
24	0,087887	0,112142	0,100309	0	0	0
25	0,095818	0,012142	0,096781	0	0	0
26	0,116549	0,087858	0,072131	0	0	0
27	0,930017	0,930017	0,943347	1	1	1
28	0,999397	0,999397	0,999684	1	1	1

Fig. 7. Output values from the glial controller.

2. The second one only uses the data that the glial controller considered relevant in each variant,
3. The third glial controller rejected input in all cases,
4. The fourth glial controller considered data relevant in at least two cases.

Tables 1 and 2 have averaged the values after repeating each experiment ten times.

Based on the data collected in Tables 1 and 2, we can observe a minimal difference in favor of the solution with a smaller number of input features. The biggest difference is when learning the entire neural network. Thanks to the selection of unique features, the learning process can be shortened in the case of LSTM by five minutes, and in the GRU variant by three minutes. This time is counted the entire network training together with data loading. The Nvidia

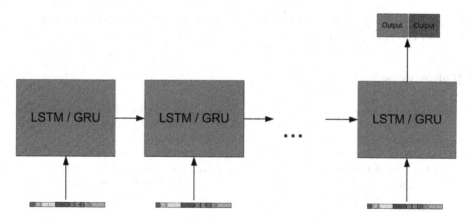

Fig. 8. A recurrent network used for classification.

Table 1. Comparison of the classification quality of the LSTM recurrent cells

Approach	Input	LSTM accuracy	LSTM precision	LSTM recall	Training time [min]
1	Input 28	0.741	0.666	0.784	26
2	Input 24	0.736	0.650	0.786	24
3	Input 19	0.746	0.663	0.796	21
4	**Input 21**	**0.749**	**0.714**	**0.785**	**21**

Table 2. Comparing the quality of the GRU recurrent cell classification

Approach	Input	GRU accuracy	GRU precision	GRU recall	Training time [min]
1	Input 28	0.732	0.685	0.756	25
2	Input 24	0.733	0.686	0.757	23
3	**Input 19**	**0.743**	**0.686**	**0.774**	**22**
4	Input 21	0.741	0.703	0.761	22

1080TI GPU was used for training. A constant learning coefficient for the whole process was used was 0.001 and the momentum value was 0.9.

5   Conclusions

The selection of parameters for training neural networks is a very difficult task. This problem is all the more difficult for analysts when dealing with completely new data that has never been studied in the past. By using the glial network, it is possible to establish sufficient parameters for more efficient training. Thus, it allows to improve the quality of neural network models and shorten the learning time of ready-made models. Attention should also be paid to the main research problem addressed by the article, i.e. the detection of redundancies among employees. Thanks to the use of the recurrent neural network, it has become possible to detect over 74% of cases whether an employee will work or leave in the next month.

References

1. Bellare, M., Canetti, R., Krawczyk, H.: Keying hash functions for message authentication. In: Koblitz, N. (ed.) CRYPTO 1996. LNCS, vol. 1109, pp. 1–15. Springer, Heidelberg (1996). https://doi.org/10.1007/3-540-68697-5_1

2. Bureau of Labor Statistics: Job openings and labor turnover summary (2022). https://www.bls.gov/news.release/jolts.nr0.htm. (Accessed 28 Mar 2022)

3. Cacioppo, J.T., Fowler, J.H., Christakis, N.A.: Alone in the crowd: the structure and spread of loneliness in a large social network. J. Pers. Soc. Psychol. **97**(6), 977 (2009)

4. Christakis, N.A., Fowler, J.H.: The spread of obesity in a large social network over 32 years. N. Engl. J. Med. **357**(4), 370–379 (2007)

5. Christakis, N.A., Fowler, J.H.: The collective dynamics of smoking in a large social network. N. Engl. J. Med. **358**(21), 2249–2258 (2008)

6. Cohen, A.: How to quit your job in the great post-pandemic resignation boom. Bloomberg Businessweek (2021)

7. Dosovitskiy, A., Springenberg, J.T., Riedmiller, M., Brox, T.: Discriminative unsupervised feature learning with convolutional neural networks. In: Advances in Neural Information Processing Systems 27 (2014)

8. Feeley, T.H., Barnett, G.A.: Predicting employee turnover from communication networks. Hum. Commun. Res. **23**(3), 370–387 (1997)

9. Feeley, T.H., Hwang, J., Barnett, G.A.: Predicting employee turnover from friendship networks. J. Appl. Commun. Res. **36**(1), 56–73 (2008)

10. Fields, R.D.: The other brain: From dementia to schizophrenia, how new discoveries about the brain are revolutionizing medicine and science. Simon and Schuster (2009)

11. Fowler, J.H., Christakis, N.A.: Dynamic spread of happiness in a large social network: longitudinal analysis over 20 years in the framingham heart study. Bmj **337** (2008)

12. Fu, R., Zhang, Z., Li, L.: Using lstm and gru neural network methods for traffic flow prediction. In: 2016 31st Youth Academic Annual Conference of Chinese Association of Automation (YAC), pp. 324–328. IEEE (2016)

13. Gallup: How to prevent employee burnout (2021). https://www.gallup.com/workplace/313160/preventing-and-dealing-with-employee-burnout.aspx. (Accessed 28 Mar 2022)

14. Hamidi, M.S., et al.: Estimating institutional physician turnover attributable to self-reported burnout and associated financial burden: a case study. BMC Health Serv. Res. **18**(1), 1–8 (2018)

15. Jankowski, J., Ciuberek, S., Zbieg, A., Michalski, R.: Studying paths of participation in viral diffusion process. In: Aberer, K., Flache, A., Jager, W., Liu, L., Tang, J., Guéret, C. (eds.) SocInfo 2012. LNCS, vol. 7710, pp. 503–516. Springer, Heidelberg (2012). https://doi.org/10.1007/978-3-642-35386-4_37

16. Klambauer, G., Unterthiner, T., Mayr, A., Hochreiter, S.: Self-normalizing neural networks. In: Advances in Neural Information Processing Systems 30 (2017)

17. Mikołajczyk, A., Grochowski, M., Kwasigroch, A.: Towards explainable classifiers using the counterfactual approach - global explanations for discovering bias in data. J. Artif. Intell. Soft Comput. Res. **11**(1), 51–67 (2021). https://doi.org/10.2478/jaiscr-2021-0004

18. Mossholder, K.W., Settoon, R.P., Henagan, S.C.: A relational perspective on turnover: Examining structural, attitudinal, and behavioral predictors. Acad. Manag. J. **48**(4), 607–618 (2005)

19. Porter, C.M., Rigby, J.R.: The turnover contagion process: An integrative review of theoretical and empirical research. J. Organ. Behav. **42**(2), 212–228 (2021)

20. Sak, H., Senior, A., Beaufays, F.: Long short-term memory based recurrent neural network architectures for large vocabulary speech recognition. arXiv preprint arXiv:1402.1128 (2014)

21. Spataro, J.: A pulse on employees' wellbeing, six months into the pandemic. Retrieved from Microsoft Work Trend Index (2020). https://www.microsoft.com/en

22. Srivastava, N., Hinton, G., Krizhevsky, A., Sutskever, I., Salakhutdinov, R.: Dropout: a simple way to prevent neural networks from overfitting. J. Mach. Learn. Res. **15**(1), 1929–1958 (2014)

23. Sunder, S., Kumar, V., Goreczny, A., Maurer, T.: Why do salespeople quit? an empirical examination of own and peer effects on salesperson turnover behavior. J. Mark. Res. **54**(3), 381–397 (2017)

24. Tabakow, P., Weiser, A., Chmielak, K., Blauciak, P., Bladowska, J., Czyz, M.: Navigated neuroendoscopy combined with intraoperative magnetic resonance cysternography for treatment of arachnoid cysts. Neurosurg. Rev. **43**(4), 1151–1161 (2020)

25. Yuan, J., et al.: Promotion and resignation in employee networks. Phys. A **444**, 442–447 (2016)

26. Zbieg, A., Zak, B., Jankowski, J., Michalski, R., Ciuberek, S.: Studying diffusion of viral content at dyadic level. In: 2012 IEEE/ACM International Conference on Advances in Social Networks Analysis and Mining, pp. 1259–1265. IEEE (2012)

27. Zini, J.E., Rizk, Y., Awad, M.: An optimized parallel implementation of non-iteratively trained recurrent neural networks. J. Artifi. Intell. Soft Comput. Res. **11**(1), 33–50 (2021). https://doi.org/10.2478/jaiscr-2021-0003

Buggy Pinball: A Novel Single-point Meta-heuristic for Global Continuous Optimization

Vasileios Lymperakis[1]($\boxtimes$) and Athanasios Aris Panagopoulos[2]

[1] Technical University of Crete, Chania, Greece
vasilis@lyberakis.gr
[2] California State University, Fresno, CA, USA

Abstract. In this work, we propose a fundamentally novel single-point meta-heuristic designed for continuous optimization. Our algorithm continuously improves on a solution via a trajectory-based search inspired by the pinball arcade game in an anytime optimization manner. We evaluate our algorithm against widely employed meta-heuristics on several standard test-bed functions and various dimensions. Our algorithm exhibits high precision, and superior accuracy compared to the benchmark, especially when complex configuration spaces are considered.

1 Introduction

Continuous optimization problems rely on optimization variables that draw their values from a non-countable set—typically a range of real numbers [20]. This is in contrast to discrete (combinatorial) optimization where the optimization variables draw values from a countable set. Many problems in various domains can be formulated and tackled as continuous optimization tasks. These range from image processing [4], and chemical engineering [27] to finance [24], and biology [26]. Additionally, machine learning techniques heavily rely on continuous optimization to optimize model parameters in order to, typically, minimize an error function [22]. Continuous optimization has drawn considerable attention [13].

Continuous optimization methods can be classified into either local or global ones. Local methods, such as naive Gradient Descent and Continuous Hill Climbing, move locally over the optimization space and aim to precisely locate the locally optimal solution. As such, they tend to be quite fast and have been used widely in numerous applications (e.g.,[3,12]). However, despite the aforementioned advantages, such methods converge to local optima when operating on non-convex configuration spaces. On the other hand, global methods aim to find the global solution, typically by moving in a less restricted manner over the optimization space. Such approaches range from meta-heuristics to modified local search, such as gradient descent using momentum [21] and adaptive subgradient methods such as Adagrad [6]. In the absence of analytical solutions, such methods approximate the global optimum—in contrast to local methods

© The Author(s), under exclusive license to Springer Nature Switzerland AG 2023
L. Rutkowski et al. (Eds.): ICAISC 2022, LNAI 13589, pp. 264–276, 2023.
https://doi.org/10.1007/978-3-031-23480-4_22

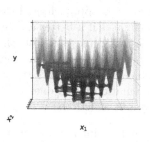

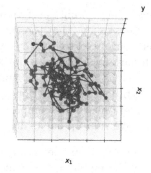

Fig. 1. Search trajectory (blue line) and coalitions (red dots) of BP on Rastrigin 3D (Color figure online)

that precisely locate a local one. Given that the objective space is usually non-convex and that one is typically interested in an approximation of the optimal solution, global methods are of great interest and widely used in practice [2].

Meta-heuristics have long been used for global continuous optimization tasks (e.g.,[5,25]). They can avoid convergence to local optima and scale to multidimensional problems, while typically not requiring a derivative of the objective function, which is a restrictive requirement of gradient-based optimization. Meta-heuristics can be broadly classified into single-point or population-based ones. The former focus on improving a single solution, while the latter on improving a collection of points based on population characteristics. Single-point meta-heuristics are generally regarded as less-fit for continuous optimization and related research is limited when compared to population-based approaches. Most single-point meta-heuristics—such as simulated annealing (SA) [15] and threshold accepting (TA) [7]—have been developed for discrete optimization problems. As such, their deployment in continuous optimization often requires non-straightforward tuning. On the other hand, population-based approaches have received considerably more attention for continuous optimization. Many such approaches have been proposed for continuous optimization over the past years, such as grey wolf optimizer [18], whale optimization [17] and particle swarm optimization (PSO) [14]. Such approaches typically demonstrate fast convergence and an ability to converge to the global optima with high precision. Nevertheless, they can still fall into local optima especially in complex configuration spaces.

Against this background, we propose a fundamentally novel single-point meta-heuristic algorithm, namely *buggy pinball* (BP), designed specifically for global continuous optimization. Our algorithm is inspired by the movement of a ball's collision and descent in the well-known pinball arcade game. Importantly, BP, is an *anytime* algorithm: it always improves over the solution while ensuring exploration. We evaluate our approach against three well-known single-point and population-based meta-heuristics on several commonly employed optimization test-bed functions and a number of dimensions. We show that BP is able to find the global optima with high accuracy and precision and in a shorter time than the benchmark. Especially, when more complex functions are considered, we show consistency on high performance, unlike any other meta-heuristic in our experiments. We believe that the superiority of our approach, and the fact that a solution is guaranteed to improve over time makes it a better choice compared to the benchmark for continuous optimization tasks, and especially those of high complexity, such as the ones that typically emerge in machine learning optimization tasks. Notably, BP's superior performance also highlights the potential of single-point meta-heuristics compared to population-based ones for continuous optimization. We sum up our main contributions as follows:

- We propose BP, a fundamentally novel single-point anytime meta-heuristic, inspired by the pinball game, which is tailored for continuous optimization.
- We experimentally show that BP ensures exploration without having to accept worse solutions, which is a common practice in meta-heuristic search.
- We evaluate our approach against widely employed meta-heuristics, on several test-bed functions and a number of dimensions.
- We show that BP performs better than both the single-point and population-based approaches considered, especially in complex configuration spaces—which proves the efficiency of our novel way of searching.

The rest of the paper is structured as follows. We first discuss background material and related work. Then, we detail our approach and discuss core motivational aspects. Subsequently, we conduct a systematic evaluation, discuss the evaluation results, and finally, conclude and present directions for future work.

2   The Buggy Pinball (BP) Algorithm

Our work is motivated by the ball's movement in the pinball game. In this game, a ball is thrown to the highest point, and by moving inside a glass-covered cabinet, it heads towards the lowest point. The player uses paddles to evade the ball from falling at the lowest point and collects points by hitting various targets. The main challenge originates from the multiple collisions of the ball, which eventually lead the ball to the lowest point. We imitated this movement by creating a trajectory-based search method, where the "ball" is moving until a collision with the objective function takes place, which occurs at the common point between the objective function and the trajectory segment—see Fig. 1. The trajectory segments start almost horizontally and become steeper with time. A trajectory segment corresponds to one round of our search algorithm. The intuition behind this movement is that when the ball is moving almost horizontally, the probability of getting into a local optimum is small, as shown in Fig. 2a, while steeper segments speed up convergence to an optimum as time progresses. Anytime algorithms are algorithms that increase the quality of the output as time progresses [10]. In BP, the trajectory segments that are progressively created, direct the search towards values that can only better optimize the cost axis. So, every new point that is detected, is guaranteed to be better than the current. Thus, Buggy Pinball is anytime, since every new segment can only provide a better solution.

(a) Trajectory Movement   (b) Entering local minimum (c) Avoiding local minimum

Fig. 2. Description of trajectory movement

Why Buggy? Even though there is a chance of getting into a local optimum solution, it is existent. To significantly decrease the possibility of converging to a local optimum, we identified the need for the pinball game to be "buggy". Instead of bouncing away, as in the real pinball, the "ball" in our algorithm is capable of continuing the search underneath the configuration space. This way, not only are local optima effectively escaped, but there is also no need to accept worse solutions to ensure exploration, unlike most meta-heuristic approaches. Thus, BP is an any-time algorithm. An illustration of being trapped into a local minimum is shown in Fig. 2b. In Fig. 2c, we can also see how BP creates trajectories underneath the configuration space and avoids convergence to a local minimum. This behavior allows BP to reduce significantly the probability of getting stuck into local optima. The BP effectiveness in avoiding local optima is verified empirically by its increased accuracy, demonstrated in our results.

BP Overview. As discussed, BP searches the space by creating trajectories. These progress via steps in a continuous simulation manner until a collision between the function and the ball occurs. Once we find that a collision occurs between two steps, a routine to precisely locate the point of collision begins. When the exact point is identified, we create a new trajectory segment starting from that point and repeat the procedure. The trajectories are created with a random direction to ensure that the configuration space is searched adequately. The elevation angle though, (i.e., the angle of descent or ascent of the segment, depending on whether we minimize or maximize a function) follows a predefined schedule: it starts at an almost horizontal level, in order to avoid local optima and becomes steeper over time to achieve faster convergence rates. By contrast, the step size is reduced over time, to achieve higher precision. As we get to bigger configuration spaces, we choose higher starting step values and smaller elevation angles. As mentioned, the trajectories are even able to advance the ball underneath the configuration space of the function. An example of a BP search on the 3D version of the Rastrigin function (Fig. 1.h in the Appendix) can be seen in Fig. 1. Each blue line represents one trajectory segment, while each red point is the common point between the function and the segment.

The BP in Detail. BP (Algorithm 1) is composed of five main parts: (1) initialization, (2) trajectory segment creation, (3) stepping forward, (4) recursive refining, and (5) cooling schedules. We note that each round corresponds to one trajectory segment, which progresses in steps. Also, an objective function is characterized by its cost (i.e. the y values) and its variables (i.e. the x_i values). Every x_i is an axis in the configuration space (see Fig. 1 for an example with two x_i variables).

Initialization. The first part (i.e., Algorithm 1, lines 1-2) is that of initializing our hyper-parameters. First, we set the original step and elevation angle, a. The original step size should be set in such a way, that we do not make too large steps in the configuration space at the beginning of the process. We have empirically found that the choice of an original step size at ~10% of the average variable range performs well (in general, the step size should be bigger as the configuration space becomes bigger). We note here that the sign of the step (variable *stepSize* in Algorithm 1) should be negative for minimization tasks and positive for maximization.

The elevation angle is defined with respect to the plane perpendicular to the y axis. For the elevation angle we always begin with a value close to 0 (e.g., 0.1) in order to perform a near-horizontal movement.[1] We have also identified that its value should be smaller in large spaces (in contrast to that for step) since a more horizontal movement is required, to effectively avoid local optima. The same "smaller values" rule applies as the number of dimensions increases.

In line 1, we also set the number of rounds, which corresponds to the number of trajectory segments to be created as well as the number of steps to be executed in each segment. The number of rounds should be set as high as possible, considering the optimization time constraints. With respect to the number of steps, we have found that a reasonable choice is one such as the product of the number of steps and the step size is three to five times bigger than the average variable range. Finally, the initialization of a random starting point takes place in line 2, and the procedure of the algorithm begins from line 3. As BP is able to escape local optima the algorithm is not very sensitive on the initial points selected regarding the convergence to a global solution. Nevertheless, a convenient initial solution can still speed up the algorithm, as further discussed in Results.

Trajectory Segment Creation. Each round corresponds to the creation and execution of a trajectory segment, starting from the current point and ending when the maximum number of steps is reached or a collision is detected. At the beginning of each round, the segment's direction is set randomly,[2] in order to ensure the best exploration of the configuration space with only the elevation angle being fixed and following a predetermined schedule. Thus, the step component for each variable is set randomly to a value within [-1, 1]. The step-towards-optimum component for the y axis that respects the elevation angle can be computed as:

$$y_{step} = \sqrt{\frac{\sin^2 a \sum_{j=1}^{d} x_{step,j}^2}{1-\sin^2 a}}$$ where d is the number of variables—i.e., the problem's

---

[1] The sign of the elevation angle is irrelevant as the square of its sin value is considered.

[2] We experimented with trajectories alternating from one direction to another; however we found this took a toll in exploration. Thus, the algorithm does not behave exactly like a pinball, however, the final trajectories do resemble a pinball movement.

Algorithm 1. Buggy Pinball

1: set $stepSize = step_{max}$; $a = a_{min}$; $\#rounds, \#steps$
2: $\boldsymbol{x}, y \leftarrow$ initialize randomly
3: **while** i in $\#rounds$ **do**
4:     $\boldsymbol{x}_{step} = \text{random}(-1,1)$
5:     $y_{step} = \sqrt{\dfrac{\sin^2 a \sum_{j=1}^{d} x_{step,j}^2}{1-\sin^2(a)}}$
6:     $\boldsymbol{x}_{step} = z\boldsymbol{x}_{step}$; $y_{step} = zy_{step}$
7:     **while** j in $\#steps$ **do**
8:        **if** $\text{crossing\_detected}(\boldsymbol{x}, y, \boldsymbol{x}_{step}, y_{step}, j)$ **then**
9:           $\boldsymbol{x}, y = \text{recursive\_refining}(\boldsymbol{x}, y, \boldsymbol{x}_{step}, y_{step}, j)$
10:           **break**
11:     $a =\text{elevation\_cooling}(a_{min}, i, \#rounds)$
12:     $stepSize =\text{stepSize\_cooling}(step_{max}, i, \#rounds)$
13: **return** $\boldsymbol{x}, y$

Algorithm 2. crossing_detected $(\boldsymbol{x}, y, \boldsymbol{x}_{step}, y_{step}, j)$

1: $A = y + (j-1)y_{step} - f(\boldsymbol{x} + (j-1)\boldsymbol{x}_{step})$
2: $B = y + jy_{step} - f(\boldsymbol{x} + j\boldsymbol{x}_{step})$
3: **return** $(A > 0 \wedge B < 0) \vee (A < 0 \wedge B > 0)$

---

dimensions. It is easy to derive this equation with the following procedure. We know for the elevation angle that: $\sin a = \frac{|\mathbf{nu}|}{|\mathbf{n}||\mathbf{u}|}$, where n is the vector perpendicular to the fundamental plane, i.e. parallel to the y axis $(0, 0, ...0, 1)$. Vector u is the trajectory's segment vector $(x_0, x_1, ..., x_{d-1}, y)$. Values $x_0, x_1, ..., x_{d-1}$ and the elevation angle are known. Solving for y gives us the y_{step} equation above.

With the above procedure, we have set the direction of the segment. What remains is to readjust the dimension-wise step components to respect the predetermined overall step size. In order to achieve this, we multiply all step components with a common factor, z, calculated as $z = \frac{stepSize}{\sqrt{x_0^2+x_1^2...+x_{d-1}^2+y_{step}^2}}$ (used in line 6 of Algorithm 1). Once we have completed this procedure, our step for the current round is ready. We then apply it for the number of steps stated or until a crossing of the objective function is detected.[3]

Stepping Forward. In this part (i.e. Algorithm 1, lines 7-10 and Algorithm 2 and 3), we start our trajectory segment search by applying the step's values on each axis, and we continue until one of the two conditions of stopping is met. As the segment proceeds in the configuration space, it moves downwards concerning the y axis, as the segment value of y decreases in every step. This results in accepting only better values (i.e. closer to the global optimum) of the current position. The crossing of the objective function by the segment is determined by checking the last two steps taken. As shown in the crossing detected function (Algorithm 2), if

---

[3] Crossing of the objective function means that a common point of the objective function and the current trajectory segment has been detected.

Algorithm 3. recursive_refining $(x, y, x_{step}, y_{step}, j)$

1: $x = x + x_{step}j$; $y = y + y_{step}j$
2: **if** $y - f(x) \approx 0$ **then**
3:     **return** $x, f(x)$
4: **else**
5:     $x_{step} = \frac{x_{step}}{2}$; $y_{step} = \frac{y_{step}}{2}$
6:     $x = x - x_{step}$; $y = y - y_{step}$
7:     **if** crossing_detected$(x, y, x_{step}, y_{step}, 0)$ **then**
8:         **return** recursive_refining$(x, y, x_{step}, y_{step}, 0)$
9:     **else**
10:         **return** recursive_refining$(x, y, x_{step}, y_{step}, 1)$

---

the difference between the y value and the function evaluation is of different sign between these steps, we know that a point on the objective function is "internal" to the last trajectory segment drawn—and "recursive refining" is triggered.

Recursive Refining. This is a process of iterative refinement, shown also in pseudo-code in Algorithm 3, used to locate the exact point where a trajectory segment crosses the objective function. It is activated only when a crossing of the configuration space is detected, otherwise, that part is skipped. In that case, a loop begins, where the point in the middle between the last two steps is examined, to determine whether this is the common point between the segment and the function (or a very good approximation). If it is not, then we choose whether we continue the loop on the upper or the lower half of the examined part, depending on where the crossing is identified, according to the signs of the points. We continue by examining the point in the middle of that part as before, and the same process is repeated until the exact location of the common point is found. Once we find this point, we stop the iteration of the current round, as we have reached the closest point to the global optimum so far.

Cooling Schedules. The last part takes place at the end of each round (i.e. Algorithm 1, lines 11–12). It determines the values of the desired step size and the angle for the upcoming round. It simply applies the cooling schedule function determined for each of the parameters. In our experiments, we used a simple linear cooling schedule, where the final values are a fraction of 1 for the step size, so we have increased precision no matter the structure of the problem, and from $0.1°$ for highly complex many-local-optima functions up to $89°$ for simple slope no-local-optima functions. This seems to work satisfactorily for each problem tested so far, but further research on the topic is desirable for future work.

3   Experiments

Background and Related Work. As discussed, in our experiments we consider various widely-employed meta-heuristics, both single-point and population-based,

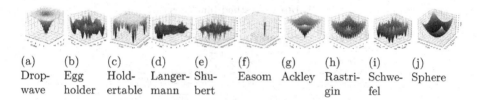

| (a) | (b) | (c) | (d) | (e) | (f) | (g) | (h) | (i) | (j) |
| Drop-wave | Egg holder | Hold-ertable | Langer-mann | Shu-bert | Easom | Ackley | Rastri-gin | Schwe-fel | Sphere |

Fig. 3. Objective functions' graphs

in order to thoroughly evaluate our approach. Simulated annealing, SA, [15] is a famous single-point, commonly employed meta-heuristic (e.g., [1,16]). Numerous SA variants have also been proposed [23]. Threshold accepting (TA) [7] is proposed as a variant of SA that is also receiving recent attention (e.g., [8,9]). One of the most famous population-based meta-heuristic approaches is particle swarm optimization, PSO [14], initially designed for continuous optimization.

To evaluate our approach, we compare BP against SA, TA, and PSO into the minimization of several benchmark optimization functions that are commonly employed in optimization evaluation [19], and on different dimensions. In more detail, the benchmark functions consider a range that spans from relatively simple uni-modal ones (i.e., Sphere and Easom) to more complex multi-modal ones (i.e., Rastrigin, Ackley, Eggholder, Schwefel, Shubert, Holdertable, Langermann, and Dropwave). As such, all algorithms are evaluated on a diverse range of optimization spaces. The function graphs and equations are reported in Fig. 3 and Table 1 of the Appendix, respectively, for concreteness. All benchmark functions are considered in three dimensions to support straightforward visualization of the results. The benchmark functions, which are also directly defined on a higher number of dimensions (i.e., Sphere, Rastrigin, Ackley, and Schwefel), have also been considered in two, four, five, and six dimensions. This range allowed us to evaluate the scalability of all approaches, while still ensuring fairness and statistical significance. To ensure statistical significance, each algorithm was executed for each function and dimension one hundred times.

In order to ensure a fair comparison, a predefined time allowance was selected and was made available for all approaches, while all experiments were run on the same machine (a 40-CPU Intel(R) Xeon(R) CPU E5-2680 v2 @ 2.80 GHz processor, with 64GB RAM). The predefined time allowance was selected to be one second, five seconds, one minute, five minutes and twenty minutes for two, three, four, five and six dimensions respectively. This time allowance was selected before the experiments were executed to avoid favoring any approach.

Ensuring fairness is a particularly prominent challenge when comparing different meta-heuristics, as they operate differently and typically depend on a number of different hyper-parameters [11]. A fair comparison should consider the "same level" of optimization/calibration with respect to the hyper-parameters for all meta-heuristics considered (e.g., the cooling schedule for SA and TA). In order to ensure fairness among all approaches, we performed a thorough exhaustive grid search—for *each algorithm, on every benchmark function, and on every dimension*—to identify the best hyper-parameters for every setting (i.e., the

Table 1. Testbed functions

Function	Equation
Dropwave	$-\dfrac{1+\cos\left(12\sqrt{x_1^2+x_2^2}\right)}{0.5(x_1^2+x_2^2)+2}$
Eggholder	$-(x_2+47)\sin\left(\sqrt{\left|x_2+\frac{x_1}{2}+47\right|}\right)$ $-x_1\sin\left(\sqrt{\left|x_1-x_2-47\right|}\right)$
Holdertable	$-\left\|\sin(x_1)\cos(x_2)e^{\left\|1-\frac{\sqrt{x_1^2+x_2^2}}{\pi}\right\|}\right\|$
Langermann*	$\sum_{i=1}^{5} c_i e^{-\frac{1}{\pi}\sum_{j=1}^{d}(x_j-A_{ij})^2}$ $\cos\left(\pi\sum_{j=1}^{d}(x_j-A_{ij})^2\right)$
Shubert	$\left(\sum_{i=1}^{5} i\cos\left((i+1)x_1+i\right)\right)$ $\left(\sum_{i=1}^{5} i\cos\left((i+1)x_2+i\right)\right)$
Easom	$-\cos(x_1)\cos(x_2)e^{-(x_1-\pi)^2-(x_2-\pi)^2}$
Ackley	$-20e^{-0.2\sqrt{\frac{1}{d}\sum_{i=1}^{d}x_i^2}}$ $-e^{\sqrt{\frac{1}{d}\sum_{i=1}^{d}\cos(2\pi x_i)}}+20+e^1$
Rastrigin	$10d+\sum_{i=1}^{d}(x_i^2-10\cos(2\pi x_i))$
Schwefel	$418.9829d-\sum_{i=1}^{d}x_i\sin\left(\sqrt{\left|x_i\right|}\right)$
Sphere	$\sum_{i=1}^{d}x_i^2$

*where $c=(1,2,5,2,3)$ and $A^T=\begin{pmatrix}3\ 5\ 2\ 1\ 7\\5\ 2\ 1\ 4\ 9\end{pmatrix}$

hyper-parameters that lead to the best performance within the predefined time allowance). The optimal hyper-parameters were used for our evaluations, ensuring a fair comparison among all approaches. The dimension range used in our evaluation enabled us to evaluate the scalability of all approaches while performing this demanding search in feasible time.

When comparing the performance of meta-heuristic approaches, it is crucial to use appropriate metrics. In this work, we use both a precision and accuracy. We calculate the accuracy percentage as the ratio of the trials, where the algorithm converged to an approximation of the global minimum over the total number of trials. We consider an algorithm to have converged to an approximation of the global minimum, if the solution discovered is better than the second-best (local) minimum. To evaluate precision, for those solutions that have converged to an approximation of the global optimum, we calculated the difference between that approximation and the global optimum itself. That is, we calculated the Mean Absolute Error (MAE) as: $MAE=\frac{\sum_{i=1}^{n}|y-x_i|}{n}$—where n is the number of trials, y the global minimum, and x_i the proposed solution on a given trial.

4   Results

Our results are shown in Table 2 for each function, algorithm, and dimension considered, with the best results in each occasion noted in bold. A higher accuracy

Table 2. Evaluation Results

Function		Accuracy (%)				Precision (MAE)			
		BP	SA	TA	PSO	BP	SA	TA	PSO
Dropwave	3D	**100%**	84%	**100%**	**100%**	1e-4	0.008	0.002	**1e-17**
Eggholder	3D	**100%**	73%	77%	48%	0.872	0.454	4.634	**1e-5**
Holdertable	3D	**100%**	**100%**	**100%**	56%	0.006	0.01	0.008	**1e-6**
Langermann	3D	**100%**	5%	59%	41%	2e-6	0.001	0.007	**2e-15**
Shubert	3D	**100%**	**100%**	**100%**	88%	0.032	0.021	0.274	**8e-6**
Easom	3D	**100%**	**100%**	**100%**	99%	0.002	0.001	7e-4	**5e-17**
Ackley	2D	**100%**	**100%**	**100%**	**100%**	2e-4	0.009	0.002	**4e-16**
	3D	**100%**	**100%**	**100%**	**100%**	7e-5	0.017	0.013	**4e-16**
	4D	**100%**	**100%**	**100%**	**100%**	**8e-5**	0.03	0.026	0.021
	5D	**100%**	**100%**	**100%**	**100%**	9e-5	0.044	0.07	**2e-15**
	6D	**100%**	**100%**	**100%**	**100%**	**3e-4**	0.062	0.115	0.063
Rastrigin	2D	**100%**	**100%**	**100%**	**100%**	7e-7	0.005	5e-4	**∼0**
	3D	**100%**	**100%**	**100%**	**100%**	8e-7	0.01	0.01	**∼0**
	4D	**100%**	**100%**	**100%**	99%	6e-7	0.035	0.044	**∼0**
	5D	**100%**	**100%**	**100%**	82%	7e-7	0.116	0.139	**∼0**
	6D	**100%**	**100%**	**100%**	46%	7e-7	0.331	0.36	**∼0**
Schwefel	2D	**100%**	95%	**100%**	95%	2e-4	6e-3	0.11	**1e-5**
	3D	**100%**	90%	93%	91%	3e-4	0.02	0.9	**2e-5**
	4D	**100%**	81%	91%	84%	3e-4	0.142	3.634	**3e-5**
	5D	**100%**	72%	86%	62%	4e-4	0.875	8.69	**3e-5**
	6D	**94%**	71%	77%	60%	3e-4	2.68	16.72	**4e-5**
Sphere	2D	**100%**	**100%**	**100%**	**100%**	0.036	5e-6	2e-12	**1e-117**
	3D	**100%**	**100%**	**100%**	**100%**	0.003	1e-6	2e-7	**4e-16**
	4D	**100%**	**100%**	**100%**	**100%**	8e-4	2e-5	9e-6	**∼0**
	5D	**100%**	**100%**	**100%**	**100%**	7e-4	1e-4	6e-5	**∼0**
	6D	**100%**	**100%**	**100%**	**100%**	7e-4	3e-4	2e-4	**∼0**

percentage indicates better performance (the algorithm discovers and approximates the global optimum more often compared to the rest of the algorithms evaluated), while lower MAE indicates better precision. An MAE of ∼0 is practically zero. Visual inspection indicated that the evaluation results are not normally distributed. A Shapiro Wilk Test with a p-value of 0.05 confirmed that we cannot assume a normal distribution. The statistical significance of all results is tested using a non-parametric Kruskal-Wallis H test and follow-up Conover's tests (along with the step-down method using Bonferroni adjustment for p-value adjustment). A p-value threshold of 0.05 is used for statistical significance. All statistical significance results are included in the Appendix.

As seen in Table 2, BP shows higher or equal accuracy rates compared to all other algorithms in all settings. The starting point for each trial and each algorithm is random. BP has no trouble approximating the global minimum from any possible starting position. Notably, it is the algorithm with the most times to achieve a 100% accuracy ratio. There are some cases, however, where all algorithms achieve almost 100% accuracy. This occurs for the "less complex" functions Ackley, Easom, Rastrigin, Schwefel (2D), Sphere, Holdertable, Shubert, and Dropwave, where all or almost all algorithms exhibit almost 100% accuracy (we elaborate below). The more "complex" functions are Schwefel (3D, 4D, 5D, 6D), Eggholder, Langermann. In those, BP always outperforms its competitors. BP reaches 100% accuracy in all these results, and its superiority to others is statistically significant, except results against TA for the Schwefel 3D, 4D cases.

Now, regarding precision, we clarify that it is calculated only for the points that have converged to an approximation of the global minimum. Thus, high precision demonstrates how well the global optimum is approximated, if the algorithm did not get stuck to a local optimum in the first place. As such, a *high precision-low accuracy* performance is *not* suitable for global optimization—since, although precise, the algorithm is not discovering the global optima often enough. That said, an adequate performance concerning precision is definitely required for global optimization algorithms. As can be seen, BP's precision is high; and it is higher than that of the other two single-point algorithms considered (SA, TA) in most cases. Follow-up tests confirm statistically significant better BP precision against SA and TA in all cases. The precision of BP seems to be lower than that of PSO (which is also developed for continuous optimization tasks), and follow-up tests indicate that this difference is statistically significant. Notably, however, PSO frequently has the worst accuracy among all algorithms, and thus is a poor choice for global optimization in these cases.

As noted already, the advantage of BP becomes greater when the more complex functions are considered, i.e., Eggholder, Schwefel and Langermann, and when we move to higher dimensions. These functions have many and deep local minima, but only a single global one. The Eggholder and Langermann are also non-symmetric. These facts make them harder to optimize in a global optimization manner, while not getting stuck in a local minimum. The higher dimensionality introduces further challenges for global optimization. That said, BP manages to achieve a 100% accuracy in all occasions except Schwefel for 6D (where an 94% is achieved). The accuracy results are also always better compared to the rest of the algorithms considered and the improvement ranges to up to 20 times better (i.e., compared to SA for Langermann 3D). When simpler functions and lower dimensions are considered the differences between the algorithms become less prominent. For instance, the Dropwave, Ackley, and Rastrigin have few local minima that are relatively shallow, while Shubert and Holdertable have many global optima. As such, most algorithms reach 100% accuracy except

SA in Dropwave 3D, PSO in Rastrigin 4D, 5D and 6D, and PSO in Holdertable 3D and Shubert 3D. Finally, when the unimodal functions, Sphere and Easom are considered, not surprisingly, all algorithms achieve a 100% accuracy.

5 Conclusions and Future Work

In this paper, we introduced a fundamentally novel single-point meta-heuristic tailored for global continuous optimization problems. Our algorithm, *buggy pinball*, is inspired by the pinball arcade game and is able to discover the global optimum in an any-time optimization manner. We evaluated our algorithm against widely-employed meta-heuristics on standard test-beds. We showed that it has a better performance compared to all benchmark approaches, especially when complex optimization functions and multiple dimensions are considered.

Future and ongoing work includes various extensions of the Buggy Pinball algorithm. For instance, investigating different cooling schedules for the step length and elevation angle parameters could improve the algorithm's effectiveness. Another valuable extension could be adapting BP for discrete optimization problems, to benefit from the perquisites of the algorithm in that domain as well.

References

1. Abdel-Basset, M., Ding, W., El-Shahat, D.: A hybrid harris hawks optimization algorithm with simulated annealing for feature selection. Artif. Intell. Rev. **54**(1), 593–637 (2021)
2. Ali, M.M., Khompatraporn, C., Zabinsky, Z.B.: A numerical evaluation of several stochastic algorithms on selected continuous global optimization test problems. J. Global Optim. **31**(4), 635–672 (2005)
3. Biehl, M., Schwarze, H.: Learning by on-line gradient descent. J. Phys. A: Math. Gen. **28**(3), 643 (1995)
4. Chambolle, A., Pock, T.: An introduction to continuous optimization for imaging. Acta Numer **25**, 161–319 (2016)
5. Dhouib, S., Kharrat, A., Chabchoub, H.: A multi-start threshold accepting algorithm for multiple objective continuous optimization problems. Int. J. Numer. Meth. Eng. **83**(11), 1498–1517 (2010)
6. Duchi, J., Hazan, E., Singer, Y.: Adaptive subgradient methods for online learning and stochastic optimization. J. Mach. Learn. Res. **12**(7) (2011)
7. Dueck, G., Scheuer, T.: Threshold accepting: A general purpose optimization algorithm appearing superior to simulated annealing. J. Comput. Phys. **90**(1), 161–175 (1990)
8. Frausto-Solis, J., Hernández-Ramírez, L., Castilla-Valdez, G., González-Barbosa, J.J., Sánchez-Hernández, J.P.: Chaotic multi-objective simulated annealing and threshold accepting for job shop scheduling problem. Math. Comput. Appli. **26**(1), 8 (2021)
9. Geiger, M.J.: Pace solver description: A simplified threshold accepting approach for the cluster editing problem. In: 16th International Symposium on Parameterized and Exact Computation (IPEC 2021). Schloss Dagstuhl-Leibniz-Zentrum für Informatik (2021)

10. Grass, J., Zilberstein, S.: Anytime algorithm development tools. ACM SIGART Bulletin **7**(2), 20–27 (1996)
11. Halim, A.H., Ismail, I., Das, S.: Performance assessment of the metaheuristic optimization algorithms: an exhaustive review. Artif. Intell. Rev. **54**(3), 2323–2409 (2021)
12. Hochreiter, S., Younger, A.S., Conwell, P.R.: Learning to learn using gradient descent. In: Dorffner, G., Bischof, H., Hornik, K. (eds.) ICANN 2001. LNCS, vol. 2130, pp. 87–94. Springer, Heidelberg (2001). https://doi.org/10.1007/3-540-44668-0_13
13. Jeyakumar, V., Rubinov, A.M.: Continuous Optimization: Current Trends and Modern Applications, vol. 99. Springer Science & Business Media (2006). https://doi.org/10.1007/b137941
14. Kennedy, J., Eberhart, R.: Particle swarm optimization. In: Proceedings of ICNN 1995-International Conference On Neural Networks, vol. 4, pp. 1942–1948. IEEE (1995)
15. Kirkpatrick, S., Gelatt, C.D., Vecchi, M.P.: Optimization by simulated annealing. Science **220**(4598), 671–680 (1983)
16. Lin, S.W., Cheng, C.Y., Pourhejazy, P., Ying, K.C.: Multi-temperature simulated annealing for optimizing mixed-blocking permutation flowshop scheduling problems. Expert Syst. Appl. **165**, 113837 (2021)
17. Mirjalili, S., Lewis, A.: The whale optimization algorithm. Adv. Eng. Softw. **95**, 51–67 (2016)
18. Mirjalili, S., Mirjalili, S.M., Lewis, A.: Grey wolf optimizer. Adv. Eng. Softw. **69**, 46–61 (2014)
19. Molga, M., Smutnicki, C.: Test functions for optimization needs. Test Funct. Optim. Needs **101**, 48 (2005)
20. Munoz, M.A., Kirley, M., Halgamuge, S.K.: The algorithm selection problem on the continuous optimization domain. In: Computational Intelligence In Intelligent Data Analysis, pp. 75–89. Springer (2013). https://doi.org/10.1007/978-3-642-32378-2_6
21. Qian, N.: On the momentum term in gradient descent learning algorithms. Neural Netw. **12**(1), 145–151 (1999)
22. Shalev-Shwartz, S., Ben-David, S.: Understanding machine learning: From theory to algorithms. Cambridge University Press (2014)
23. Siddique, N., Adeli, H.: Simulated annealing, its variants and engineering applications. Int. J. Artif. Intell. Tools **25**(06), 1630001 (2016)
24. Taylan, P., Weber, G.W., Yerlikaya, F.: Continuous optimization applied in mars for modern applications in finance, science and technology. In: ISI Proceedings of 20th Mini-euro Conference Continuous Optimization and Knowledge-based Technologies, pp. 317–322. Citeseer (2008)
25. Vanderbilt, D., Louie, S.G.: A monte carlo simulated annealing approach to optimization over continuous variables. J. Comput. Phys. **56**(2), 259–271 (1984)
26. Weber, G.W., Özöğür-Akyüz, S., Kropat, E.: A review on data mining and continuous optimization applications in computational biology and medicine. Birth Defects Res. C Embryo Today **87**(2), 165–181 (2009)
27. Xiong, Q., Jutan, A.: Continuous optimization using a dynamic simplex method. Chem. Eng. Sci. **58**(16), 3817–3828 (2003)

Privacy Preserving by Removing Sensitive Data from Documents with Fully Convolutional Networks

Marcin Korytkowski[1]🆔, Jakub Nowak[1]🆔, Rafał Scherer[1(✉)]🆔, and Wei Wei[2]🆔

[1] Czestochowa University of Technology, al. Armii Krajowej 36, Czestochowa, Poland
{marcin.korytkowski,jakub.nowak,rafal.scherer}@pcz.pl
[2] Shaanxi Key Laboratory for Network Computing and Security Technology,
School of Computer Science and Engineering, Xi'an University of Technology,
Xi'an 710048, China
weiwei@xaut.edu.cn

Abstract. We present a new approach to anonymizing personal data in text files. In the conducted research, an approach was applied that enables the analysis of sentence sentences with the use of neural networks. Contrary to other currently proposed methods, the presented work analyzes the context of a fragment of the text, which enables the detection of sensitive information not only on the basis of specific words but on the basis of "understanding" the context, such as "mayor of Paris", "son of the CEO" of a specific company. We present a proprietary solution using convolutional networks connected with glial cells, enabling the selection of the optimal size of the CNN network structure.

Keywords: Privacy preserving · Anonymization · Text analysis

1 Introduction

Nowadays, many enterprises and private persons strive for protecting their data against leakage. It can be information about both health and company secrets, e.g. research works. The subject of the processing of sensitive data is also extremely important in the context of EU regulations, e.g. Directive 95/46/EC of the European Parliament and of the Council of 24 October 1995 on the protection of individuals with regard to the processing of personal data and on the free movement of such data and the criminal and financial liability of persons creating and processing such collections of information. It is also worth noting that the theft of sensitive data may be used to assess the health condition of politicians or other decision-makers. Automatic anonymization of unstructured text it still a challenge. Paper [6] proposed a solution to text anonymization based on word embedding. The idea was to represent all the entities appearing in the document as word vectors that

The work was supported by The National Centre for Research and Development (NCBR), the project no POIR.01.01.01-00-1431/19.

L. Rutkowski et al. (Eds.): ICAISC 2022, LNAI 13589, pp. 277–285, 2023.
https://doi.org/10.1007/978-3-031-23480-4_23

capture their semantic relationships. The author of [4] deals with embedded privacy detection in complex document formats such as PDF and Microsoft's Compound File Binary Format.

In this paper, we present a system for detecting documents containing sensitive data, also in cases where they are intentionally hidden there. For obvious reasons, the classification of data into one of two classes: contain sensitive data and without this data must be done automatically. In a situation where nowadays even small entities process gigabytes of information daily, a human is not able to manually verify the content of processed files. We propose a solution that fully automates this process based on AI techniques. The task facing the system is to detect and remove sensitive data from documents entered at its entrance. The second, extremely important aspect discussed in this paper is the selection of the optimal structure of the weave network. When using deep learning techniques, optimization of the structure of the weaving networks is of great importance. While the learning process itself must take place on highly efficient devices, the same operation of the learned network should not require such tools. Cleansing structures from unnecessary connections/weights or neurons allows for a significant reduction in the number of calculations and may allow the execution of sentences on e.g. ordinary mobile devices. For this purpose, the paper proposes the use of a technique based on the observation of the activity of human brain glial cells. In [16] a concept of PrivacyBot is proposed that detects sensitive information in user-generated unstructured texts. It also recognizes categories of sensitive information types defined based on existing work and Art. 9 of the European Union (EU) General Data Protection Regulation (GDPR).

The rest of the paper is organized as follows. In Sect. 2 we describe the datasets we created from various sources and in Sect. 3 a proposed model for data classification. Section 4 presents a way to encode the data for the network and Sect. 5 describes training of the system. Section 6 concludes the paper.

2   Data

Training and testing sets for the detection of sensitive data in files containing plain text embedded in various formats, e.g. DOCX, PDF, were generated as follows. In the collections of biographies of famous characters, Wikipedia articles, databases available at:

> https://opus.nlpl.eu/OpenSubtitles-v2018.php
> https://opus.nlpl.eu/ParaCrawl.php
> https://dumps.wikimedia.org/plwiki/
> https://dumps.wikimedia.org/

and other documents (e.g. EZD RP – Electronic Document Management in Polish Public Administration [10] documents) in the possession of the authors, sets of training and testing data containing and not including sensitive data were built. It should be noted that a significant part of the data consisted of press reports, selected manually by us, which indicated sensitive content, e.g. "mayor of Paris", "son of the mayor" of a city. The content selected in this way was fed to a tokenizer, built in accordance with the BERT (PolBERT) algorithm [8]. As a result of this operation, one-hot vector was obtained for each word. The tensors generated in this way were then fed to the input of the BERT [1] network, and as a result of its operation, hashes (hashs) of 768 real numbers were obtained. Example sentences containing sensitive information:

1. The son of the CEO of Abbea Inc. caused a car accident under the influence of alcohol.
2. Mr. John from Baker Street 2233 in London points out that the building is after extension.

Example sentences not containing sensitive information:

1. All reports have been sent to the e-mail address,
2. All people from Baker Street in London were evacuated.

3   Convolutions Glial Neural Network

The innovative discovery presented in the article is the so-called glial network. One of the properties it has is the ability to adjust its own size (number of neurons) without worsening the performance of the classifier. Artificial neural networks are the foundation of the most commonly used data analysis algorithms. However, they are treated as black boxes with a specific task to perform. The evaluation of the quality of their operation can be checked on the basis of statistical measures based on testing samples. Although the method of operation of neural networks is known and algorithms adjusting structures to properly operate on specific data sets are known, the current state of knowledge does not allow to describe (e.g. in the form of rules) the operation of the processes embedded in the trained model. There are also unknown methods for transferring freely knowledge between networks apart from transfer learning in CNNs [7,12]. The main research thread that has been undertaken here is the verification of the possibility of using the latest discoveries regarding the structure and operation of the human brain to optimize the construction of artificial neural networks. In the latest world medical literature, there is some information about the impact of the so-called glial cells for building connections between neurons in the brain and for the flow of information. It should also be noted that in the last two years a lot of research has been undertaken in the field of the use of glial cells in the treatment of oncology, cardiology and dentistry. The pioneers in this area are Polish doctors from Wrocław, who, based on their knowledge of glial cells, reconstructed a torn spinal cord [15]. How important is the influence of these cells on the thought process can be found in the book "The other brain" by R. Douglas Fields [2]. Based on this information, we created a new neural network architecture, the purpose of which will be to supervise the process of building and training convolutional networks. This approach is analogous to the existing medical knowledge regarding the way the human brain works. It clearly shows that glial cells protect neuronal cells from damage and are responsible for the construction of connections between them. Therefore, as part of the work, new neural structures of neural networks were developed, which in the training by regulating the weight values turn on or off the given connections between neurons, and thus eliminate unnecessary neurons or groups of neurons. The obvious fact is that reducing the structure of the neural network results in a significant reduction in the number of calculations and thus the response time to the given input signal. In this paper, it was decided to investigate the processes accompanying the learning and operation of convolutional networks, with particular emphasis on the interpretation of the knowledge stored in them, using the proprietary model inspired by biological glial cells. In order to present the possibilities of the proposed new convolutional network structures in the field of model simplification (pruning), the developed method can be successfully used to simplify the construction of convolutional networks, interpret filters in these networks and to improve the quality of convolutional networks. The current applications of this innovative concept have mainly concerned the issues of improving the security of computer networks by identifying users on the basis of generated traffic and detecting phishing attacks. The glial cells in these applications were fused with convolutional layers. The network model was made on the basis of glial cells that were added to the dense networks. The inputs to the structure designed in this way were hashes obtained on the basis of the outputs of the PolBERT network. It should be noted here that the convolutional network receives its input signal in a highly compressed form. The diagram of the glial network is presented in Fig. 1. According to Fig. 1, the so-called glial driver. This driver is a dense network with three layers. The driver input is the same as the convolutional network input. The driver outputs through the multiplication operator are designed to

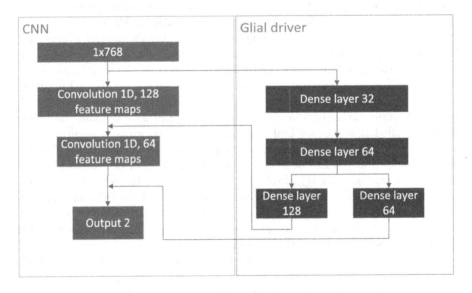

Fig. 1. Neural network model with glial cells.

strengthen or weaken the outputs of the filters in the convolutional network. They are therefore the equivalent of the action of glial cells in the human brain. The operation of the controller can also be compared to the dropout operator [11,14] in the network training process. However, in the proposed case, it has the advantage of being deterministic. Training a network designed in such a way consists of two stages. In the first one, we block the weights of the controller neurons, while the neurons of the convolutional network are trained. In the second stage, only the neurons in the controller are adapted, with the total lockout of the weight value of the convolutional network. There, however, the input was vectors describing several different characteristics. In the case of the problem considered in the paper, the input data are vectors containing encoded text.

To detect documents containing widely understood sensitive data, we decided to use the latest discoveries in the field of text processing, i.e. the aforementioned PolBERT model. We combined the original concept of glial structures with data coding using methods related to the PolBERT models and we conducted research on the system designed in this way. One of the most important elements of the PolBERT method is Tokenizer. Input vectors for the BERT network were built on the basis of encoded word indexes (Fig. 2).

4   Encoding of Input Vectors

The input of the neural network is the output vector with the dimension 768. This dimension is related to the output of the PolBERT network [8]. We encode only one sentence at a time. A sentence is a fragment of text that appears between a capital letter and a full stop. Each of the sentences is divided into individual words from the dictionary with the help of the tokenizer. This division is done by greedy string matching.

After processing the text fragments in this way, the obtained data was encoded in vectors with a size of 768 real values. Only this information was obtained at the input of the glial network. Its task was to classify documents into one of two classes: containing sensitive data and not containing sensitive information.

5   Network Optimization and Training with Glial Cells

Training such a defined structure required the development of a dedicated learning method. On the basis of numerous experiments, it was found that in order to obtain optimal results, the learning process should be carried out in an alternating manner—we should repeatedly train the CNN block and then the glial controller. In the presented model, two activation functions were used: sigmoidal (values in the range $< 0, 1 >$) for the glial network outputs and ReLU [3] in other cases.

The results of the simulations show that thanks to the proposed concept, it is possible to remove entire feature maps without losing the quality of operation. An exemplary scheme of removing a single feature map is shown in Fig. 3. Green

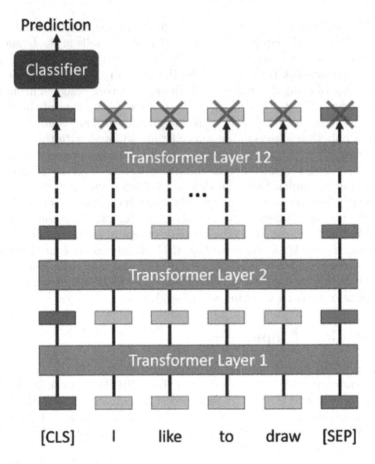

Fig. 2. Classification in BERT transformer-based architecture [9].

color is used to mark filters used to create the map. Blue color denotes filters using the same map. After training is complete and inactive feature maps are deleted, the glial controller can be disconnected. However, this requires additional training of the remaining structure. An additional 10 epochs were enough in the simulations. After that, the CNN network can be successfully used to anonymize documents.

The learning parameters were selected empirically on the basis of the experiments and they are shown in Table 1.

The obtained results showed that the use of glial cells in combination with the baseline values from the PolBERT structure did not bring a significant improvement in the classification results (an average improvement in efficiency by 1% for 10-fold repeated tests). It is worth emphasizing, however, that the obtained results were at the level of 97.7% (glial network) and 96.6% (without glial cells). An interesting fact, however, is the possibility of using the glial network to simplify the structure. The conducted research showed that glial cells identified

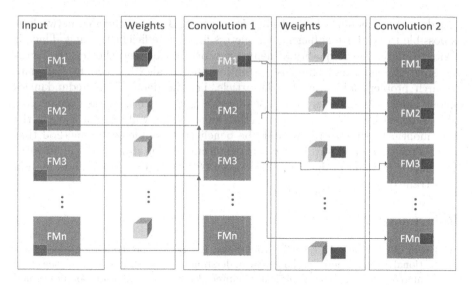

Fig. 3. Glial network to determine the significance of the necessary parameters.

Table 1. Learning parameters and results of the glial network

The size of the minibatch	36
Momentum	0.9
Learning factor	0.0005 to 0.001 decreased during learning
Number of epochs	200 glial controller, 200 CNN network + 10 after switching off the glial controller
Learning algorithm	the SGD momentum, Pytorch library
Number of simulation repetitions	200
Best result for the test dataset	98.2%
Average result from 50 independent training trials for the test dataset	97.7%
Processing time for 25 examples (input in the form of files)	1.2 [seconds]
Input vector size	$1 \times 768 \times 1$

those neurons that can be removed from the network without losing the quality of operation. In this way, the pruning process was carried out at the level of 13% (i.e. as much as 13% of neurons could be removed from the network without losing the quality of operation).

6   Conclusions

The selection of parameters for training neural networks is a very difficult task. This problem is all the more difficult for analysts when dealing with completely new data that has never been studied in the past. By using the glial network, it is possible to establish sufficient parameters for more efficient training. Thus, it allows to improve the quality of neural network models and shorten the learning time of ready-made models. Attention should also be paid to the main research problem addressed by the article, i.e. the detection of redundancies among employees. Thanks to the use of a recursive network, it has become possible to detect over 74 % of cases whether an employee will work or leave in the next month. In the future, it would be beneficial to implement a method that enables multiple parties to jointly learn an accurate neural-network model without sharing their input datasets [13] or even try to remove sensitive information from trained models [5].

References

1. Devin, J., Chang, M., Lee, K., Toutanova, K.: BERT: pre-training of deep bidirectional transformers for language onderstanding. In: Proceedings of the 2019 Conference of the North American Chapter of the Association for Computational Linguistics: Human Language Technologies, vol. 1, pp. 4171–4186 (2019)
2. Fields, R.D.: The other brain: From dementia to schizophrenia, how new discoveries about the brain are revolutionizing medicine and science. Simon and Schuster (2009)
3. Fukushima, K.: Cognitron: a self-organizing multilayered neural network. Biol. Cybern. **20**(3), 121–136 (1975)
4. Garfinkel, S.L.: Leaking sensitive information in complex document files-and how to prevent it. IEEE Secur. Priv. **12**(1), 20–27 (2013)
5. Guo, T., Guo, S., Zhang, J., Xu, W., Wang, J.: Vertical machine unlearning: Selectively removing sensitive information from latent feature space. arXiv preprint arXiv:2202.13295 (2022)
6. Hassan, F., Sánchez, D., Soria-Comas, J., Domingo-Ferrer, J.: Automatic anonymization of textual documents: detecting sensitive information via word embeddings. In: 2019 18th IEEE International Conference On Trust, Security And Privacy In Computing And Communications/13th IEEE International Conference On Big Data Science And Engineering (TrustCom/BigDataSE), pp. 358–365. IEEE (2019)
7. Karam, C., Zini, J.E., Awad, M., Saade, C., Naffaa, L., Amine, M.E.: A progressive and cross-domain deep transfer learning framework for wrist fracture detection. J. Artif. Intell. Soft Comput. Res, **12**(2), 101–120 (2022). https://doi.org/10.2478/jaiscr-2022-0007
8. Kłeczek, D.: PolBERT: attacking polish NLP tasks with transformers. In: Proceedings of the PolEval 2020 Workshop, pp. 79–88 (2020)
9. McCormick, C., Ryan, N.: BERT fine-tuning tutorial with pytorch. Accessed 24 Jan 2021 (2019)
10. NASK: EZD RP - Electronic Document Management in Public Administration (2019). https://en.nask.pl/eng/activities/digitisation-of-poland/ezd-rp-electronic-docum/3312,EZD-RP-Electronic-Document-Management-System.html. Accessed 04 April 2022
11. Shi, L., Copot, C., Vanlanduit, S.: Evaluating dropout placements in Bayesian regression Resnet. J. Artif. Intell. Soft Comput. Res. **12**(1), 61–73 (2022). https://doi.org/10.2478/jaiscr-2022-0005

12. Shin, H.C., et al.: Deep convolutional neural networks for computer-aided detection: CNN architectures, dataset characteristics and transfer learning. IEEE Trans. Med. Imaging **35**(5), 1285–1298 (2016)

13. Shokri, R., Shmatikov, V.: Privacy-preserving deep learning. In: Proceedings of the 22nd ACM SIGSAC Conference on Computer and Communications Security, pp. 1310–1321 (2015)

14. Srivastava, N., Hinton, G., Krizhevsky, A., Sutskever, I., Salakhutdinov, R.: Dropout: a simple way to prevent neural networks from overfitting. J. Mach. Learn. Res. **15**(1), 1929–1958 (2014)

15. Tabakow, P., Weiser, A., Chmielak, K., Blauciak, P., Bladowska, J., Czyz, M.: Navigated neuroendoscopy combined with intraoperative magnetic resonance cysternography for treatment of arachnoid cysts. Neurosurg. Rev. **43**(4), 1151–1161 (2020)

16. Tesfay, W.B., Serna, J., Rannenberg, K.: PrivacyBOT: detecting privacy sensitive information in unstructured texts. In: 2019 Sixth International Conference on Social Networks Analysis, Management and Security (SNAMS), pp. 53–60. IEEE (2019)

A New Approach to Statistical Iterative Reconstruction Algorithm for a CT Scanner with Flying Focal Spot Using a Rebinning Method

Piotr Pluta[✉]

Department of Intelligent Computer Systems, Czestochowa University of Technology,
Armii Krajowej 36, 42-200 Czestochowa, Poland
piotr.pluta@pcz.pl
http://www.kisi.pcz.pl/

Abstract. This work is related to our original statistical model-based iterative reconstruction conception for medical computed tomography with the flying focal spot. This new reconstruction approach is based on a continuous-to-continuous data model and the forward model formulated as a shift-invariant system. The proposed reconstruction methodology resembles the single slice rebinning concept, belonging to the so-called "nutating" reconstruction algorithms. Our algorithm is classified as an iterative reconstruction method. The proposed forward model is derived as a shift-invariant system. Thanks to this fact, it is possible to use an FFT algorithm to reduce the computational complexity of the reconstruction problem. Because of this, we can obtain reconstructed images in a time comparable to that of FBP methods, which is especially important for ambulatory purposes. The performed by us computer simulations have shown that the statistical reconstruction conception presented here outperforms the referential traditional FBP method concerning the image quality obtained and can be competitive regarding the time of the reconstruction performance.

Keywords: Image reconstruction from projections · Computed tomography · Iterative reconstruction algorithm · Flying focal spot

1 Introduction

Despite many years since the introduction of the first CT device, the search for new designs continues. At the beginning of the XXIth century, one of these designs was for a medical spiral scanner with a flying focal spot [1]. The main aim of this new technique was to increase the sampling density of the integral lines in reconstruction plane, and in the z-direction, to improve the sampling density in just the z-direction. Of course, this has meant that new reconstruction methods have had to

The authors thank Dr. Cynthia McCoullough and the American Association of Physicists in Medicine for providing the Low-Dose CT Grand Challenge dataset.

L. Rutkowski et al. (Eds.): ICAISC 2022, LNAI 13589, pp. 286–299, 2023.
https://doi.org/10.1007/978-3-031-23480-4_24

be formulated to allow for using of projections obtained from such scanners. From a practical point of view, one of the most important among these reconstruction methods is the adaptive multiple plane reconstruction (AMPR) method, (see e.g. [2]). The AMPR conception, which can be classified as a nutating reconstruction method, is a development of the advanced single slice rebinning (ASSR) method (see e.g. [1,3,11]). However, this rebinning approach has several drawbacks. One of these is its limited ability to suppress noise, caused by the linear nature of the signal processing. That means that it cannot be considered for systems that aim to reduce the dose of X-ray radiation absorbed by patients during examinations. In turn, the most interesting research in this area are statistical approaches, especially those belonging to the model-based iterative reconstruction (MBIR) group of methods [4]. In these conceptions, a probabilistic model of the measurement signals is formulated (a methodology based on the D-D data model). Unfortunately, up to now, the MBIR methods used commercially have several serious drawbacks: the calculation complexity of the reconstruction problem is problematic (proportional to N^4, where N is the image resolution), and the statistical reconstruction procedure based on this methodology necessitates simultaneous calculations for all the voxels in the range of the reconstructed 3D image, the size of the forward model matrix $\mathbf{A}$ is vast, and the reconstruction problem is extremely ill-conditioned [8]. These drawbacks can be reduced by using an approach that is based on a continuous-to-continuous (C-C) data model formulated as a shift-invariant system (this type of system is also used in other applications [5]).

It has been previously proposed formulations of the reconstruction problems for parallel scanner geometry [6], and for the spiral cone-beam scanner [7]. In this paper, we present the concept of a statistical reconstruction algorithm that uses helical cone-beam projections with the flying focal spot of the x-ray source. This approach has originally formulated a rebinning method with an iterative reconstruction method.

2   Scanner Geometry

In our work, we have taken into consideration the CT scanner with a flying focal spot. A general view of this system, with fundamental parameters, is shown in Fig. 1. The measurement system is rotated around the z axis, and in each rotation, the view angle α is incremented by a value Δ_α (for every focal spot position separately). At a given angle α_θ, all the detectors placed on the screen take measurements, and in this way projections $p\left(\alpha_\theta, \beta_\psi, \dot{z}_k\right)$ are obtained, where the integral line runs from the focus of the screen to a given detector identified by the pair of indexes ψ and k, as depicted in Fig. 1. Other important nominal parameters of this system are the focus-isocenter distance R_f and the normal focus-detector distance R_{fd}.

In the case we analyzed, a measurement system was used that allows for deflections of the focal spot both in the reconstruction plane (αFFS) and in the z-direction (zFFS). The flying focal spot in the α-direction aims to improve the resolution of the sampling of the integral lines in the reconstruction plane, and

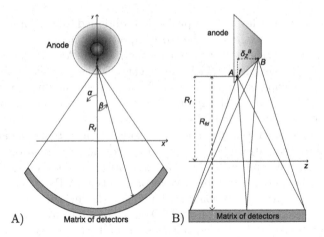

Fig. 1. General view of the measurement system of the scanner: in the reconstruction plane (A); in the z-direction (B).

in the z-direction aims to improve the resulting resolution of the reconstruction planes in just the z-direction. In this technique, this is possible by switching the focal spot between different places on the anode of the X-ray tube. These specific points on the anode are pointed out in Fig. 2.

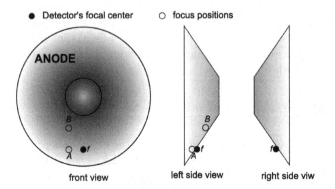

Fig. 2. Topology of the focal spots on the anode. Orientation in the reconstruction plane (αFFS).

Changing the focal spot during the measurement process also affects the parameters of the projection system, i.e. the focus-isocenter distance R_f, the rotation angle α, and the z position of the focal spot. In the case of focus position No. 1, only an increment in α is specified, namely δ'_α, and in the case of focus position No. 2, all three increments have to be taken into account: δ''_α, δ''_{R_f} and δ''_z. The adjusted angles of rotation can be determined using the simple relations

$\alpha' = \alpha + \delta'_\alpha$ and $\alpha'' = \alpha + \delta''_\alpha$ for the focal spots No. 1 and 2, respectively. The adjusted positions of the detector with index k can be calculated using the relation $\dot{z}''_k = \dot{z}_k + \delta''_{\dot{z}}$ (for the focal spot No. 2), where $\dot{z}_k$ describes the positions of these detectors relative to the nominal situation. As for the focal spot position, the nominal focus-isocenter distance is increased easily, i.e. $R''_f = R_f + \delta''_{R_f}$. It should be noted that the integral lines are no longer equi-angularly distributed, for both focal spot positions.

3   Problem of Equi-Angularity in Projections in FFS Technique

The problem of equi-angularity in the performed projections with FFS is well known. This paper presents the completely different approach to solving this problem with a minimum complications in operations performed during the reconstruction process.

If we consider only points in a 3D space without an angle between rays, it is possible to describe an x-ray in the following way:

$$xray(F'_x, F'_y, F'_z, F''_x, F''_y, F''_z, Q_x, Q_y, Q_z, D_x, D_y, D_z, p, s_m, \alpha_\psi), \qquad (1)$$

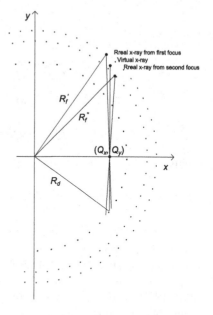

where F'_x, F'_y, F'_z are positions of first focus, F''_x, F''_y, F''_z are positions of second focus, D_x, D_y, D_z are positions of detector, Q_x, Q_y, Q_z are positions of semi-isocenter, p is value of projection, s_m and α_ψ are auxiliary values for future parallel back-projection. All positions of the first focus, the second focus, detector, and semi-isocenter are considered in the 3D space $(x - y - z)$.

The semi-isocenter is a referential point in this approach, and relating to this fixed point we can calculate new positions of the virtual detector based on an equation of three-point commonality. The placements of this and other points are depicted in Fig. 3.

Determination of parameters of virtual $xray()$ for focus positions is as follows:

$$F^T_{x_k} = F^T_{x_0} \cdot \cos\alpha - F^T_{y_0} \cdot \sin\alpha; \qquad (2)$$

$$F^T_{y_k} = F^T_{x_0} \cdot \sin\alpha + F^T_{y_0} \cdot \cos\alpha; \qquad (3)$$

Fig. 3. Location of focus, detector, and semi-isocenter.

$$F^T_{z_k} = F^T_{z_0} + (F^T_{y_k} \cdot \sin \alpha_0 + F^T_{x_k} \cdot \cos \alpha_0) \cdot \frac{1}{a}; \tag{4}$$

where:

$$F^T_{x_0} = s; \tag{5}$$

$$F^T_{y_0} = \sqrt{s^2 + (R_{f_0} + \delta^T_{R_f})^2}; \tag{6}$$

$$F^T_{z_0} = z_0; \tag{7}$$

where T is a first(') or second(") focus positions, and z_0 is places of reconstruction in z axis. For detectors position:

$$D_{x_k} = D_{x_0} \cdot \cos \alpha - D_{y_0} \cdot \sin \alpha; \tag{8}$$

$$D_{y_k} = D_{x_0} \cdot \sin \alpha + D_{y_0} \cdot \cos \alpha; \tag{9}$$

$$D_{z_k} = D_{z_0} + (D_{y_k} \cdot \sin \alpha_0 + D_{x_k} \cdot \cos \alpha_0) \cdot \frac{1}{a}; \tag{10}$$

where

$$D_{x_0} = s; \tag{11}$$

$$D_{y_0} = \sqrt{s^2 + R^2_{d_0}}; \tag{12}$$

$$D_{z_0} = z_0. \tag{13}$$

For semi-isocenter position:

$$Q_{x_k} = Q_{x_0} \cdot \cos \alpha - Q_{y_0} \cdot \sin \alpha; \tag{14}$$

$$Q_{y_k} = Q_{x_0} \cdot \sin \alpha + Q_{y_0} \cdot \cos \alpha; \tag{15}$$

$$Q_{z_k} = Q_{z_0} + (Q_{y_k} \cdot \sin \alpha_0 + Q_{x_k} \cdot \cos \alpha_0) \cdot \frac{1}{a}; \tag{16}$$

where

$$Q_{x_0} = s; \tag{17}$$

$$Q_{y_0} = 0; \tag{18}$$

$$Q_{z_0} = z_0; \tag{19}$$

where s is the distance between two nearest parallel virtual x-rays, and coefficient a is the tangent of the spiral pitch. Thus determining virtual parameters could help us to compare and find the nearest real x-ray. First, we need to calculate ζ angle between real and virtual focus and semi-isocenter that can be calculated using following relations:

$$\zeta = \arccos \left(\frac{\hat{w}_x \cdot \hat{v}_x + \hat{w}_y \cdot \hat{v}_y}{\sqrt{\hat{w}_x^2 + \hat{w}_y^2} \cdot \sqrt{\hat{v}_x^2 + \hat{v}_y^2}} \right), \tag{20}$$

where

$$\hat{v}_x = f_x^T - Q_x;$$
$$\hat{v}_y = f_y^T - Q_y;$$
$$\hat{w}_x = F_x^T - Q_x;$$
$$\hat{w}_y = F_y^T - Q_y.$$

(21)

where f is the real focus position, and F is the virtual focus position. Additional we need to calculate:

$$\xi = \hat{w}_x \cdot \hat{v}_y - \hat{v}_x \cdot \hat{w}_y.$$

(22)

Now, we can choose the two most appropriate real focuses, based on the below requirements:

$$f \uparrow = \begin{cases} \min \zeta \\ \xi > 0 \end{cases},$$

(23)

$$f \downarrow = \begin{cases} \min \zeta \\ \xi < 0 \end{cases}.$$

(24)

The next step is a choosing of the most suitable detectors. For each best-chosen focus, we need to find the four best-chosen detectors. First, we select these detectors in $x - y$ space and later in z direction. Bearing in mind, how important is a semi-isocenter position we attempt to calculate the new placement of the virtual detectors based on a coordinates of the real focus. It is necessary because it is highly probable that real and virtual detectors are places the some point of 3D space. Therefore, it is needed to recalculate these positions, according to the following relations:

$$\begin{cases} (D_x - f_{x0})^2 + (D_y - f_{y0})^2 = r_{fd}^2 \\ (Q_x - f_x)(D_y - f_y) - (Q_y - f_y)(D_x - f_x) = 0 \end{cases},$$

(25)

where the looking for virtual detector have coordinates D_x, D_y, and know localization of semi-isocenter is Q_x, Q_y, and knew real focus positions f_x, f_y, and real detector's focal center f_{x0}, f_{y0}.

It is worth noting that in a case, if we obtain more then four solutions, we reject two after re-checking, which will not be colinear with the real focus and virtual semi-isocenter, and then choose the one that is the closest to the semi-isocenter.

The last two detectors regarding z-dimension are determined according to the following equations:

$$d_z = z_0 + (l - \frac{L}{2}) \cdot d_{wc},$$

(26)

z_0 is the current table position relative to the zero point, l is the row number, L is the number of all rows, d_{wc} is the width of the detector.

The chosen real x-rays can be used for 3D interpolation operation of a virtual x-ray. Of course, they are determined as the intermediate values before we start to determine parallel projections $xray_{pararell} = (p, s_m, \alpha_\psi)$. For this reason, we use these parameters to describe a position of virtual focus, detectors, and semi-isocenter.

After these operations, we use a standard back-projection procedure based on parallel projections to obtain a referential image and a starting image for the iterative method. Presented above calculations in a few separate kernels in GPU are implemented. We show the execution time of implementation of the above pointed out procedures in Table 1, using the following GPU cards: nVidia Titan V, nVidia RTX3080, nVidia GTX960M (an old mobile GPU dedicated for notebooks).

Table 1. Comparison of the computation times for the different implementations of the procedure for obtaining of the referential image.

Operations	TitanV	RTX3080	GTX960M
Preparing virtual x-ray	0.8 ms	1.2 ms	11.5 ms
Calculating virtual x-ray	24.2 ms	106 ms	966.9 ms
Back-projection	3.0 ms	2.2 ms	55.8 ms

4   Iterative Reconstruction Algorithm

In our approach, we utilize an optimization problem, based on the well-known maximum-likelihood (ML) estimation method, which is consistent with the following continuous-to-continuous data model:

$$\mu_{\min} = \arg\min_{\mu} \left(\int_x \int_y \left(\int_{\bar{x}} \int_{\bar{y}} \mu\left(\bar{x}, \bar{y}\right) \cdot h_{\Delta x, \Delta y} d\bar{x} d\bar{y} - \tilde{\mu}\left(x, y\right) \right)^2 dx dy \right), \quad (27)$$

where $\tilde{\mu}\left(x, y\right)$ is an image obtained by way of a back-projection operation, $\mu\left(x, y\right)$ is a reconstructed image, and the kernel $h_{\Delta x, \Delta y}$ is precalculated according to the formula:

$$h_{\Delta x, \Delta y} = \int_0^{2\pi} int\left(\Delta x \cos\alpha + \Delta y \sin\alpha\right) d\alpha, \quad (28)$$

and $int\left(\Delta s\right)$ is a linear interpolation function used during the back-projection operation, as will be shown later.

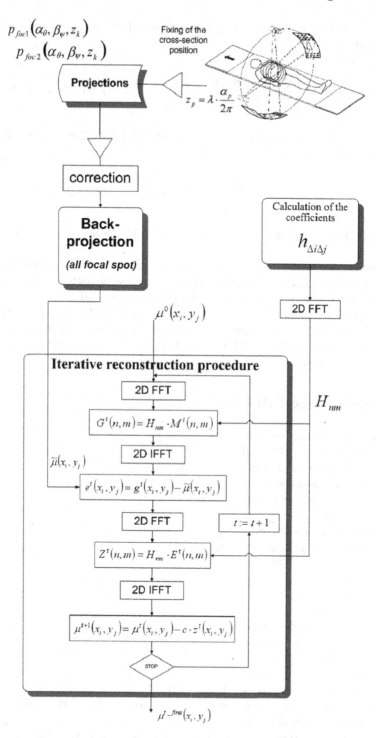

Fig. 4. Statistical iterative reconstruction algorithm.

It is clear that we have to apply the problem shown above in a discrete form, i.e. as follows:

$$\mu_{\min} = \arg\min_{\mu} \left(\sum_{i=1}^{I} \sum_{j=1}^{I} \left(\sum_{\bar{i}=1}^{I} \sum_{\bar{j}=1}^{I} \mu\left(x_{\bar{i}}, y_{\bar{j}}\right) \cdot h_{\Delta i, \Delta j} - \tilde{\mu}\left(x_i, y_j\right) \right)^2 \right), \quad (29)$$

where I is a dimension of the processed image, and the discrete kernel $h_{\Delta i, \Delta j}$ is precalculated according to the formula:

$$h_{\Delta i, \Delta j} = \frac{\Delta_\alpha}{\Delta_s^2} \sum_{\theta=0}^{\Theta-1} int\left(\Delta i \cos\theta\Delta_\alpha + \Delta j \sin\theta\Delta_\alpha\right), \quad (30)$$

and $int\left(\Delta s\right)$ is an interpolation function used during the back-projection operation.

It should be underlined that the presence of a shift-invariant system in the reconstruction problem (27) implies that this system is better conditioned than the least squares problem present in the D-D approach [8].

The proposed algorithm is described schematically in Fig. 4. It is worth noting that the convolution operation from this algorithm can be implemented in the frequency domain (using an FFT algorithm), what significantly accelerates the necessary calculations.

5   Experimental Results

In our experiments, we have used projections obtained from a Somatom Definition AS+ (helical mode) scanner with the following parameters: reference tube potential 120 kVp and quality reference effective 200 mAs, $R_{fd} = 1085.6$ mm, $R_f = 595$ mm, number of views per rotation $\Psi = 2304$, number of pixels in detector panel 736, detector dimensions were 1.09 mm × 1.28 mm. We have fixed the size of the processed image at 512 × 512 pixels. A discrete representation of the matrix $h_{\Delta x, \Delta y}$ was determined before an actual recon-

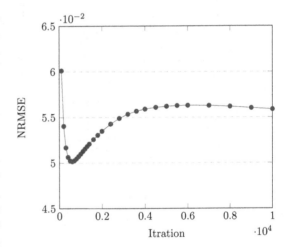

Fig. 5. NRMSE for an iterative method for the mathematical model, z: 0, noise: medium.

struction process started. These coefficients were fixed (transformed into the frequency domain) for the whole iterative reconstruction procedure. The image obtained after the back-projection operations were then subjected to a process of reconstruction (optimization) using an iterative procedure. A specially prepared result of an FBP reconstruction algorithm was chosen as the starting point of this procedure (using projections obtained from the first focal spot position. It is worth emphasise that our reconstruction procedure was performed without any regularization regarding the objective function from (27).

In our experiments, we have performed calculations necessary to realize the iterative reconstruction using hardware implementation based on the same GPU mentioned above. There are presented results using different realizations in Table 2. According to an assessment of the quality of the obtained images by a radiologist, 5000 iterations are enough to provide an acceptable image. Both hardware implementations gave almost the same results (less than 3–5s depending on the used GPU).

Table 2. Comparison of the computation times for the different realizations of the iterative reconstruction procedure.

Number of iterations	TitanV	RTX3080	GTX960M
1000	705.7 ms	588 ms	12 173 ms
3000	2 110.3 ms	1 750 ms	36 528 ms
5000	3 516.3 ms	2 914 ms	60 886 ms
7000	4 920.1 ms	4 087 ms	85 222 ms
10000	7 030.5 ms	5 833 ms	121 755 ms

We can compare the results obtained by assessing the views of the reconstructed images in Fig. 6, where the full dose projections were used, and in Fig. 7 where the quarter-dose projections were considered. In both cases, Figures (A) depict reconstructed images obtained using the standard FDK algorithm (with linear interpolation function and Shepp-Logan kernel), and Figures (B) and (C) present reconstructed images where the statistical approach presented in this paper was used: only number of iterations is different, 5000 (B) and 7000 (C).

Additionally, we compare our method to a mathematical Shepp-Logan data model with medium noise. These virtual noises are similar to noises in a quarter dose of original data. Results we present in Fig. 5, where are results of NRSME (normalized root square mean error).

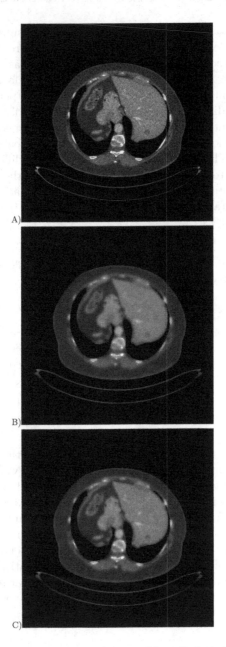

Fig. 6. View of the reconstructed image (a case with pathological changes in the liver) using full-dose projections with the application of: the standard FDK algorithm (A); the statistical method presented in this paper (5000 iterations) (B); the statistical method presented in this paper (7000 iterations) (C).

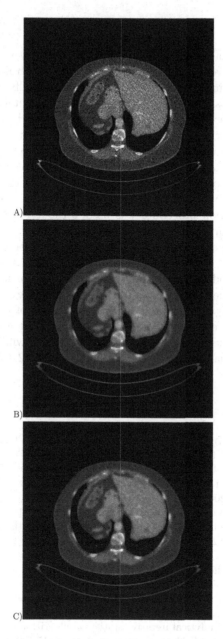

Fig. 7. View of the reconstructed image (a case with pathological changes in the liver) using quarter-dose projections with the application of: the standard FDK algorithm (A); the statistical method presented in this paper (5000 iterations) (B); the statistical method presented in this paper (7000 iterations) (C).

6   Conclusion

An original analytical preparation and acquisition x-ray value of absorption in different meaning levels can be used in all scanners with flying focal spots and even in some cases in multi-source scanners. Additionally, thanks to this approach it becomes possible to use an original statistical iterative reconstruction algorithm, which has been shown too. We have carried out experiments that have proved that our reconstruction method is very fast, mainly thanks to the utiliaztion of the individual implementation in CUDA on GPU and thanks to an original implementation of an FFT algorithm in GPU. Regarding the complexity of the iterative approach, it is worth emphasize that if the image resolution is assumed to be $N \times N$ pixels, this complexity is proportional to $N^2 \log_2 N$. The iterative reconstruction procedure was used without regularization term, carrying out only an early stopping regularization strategy. Our conception yields satisfactory results regarding the quality of the reconstructed images, and, to put into perspective, significantly reduces the dose of X-rays absorbed by the patients. Our method is easy to implement and open for using different modes of focal spot movement (both z and angle flying) and even multi-source scanners. Additionally, it is woth underline that the price of the hardware used in our experiments is relatively low (about 4000 USD for a desktop computer), and it works even in old notebooks. Further research will be devoted to integrating computational intelligence methods (e.g. [9,10,12,13]) into the approach presented here.

References

1. Kachelriess, M., Knaup, M., Penssel, C., Kalender, W.: Flying focal spot (FFS) in cone-beam CT. IEEE Trans. Nucl. Sci. **53**(3), 1238–1247 (2006)
2. Flohr, T., Stierstofer, K., Bruder, H., Simon, J., Polacin, A., Schaller, S.: Image reconstruction and image quality evaluation for a 16-slice CT scanner. Med. Phys. **30**(5), 832–845 (2003)
3. Cierniak, R., Pluta, P., Kaźmierczak, A.: A practicals Statistical approach to the reconstruction problem using a single slice rebinning method. J. Artif. Intell. Soft Comput. Res. **10**(2), 137–149 (2021)
4. Zhou, Y., Thibault, J.-B., Bouman, C.A., Hsieh, J., Sauer, K.D.: Fast model-based x-ray CT reconstruction using spatially non-homogeneous ICD optimization. IEEE Trans. Image Process. **20**, 161–175 (2011)
5. Pawlak, M., Panesar, G.S., Korytkowski, M.: A Novel Method for Invariant Image Reconstruction. J. Artif. Intell. Soft Comput. Res. **11**(1), 69–80 (2021)
6. Cierniak, R.: An analytical iterative statistical algorithm for image reconstruction from projections. Appl. Math. Comput. Sci. **24**, 7–17 (2014)
7. Cierniak, R.: Analytical statistical reconstruction algorithm with the direct use of projections performed in spiral cone-beam scanners. In: The 5th International Meeting on Image Formation in X-Ray Computed Tomography, pp. 293–296. Salt Lake City (2018)
8. Cierniak, R., Lorent, A.: Comparison of algebraic and analytical approaches to the formulation of the statistical model-based reconstruction problem for x-ray computed tomography. Comput. Med. Imaging Graph. **2**, 19–27 (2016)

9. Bilski, J., Kowalczyk, B., Marchlewska, A., Zurada, J.M.: Local Levenberg-Marquardt algorithm for learning feedforwad neural networks. J. Artif. Intell. Soft Comput. Res. **10**(4), 229–316 (2020)
10. Duda, P., Jaworski, M., Cader, A., Wang, L.: On training deep neural networks using a streaming approach. J. Artif. Intell. Soft Comput. Res. **10**(1), 15–26 (2020)
11. Cierniak, R., et al.: A new statistical reconstruction method for the computed tomography using an X-Ray tube with flying focal spot. J. Artif. Intell. Soft Comput. Res. **11**(4), 271–286 (2021)
12. Bilski, J., Kowalczyk, B., Marjański, A., Gandor, M., Zurada, J.: A novel fast feedforward neural networks training algorithm. J. Artif. Intell. Soft Comput. Res. **11**(4), 287–306 (2021)
13. Zini, J., Rizk, Y., Awad, M.: An optimized parallel implementation of non-iteratively trained recurrent neural networks. J. Artif. Intell. Soft Comput. Res. **11**(1), 33–50 (2021)

Algorithm for Solving Optimal Placement of Routers in Mines

Alan Popiel and Marcin Woźniak[(✉)]

Faculty of Applied Mathematics, Silesian University of Technology,
Kaszubska 23, 44-100 Gliwice, Poland
alanpop263@student.polsl.pl, marcin.wozniak@polsl.pl

Abstract. In this paper we present a model of optimization in a form of an algorithm which solves optimal placement of routers in N chambers with $N-1$ or less connections between them in a network system used i.e. for mining industry. The model is presented for using 2 types of routers with different signal strengths. Proposed algorithm is defined and examined on sample graph structures. Presented algorithm shall behave computational complexity of $O(n^2)$.

Keywords: Graph optimization · Wireless signal in mines · Router placement

1 Introduction

In today's mines one of very important factors is communication. As we can read in [1], wireless communication is currently used underground and gaining popularity. One of many upsides of wireless communication is its use for safety, workers can have RFID chips that enable tracking. In case of accident position of staff needed to be rescued is known.

We can find many possible applications of wireless networks in IoT systems, which control smart environments, for example at a house infrastructure [9]. In all IoT systems it is important to maintain energetic efficiency, which is also supported by configuration of networking model as presented in [6]. It is also important to optimize data transfer of information, which can be done by using various approaches. A complex study for network positioning for library unit was presented in [3]. In [2] was presented comparison between coexistence algorithms for wifi networks, while a model presented in [4] presents how to use artificial intelligence for optimal power allocation in networking systems. In [5] was presented an interesting study of access points location to solve the optimal configuration of networking systems. The study in [7] gave presentation of possible metrics to evaluate dense wifi networking systems. We can also find a variety of possible models for data analytic in different forms. Very often we can use rough set theory to ensure classification of incomplete data [10]. There are also interesting approaches in industrial informatics by using fuzzy set theory [11].

L. Rutkowski et al. (Eds.): ICAISC 2022, LNAI 13589, pp. 300–313, 2023.
https://doi.org/10.1007/978-3-031-23480-4_25

In general, optimal network configuration is related to access point placement. The network can be defined in a form of graph, so that to solve this problem we can use algorithms for graph solving. Additionally, the efficiency of the transmission is always related to the environment in which we use the networking system. At industrial sites or mines, due to thick walls, metal reinforcements, bends and corners in the tunnels its really hard environment to implement wireless signal, because it's being damped significantly or nullified altogether. So the signal from router would travel in straight lines that are tunnels, coming from the room where router is installed.

In this paper is presented a model for optimal configuration of communication nodes for wireless networking system to maintain proper signal level and optimal positions of transmitters in the network. Similar solutions for application of wifi networking models in industrial areas are presented in [8], where the model of networking was used in underground tunnels to improve communication. However in this model we work with potential improvement to number of transmitters in large networking systems. The proposed model is implemented in java and tested for optimal efficiency of networking configuration on a variety of networks.

2   Proposed Solution

To best model situation in a mine, this paper proposes a graph where nodes would be rooms, while tunnels would be edges. In this approach any big bends or corners in the tunnel would be treated as room in graph, due to signal having to do a sharp turn that can nullify it.

In tunnels longer than range of 1st type of router we will place room at the end of that range, for example router has range of 50 m and tunnel has length of 200 m. This will result in structure of 5 nodes connected in line.

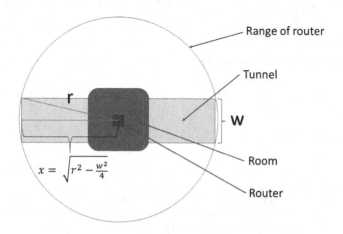

Fig. 1. Range of router in relation to varying tunnel width.

We can also take into consideration tunnel width "W". In this case we don't place rooms every "R" meters, but instead every "X" meters as presented in Fig. 1. To provide signal[1] on all tunnel length.

In this model there would be 2 types of routers placed:

- **1st type of router** - Provides signal to a room in which it's placed and all rooms that are connected to it.
  This can be achieved, because signal will travel in tunnels like in straight lines due to all things said earlier;
- **2nd type of router** - Provides signal to a room in which it's placed and all rooms directly connected to it and their neighbours.
  This can be achieved by placing one stronger router of 2nd type in room and passive reflectors[2] in rooms connected to it. Those reflectors would pass signal one room further.

While using 1st type of router our only cost of installation is said router so we assign cost 1 to it. But when we use 2nd type of router, we have to pay for passive reflectors and better device with bigger signal strength, so we assign cost 2 to it.

Packages needed for proper functioning of algorithm :

- `java.util.*` for data collections.

Our implementation of the algorithm with extra functionalities uses these additional packages :

- `java.util.concurrent.ThreadLocalRandom` for generating random graphs;
- `java.io.*` for serialization and deserialization of generated graphs or data;
- `org.graphstream` for data visualization.

In the input to the algorithm we take table of connections[3], where first value is number of nodes. As the result we get 2 lists of node IDs. Each list with different type of router.

3   Mathematical Aspects

Proposed model of node reduction for applied algorithm is presented in Fig. 2. We can define elements of this:

---

[1] Main point of the algorithm is to provide signal in "rooms" of model, there are examples where not all tunnels get signal to minimize cost of the system.

[2] They are used in mines as we can read here [1].

[3] Example : we want to show connection between node 1 and 2, plus node 2 and 3. Table of connections would look like this : [3,1,2,2,3].

- **Red node** - Node with connection to any unrestricted structure of nodes.
- **Black node** - Normal node.
- **Blue node** - Optional node[4].
- **Orange square** - Node with 1st type of router.
- **Red cross** - Node with 2nd type of router.
- **Purple node** - Node with signal.

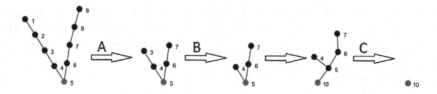

Fig. 2. Simplified image of graph's possible connections after using methods.

At start we have finite[5], number of nodes connected to node 5 in a straight line. Method "A" deletes every 3 points in this line, so in the worst case scenario we are left with 2 points in a line connected to node 5.

Method "B" deletes lines constructed from 2 nodes connected to node 5, if there are 2 or more lines like that. That leaves us with only possible structure of 2 nodes in line and 1 single point attached to node 5. Next algorithm requires more knowledge about structure, so we need to go deeper into a graph.

Now node 10 is a red point. At this stage, ends[6] can only look like that, because if we had any other arrangement, other methods would solve them. Method "C" solves structure shown on the Fig. 1, after 3rd step.

In that case we are left with few restricted structures that our last method "D" will solve.

3.1   Preface to Methods

Some structures recognized by the methods can be simplified in display, for example we can show two structures as one while changing properties of a point, but we choose this way of displaying it because algorithm uses different logic to recognize those structures.

On other times we must add some recognition of extra structures as a consequence of using table of connections instead of adjacency matrix.

---

[4] If this node is connected to given point for example point "A". Any number of single nodes can be connected to point "A" but aren't required in order for method to work.

[5] There could be more nodes connected in a straight line to node 1 and 9.

[6] Structures made using combination of normal nodes and nodes with 1 edge.

All methods start scanning graph at the ends, at the same time. Because of that we need to take into a consideration that structure that we are looking for might overlap with a part of itself thus resulting in more exception structures to consider in a method.

The reason for beginning process of solving the graph from the ends is that we can delete solved structure and make table of connections smaller. This result in enchanted speed of algorithm, because the next method will consider smaller graph.

Another reason is that we need to provide signal to all rooms. If we want to do this optimally we have to find structures in which placement of routers deeper in the graph wont change the solution already given by our method. And structures like that contained by the graph that we don't know the structure of can be localized near ends.

3.2   Method X - DeletePointsWithWifi

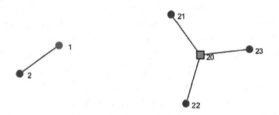

Fig. 3. Structures solved by method DeletePointsWithWifi.

Method in Fig. 3 main structure deletes nodes with 1 edge that have signal in them and are connected to other nodes. We can't simply delete all nodes with signal, because there is a chance the best possible solution will require us to insert a router in a point that already has signal.

While using this method we must look for exceptions like structure on the right. If central node (example node 20) doesn't have signal we must place router there. We must do it in this method because when we will delete all nodes connected to node 20, information about this node will be lost.

This method has inside loop that runs as long as there aren't any ends on the graph that have signal.

3.3   Method A - IsStraightLine

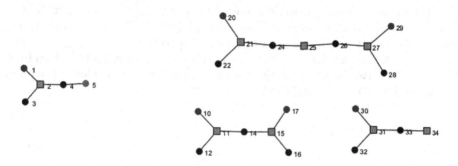

Fig. 4. Structures solved by method IsStraightLine.

Main part of the method in Fig. 4 is on the left in the pictures, exception structures are on the right. This is the best way for solving main structure because, even if we had 2nd type router at node 5, we would still need to have 1st type router in the node 2. This way we can delete all nodes from 1 to 4 and don't have to look further into graph.

There are 3 exceptions that we must look into. On graph with nodes from 10 to 17 we can see the main structure overlapping with itself.

On the other graphs we have same type of problem. While main structure would be deleted, information about the point would be lost and this node would be left without signal. So we have to check if node 5 in main structure has more than only 1 edge, or if there are other structures connected to it, besides that from method 1. For example if we had a Red point connected to node 25 or 34 there wouldn't be a router placed there.

We can have more main structures of IsStraightLine connected to node 25 in this exception.

3.4   Method B - TwoLines

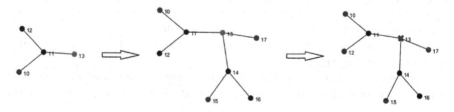

Fig. 5. Structures solved by method TwoLines.

The method in Fig. 5 is searching for at least 2 structures connected to one common point (node 13 in example). There aren't any restrictions about what other nodes can be connected to this point.

After point like that is identified method places 2nd type router in it. In next step point is deleted, so as every node with 1 edge that is connected to it, due to proprieties[7] of storing our data in table of connections.

In last step we mark nodes that we can travel in less or equal to 2 steps from the node with router. We mark those nodes as nodes with signal and leave them to method X - DeletePointsWithWifi to delete.

3.5   Method C - TwoLinesLX

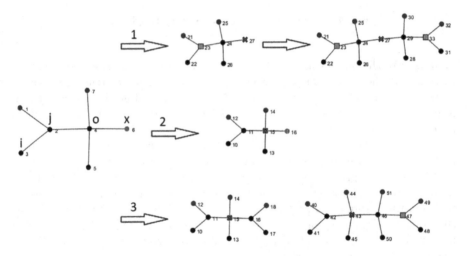

Fig. 6. Structures solved by method TwoLinesLX.

The method in Fig. 6 is searching for structure on the left. We put letters i,j,o,x on nodes that are important for logic.

In the first case we have at least 2 structures connected to the same point[8]. When this situation occurs we place 1st type routers in every point "j" of the structure. Then we must place 2nd type router in common point "x" of the structures. Node number 27 still can have unrestricted connections to it.

In the second case when only one structure is found we place 2nd type router in point "o" of the structure.

Third case are exceptions. Those structures are product of overlapping main structure. In the graph with nodes from 10 to 18 method would start in node 10

---

[7] When we delete point from the table, we delete a pair of point that we want to delete and point that its connected to. This way we break connections between nodes.

[8] In example it's node number 27.

and 17, thus resulting in placing 2 routers in point "x" without this exception. Structures here share same point "o".

Second graph has point "o" of one structure overlapping with point "x" of another structure. In this case we put 2nd type router in point "o" of one structure and 1st type of router in point "j" of second structure.

All variations of main structure will be deleted automatically, nodes in range of 2nd type of router will be handled by method DeletePointsWithWifi.

3.6   Method D - Ends

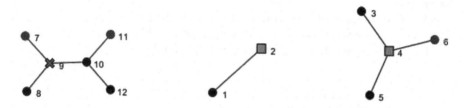

Fig. 7. Structures solved by method Ends.

The method in Fig. 7 is collection of possible structures that are left after all solving methods were run. If we have 2 nodes connected, we place 1st type of router in one of them. If we have more than 2 nodes with one edge connected to a common point, we place 1st type of router in this point. In the last structure we place 2nd type of router in node 9 to give signal to all the nodes.

3.7   Method E - CheckLX

Fig. 8. Structure solved by method CheckLX.

The model in Fig. 8 is a special method, because it runs only once after graph has been solved. It checks if structures from method TwoLinesLX that have 2nd type router in point "o", also have 2nd type router in point "x". If such structure is found it removes router from point "o" and places 1st type router in point "j" of this structure.

This can happen if we have structures from TwoLinesLX connected to part of structure from TwoLines, in the point with router and at least one other extra point. As a result method TwoLines places router after TwoLinesLX, giving us this case.

3.8 Proposed Algorithm to Solve Graph Optimization

Methods TwoLines and TwoLinesLX break graph into smaller pieces thus resulting in more ends and faster processing of the graph, because now we have more potential cases checked by the methods.

After removing nodes from graph with each method, we go to the method A. If we can't remove nodes, we go to the next method. This way we can guarantee best possible solution. In Fig. 9 we presented our implemented approach.

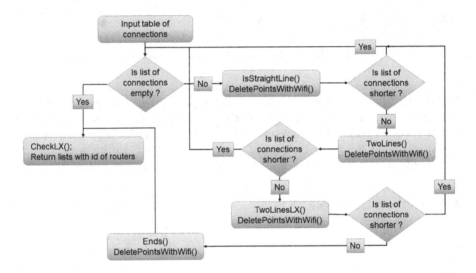

Fig. 9. Pseudo code for proposed algorithm.

4 Experimental Results

In this section we present results of numerical experiments, where for all figure we use the following assumptions:

- Green node - Node that isn't solved/deleted yet.
- Black node - Normal deleted node with signal.
- Orange square - Node with 1st type of router.
- Red cross - Node with 2nd type of router.

4.1   Solving Graph with 100 Nodes

In this subsection we will show how our algorithm solves graph step by step. Only important steps that delete part of graph will be included, but the number of step will be displayed[9], so we can keep track of how many methods already scanned graph in Figs. 10–11.

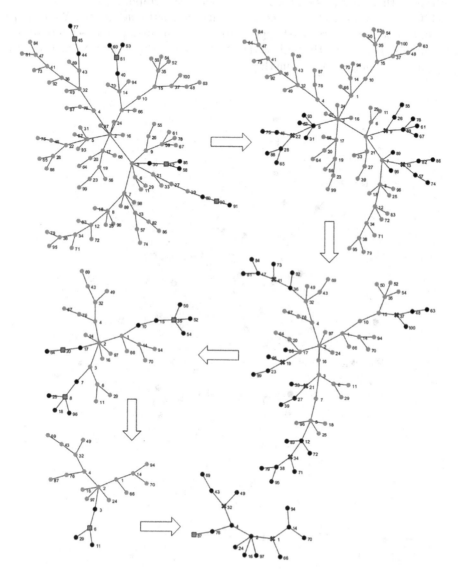

Fig. 10. Step, Method: 1A, 3B, 6C, 7A, 8A, 11C.

---

[9] Example: If 1st method doesn't solve anything, we will start from 2nd method - step 2.

Applied process of optimization follows:

- 3B - Some parts of the graph have signal. In this case, those are nodes: 2,3,6,21,7,8. But they can't be deleted yet, because of that they remain green.
- 6C - In this picture we have example of a case when we have signal in a node, but in the future there will be a router on the same spot. Node number 35 will be that point, its on the right branch of the graph.
- 11C - Now we are at last step in solving this particular graph. We can see that in node 87 there is 1st type of router and at first it may seem odd, but we must remember that all methods also run with method X. In this case router was placed by this method.

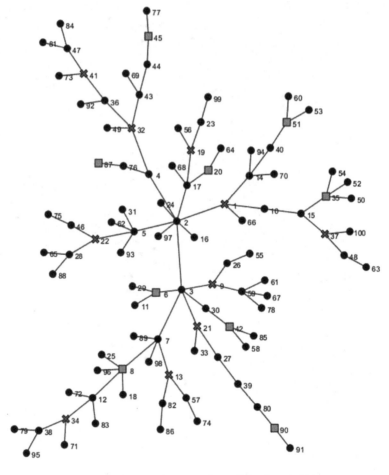

Fig. 11. Graph with 100 nodes - Solved in 0.035 s.

4.2   Computational Complexity

We run the algorithm[10] on different sizes of graphs to estimate computational complexity. As a result we have $O(n^2)$ with $R^2 = 0.99$ presented in Fig. 12.

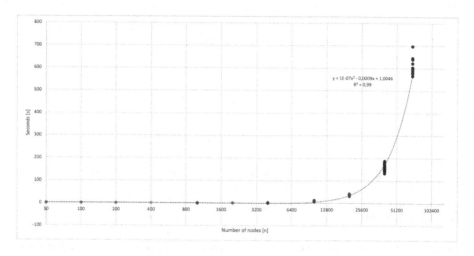

Fig. 12. Computational complexity chart.

As the number of nodes grow, in Fig. 13 we can observe less percent difference in time, because in smaller graphs we can find a solution using 1 method or remove big portion of graph and get a near instant result. We can also encounter really unfortunate node connections in which we need to check graph with all methods, thus having big impact on time.

On bigger graphs differences in time are caused mostly by the method IsStraightLine, because after deleting part of the graph this method is starting search of the structure again looking thought all ends, sometimes finding only one structure to delete, even in graphs with thousands of nodes.

Possible solution to this problem is to do a recursive variant of this method that takes as input nodes that were connected to deleted structures, but then we would also need to do a recursive variant of method DeletePointsWithWifi in order for this to work.

---

[10] Tests were run on CPU: Intel Core i5-3470.

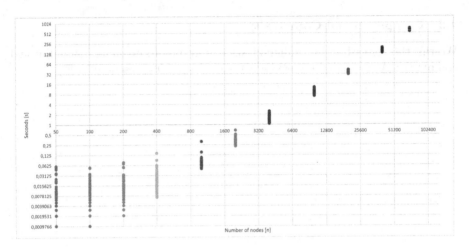

Fig. 13. Time in relation to number of nodes on log_2 scale.

5   Conclusions

In above article we presented an algorithm for solving placement of signal emitters in graphs with the minimal possible cost. For example of real world solution we used placing routers in mines, but we are sure this algorithm can be helpful in other situations as well. Proposed model is scalable, because signal emitters doesn't have fixed range but instead they operate on multiple of range. This way, if we have for example bigger mine we can assign 100 m to 1st router range and max length of edge between 2 nodes, then 200 m for 2nd type router.

Also due to many optimizations in code and logic we managed to bring down computational complexity to $O(n^2)$ so we can process even large situations in relatively short time.

References

1. Patri, A., Nayak, A., Jayanthu, S.: Wireless communication systems for underground mines-a critical appraisal. Int. J. Eng. Trends Technol. **4**(7), 3149–3153 (2013)
2. Zhang, J., et al.: Coexistence algorithms for LTE and WiFi networks in unlicensed spectrum: performance optimization and comparison. Wirel. Netw. **27**(3), 1875–1885 (2021)
3. Alemi, K., et al.: Optimization model for network coverage of SFU Bennett library. Anal. Now. **2021**, 15–31 (2021)
4. Yin, R., et al.: Distributed spectrum and power allocation for D2D-U networks: a scheme based on NN and federated learning. Mob. Netw. App. **26**(5), 2000–2013 (2021)
5. Qiu, S., et al.: Joint access point placement and power-channel-resource-unit assignment for IEEE 802.11 ax-based dense WiFi network with QoS requirements. IEEE Trans. Mob. Comput. (2021)

6. Sun, J.-S., Zhu, T., Wozniak, M.: Intelligent spacing selection model under energy-saving constraints for the selection of communication nodes in the Internet of Things. Mob. Netw. App. **628**, 1–9 (2021)
7. Kala, S.M., et al.: Evaluation of theoretical interference estimation metrics for dense WI-FI networks. In: 2021 International Conference on COMmunication Systems and NETworkS (COMSNETS). IEEE (2021)
8. Wei, W., et al.: Multi-sink distributed power control algorithm for cyber-physical-systems in coal mine tunnels. Comput. Netw. **161**, 210–219 (2019)
9. Wozniak, M., et al.: 6G-enabled IoT home environment control using fuzzy rules. IEEE Internet of Things J. **8**(7), 5442–5452 (2020)
10. Nowicki, R.K., Grzanek, K., Hayashi, Y.: Rough support vector machine for classification with interval and incomplete data. J. Artif. Intell. Soft Comput. Res. **10**, 1–10 (2020)
11. Niewiadomski, A., Kacprowicz, M.: Type-2 fuzzy logic systems in applications: managing data in selective catalytic reduction for air pollution prevention. J. Artif. Intell. Soft Comput. Res. **11**, 1–13 (2021)

An Improved Structure-Based Partial Solution Search for the Examination Timetabling Problem

Christopher Rajah[1] and Nelishia Pillay[2]($\boxtimes$) ®

[1] School of Mathematics, Statistics and Computer Science,
University of KwaZulu-Natal, KwaZulu-Natal, South Africa
`CHRISTOPHER.RAJAH@kzntreasury.gov.za`
[2] Department of Computer Science, University of Pretoria, Gauteng, South Africa
`nelishia.pillay@up.ac.za`

Abstract. The effectiveness of the Structure-Based Partial Solution Search (SBPSS) in solving the examination timetabling problem (ETP) was shown in previous work. The research presented in this paper extends this work by improving the previous version of the SBPSS. Two improvements were made, namely, additional search operators for exploitation and better control of exploration and exploitation in the deconstruction phase of the approach. The SBPSS was also evaluated on additional benchmark sets, namely the Carter and Yeditepe benchmark sets, in addition to the ITC2007 benchmark set which it was evaluated on in previous work. The improved SBPSS was found to outperform the original version. The performance of the improved SBPSS was also found to be competitive to state-of-the-art approaches applied to the examination timetabling problem, producing two best results for the Carter benchmark set, four best results for the ITC2007 benchmark set, and best results for all instances of the Yeditepe benchmark set.

Keywords: Examination timetabling · Partial solution space · Structure-based search · Multipoint search

1 Introduction

The examination timetabling problem is a well-researched problem [10] which is still receiving a lot of attention. The problem involves scheduling examinations in timetable slots so as to satisfy the hard constraints of the problem and minimize the number of soft constraints violated. Hard constraints are constraints that have to be met in order for the timetable to be feasible. Soft constraints are constraints that we would like to be satisfied, but this may not be possible as these constraints can be contradictory. Hence, we aim to minimize the number of soft constraints violated.

In previous work the effectiveness and potential of the Structure-Based Partial Solution Search (SBPSS) for solving the examination timetabling problem [11], specifically the ITC2007 benchmark set for the second international

L. Rutkowski et al. (Eds.): ICAISC 2022, LNAI 13589, pp. 314–326, 2023.
https://doi.org/10.1007/978-3-031-23480-4_26

timetabling problem, was established. This previous study also identified the following shortcomings of the approach. Firstly, in previous work three operators were employed to perform exploitation. This study will investigate further exploitation operators. The second pertains to exploration in the deconstruction phase. The aim of the deconstruction phase is to allow for further exploration after complete solutions are created to move the search in case it is stuck at a local optimum. This is achieved by removing components from the complete solutions. The number of components removed are specified by a problem dependent parameter value with the aim of the parameter value controlling the amount of exploration. However, previous work has indicated that this is not effective in controlling the exploration rate. Too low a rate results in exploitation and two high a rate prevents the algorithm from converging. This study extends the current version of SBPSS to overcome these shortcomings. The version of SBPSS with these extensions, SBPSS-I, was found to outperform SBPSS. The performance of SBPSS-I was also found to be competitive with state of the art approaches. SBPSS-I produced the best results for two problem instances for the Carter benchmark set, four problem instances for the ITC2007 benchmark set and for all problem instances for the Yeditepe benchmark set.

Hence the main contributions of this research are: 1) an extension of the SBPSS to include additional operators for exploitation, 2) an extension of the SBPSS to investigate a mechanism for better control of exploration in the deconstruction phase, 3) further evaluation of SBPSS and SBPSS-I on two additional examination timetabling problems.

The following section presents an overview of previous work aiming to solve the examination timetabling problem. Section 3 describes the extended SBPSS, SBPSS-I. The experimental setup used to assess the performance of SBPSS-I is evaluated in Sect. 4. Section 5 discusses the performance of the SBPSS-I. The conclusions of the study and future extensions of the research are reported in 6.

2   Related Work

This section reports on those studies producing one or more best results for one of the three examination timetabling benchmark sets used for evaluation of SBPSS-I in this study, namely, the Carter benchmark set, the ITC2007 benchmark set and the Yedetipe benchmark set as well as approaches that are similar to SBPSS-I. The performance of SBPSS-I is compared to that of these approaches in Sect. 5. The approach producing the best results for a number of instances of the Carter benchmark set is the greedy scheduler with backtracking implemented by Caramia et al. [4]. In the study conducted by Burke et al. [2] variable neighbourhood search is used to solve the examination timetabling problem. The approach produced a best result for the Carter benchmark set. Mandal and Kahar [7] employed great deluge to improve the quality of partial solutions, while at the same time assigning unallocated examinations, to produce examination timetables. The proposed approach performed comparatively to state-of-the-art approaches when applied to the Carter benchmark set. More

recently Bellio et al. [1] have employed a two-stage approach, employing simulated annealing on both phases, to firstly create feasible timetables and then
improve the quality of the feasible solutions produced in the first phase. This
approach produced best results for four instances of the Carter benchmark set.
The multi-phased approach by Muller [9], hybridising random solution creation,
hill-climbing, bounded great deluge and simulated annealing, on each of the
phases respectively, produced the best results for the ITC2007 benchmark set
for the second international timetabling competition. The approach also produced a best result for the Yedetipe benchmark set. In the study conducted by
Gogos et al. [5] GRASP was used to create initial timetables which were then
improved by simulated annealing followed by integer programming. The approach produced competitive results for the ITC2007 benchmark set. The new
hill-climbing method called Step Counting Hill Climbing Algorithm (SCHC) proposed by Bykov and Petrovic [3] produced best results for seven of the twelve
ITC2007 problem instances. A more recent approach applied to the ITC2007
benchmark set is the simulated annealing variation, *Fast*SA employed by Leite et
al. [6]. The approach performed comparatively to state-of-the-art-approaches on
the Carter and ITC2007 problem instances. Muklason et al. [8] have hybridized
the application of ordering heuristics, squeaky wheel optimization and a hyper-
heuristic to solve the examination timetabling problem. The approach produced
a best result for an instance of the Yedetipe benchmark set.

3   Improved Structure-Based Partial Solution Search (SBPSS-I)

The SBPSS-I algorithm is depicted in Algorithm 1. The algorithm is composed
of two phases, the construction phase (lines 5 to 12) and deconstruction phase
(lines 14 to 18). The construction phase focuses on constructing solutions by
incrementally exploring the partial solution space. On each iteration of the construction phase a component is added to the partial solution. The saturation
degree heuristic [10] is used to select the component, i.e. the examination, to
allocate to the partial timetable. The partial solutions are grouped into regions
using Algorithm 2. Similar partial solutions are grouped together. The similarity
of partial solutions is determined using Eq. 1:

$$simInd(p_i, p_j) = \frac{\sum_{i=1}^{m} c_{ij}}{m} * 100 \qquad (1)$$

where: $simInd(p_i, p_j)$ is the index indicating how similar p_i, is to p_j, m is the
number of components in partial solution p_i, c_{ij} is 1 if the component c_{ij} is in
both p_i and p_j otherwise 0

In line 7 of the Algorithm 2 $simThresh$ is used to determine the similarity
to solutions that are acceptable. Once the different regions have been established, problem specific exploitation operators are then applied to each region to
perform exploitation. These operators are applied sequentially with the period
operators applied first followed by the room operators. As application of these

---

Algorithm 1. Improved Structure-Based Partial Solution Search Algorithm

---

1: **procedure** SBPSS-I($simThresh, unscheP$)
2:     Initialize $parSolns$ partial solution points
3:     Delineate $parSolns$ into regions in the search space $curRegs =$
4:     findRegions($simThresh, parSolns$)
5:     **for** $its \leftarrow 1, exploreIts$ **do**
6:         **while** $parSolns$ points are incomplete **do**
7:             Add a new solution component to each partial solution in $parSolns$ to produce $newParSolns$
8:             **for** $lsIts \leftarrow 1, exploitIts$ **do**
9:                 Delineate $parSolns$ into regions in the search space $newRegs =$ $findRegions$(simThresh,newParSolns)
10:                 Perform search in each region in $newRegs$
11:                 Evaluate $newRegs$ and replace $curRegs$ with $newRegs$ if better
12:             **end for**
13:         **end while**
14:         The best performing complete solution becomes the new current best if it
15:         is better
16:         **if** the current best has remained unchanged for $nAttempts$ iterations **then**
17:             Remove $unschePlargeP$ number of solution components in each
18:             completed candidate solution
19:         **else**
20:             Remove $unschePsmall$ number of solution components in each
21:             completed candidate solution
22:         **end if**
23:     **end for**
24: **end procedure**

Algorithm 2. Procedure for Finding Regions in the Partial Solution Space [11]

1: **procedure** FINDREGIONS($simThresh, nParSolns$)
2:     Add the first partial solution to reg_0
3:     Add reg_0 to list of regions
4:     **while** $nParSolns$ is not empty **do**
5:         Remove $parSoln$ from $nParSolns$
6:         **for** $i \leftarrow 1, n$ **do**
7:             **if** $parSoln$ is similar to solutions in reg_i **then**
8:                 Add $parSoln$ to region reg_i
9:             **end if**
10:         **end for**
11:         **if** $parSoln$ is not added to a region **then**
12:             Create a new region reg_j
13:             Add $parSoln$ to reg_j
14:         **end if**
15:     **end while**
16: **end procedure**

---

operators will change the structure of the partial solutions, once exploitation is completed the resulting partial solutions are reorganised into regions according to similarity of structure. This process of adding components to partial solutions, organizing partial solutions into regions according structural similarity and applying exploitation operators is repeated iteratively until complete solutions are produced. The approach then enters the deconstruction phase. This phase aims at introducing a further level of exploration if the algorithm appears to be stuck in a local optimum by deallocating a larger number of components from the solutions than in the original algorithm. If this is not the case a smaller number of components are removed from the solution to perform exploration without a big impact on convergence of the algorithm. The number of components to remove in the case of the algorithm reaching a local optimum and the number of components to remove in the case it is not at a local optimum are both problem-specific parameters. This mechanism to control the amount of exploration during the deconstruction phase is an improvement on the previous version of the SBPSS.

This study builds on previous work by introducing additional move operators for exploitation performed in line 9 of the algorithm. In previous work four operators, *MovePeriodSame*, *MovePeriodRandom*, *MoveRoomSame* and *MoveRoomRandom* [11] were used. In all cases if the move operator results in one or more conflicts a Kempe chain operator is used to remove clashes [11]. The following additional exploitation operators are applied in this study:

- *PeriodChange*- This operator randomly selects two periods in a randomly selected timetable. The exams are swapped between the two selected periods. The swap is done provided there are no hard constraint violations and there is a reduction in the soft constraint cost.
- *SwapPeriodRandom*-Two periods are randomly selected. Two examinations in these periods are randomly selected. The examinations are swapped if the swap reduces the hard and/or soft constraint cost of the timetable.
- *2WaySwapPeriodRandom*-This operator is similar to the *SwapPeriodRandom* operator where three periods are considered instead of two. An exam is randomly selected from a randomly selected period in a randomly selected timetable. The cost of swapping the exam with another exam from a different period is compared to that of swapping the same exam with an exam from a third period. The swap with the largest decrease in the soft constraint cost and no hard constraint violation is performed.

SBPSS-I requires the following parameters to be set:

- *simThresh* - Is the threshold used to categorize two partial solutions as similar. This is expressed as a percentage indicating the similarity.
- *exploitIts* - The number of iterations that exploitation is performed for.
- *exploreIts* - The number of iterations that exploration is performed for.
- *nAttempts* - The number of iterations for which the *currentBest* must be the same to perform more exploration in the deconstruction phase.

The following section describes the experimental setup for the experiments to evaluate the performance of SBPSS-I.

4   Experimental Setup

This section describes the examination timetabling problems that SBPSS-I was evaluated on, the parameter values used and the statistical tests used in the performance comparison between SBPSS and SBPSS-1.

The performance of SBPSS and SBPSS-I is tested on three examination timetabling problems, namely, the Carter problem, the ITC2007 problem and the Yedetipe problem. The Carter benchmark set contains data collected for educational institutions in Canada and United States. The hard constraint for the problem is that there must be no clashes, i.e. a student must not be scheduled to write more than one examination at a time. The soft constraint for the problem is that the examinations must be well spaced for each student. The soft constraint cost is assessed using the following equation:

$$\frac{\sum w(|e_i - e_j|)N_{ij}}{S} \tag{2}$$

where: $|e_i - e_j|$ is the distance between the periods of the examination pair e_i and e_j which have students in common; S is the total number of students; N_{ij} is the number of students common to both examinations; $w(1)=16$, $w(2)=8$, $w(3)=4$, $w(4)=2$, $w(5)=1$, $w(n)=0$ for $n > 5$.

Table 1 lists the problem instances in the benchmark set. The conflict density is the ratio of the number of students that could potentially be involved in clashes to the total number of students.

Table 1. Carter benchmark set

Instance	Institution	No. of periods	No. of Exams	No. of students	Density of conflict matrix
car-f-92 I	Carleton University	32	543	18419	0.14
car-f-91 I	Carleton University	35	682	16925	0.13
ear-f-83 I	Earl Haig Collegiate Institute	24	190	1125	0.27
hec-s-92 I	Ecoles des Hautes Etudes Commerciale	18	81	2823	0.27
kfu-s-93	King Fahd University of Petroleum and Minerals	20	461	5349	0.06
lse-f-91	London School of Economics	18	381	2726	0.06
rye-s-93	Ryerson University	23	486	11483	0.08
sta-f-83 I	St Andrews Junior High School	13	139	611	0.14
tre-s-92	Trent University	23	261	4360	0.18
uta-s-92 I	Faculty of Arts, University of Toronto	35	622	21266	0.13
ute-s-92	Faculty of Engineering, University of Toronto	10	184	2749	0.08
yor-f-83 I	York Mills Collegiate Institute	21	181	941	0.29

The ITC2007 benchmark set consists of real-world problem instances and was created for the second international timetabling competition held in 2007. The hard constraints for the benchmark set are: students must not be scheduled to write more than one examination at the same time, i.e. there must be no clashes; room capacities must be satisfied; period durations must be satisfied; ordering requirements must be satisfied e.g. one examination being scheduled before another or two examinations must take place at the same time; room requirements must be satisfied, e.g. Physics practical examination has to be scheduled in the lab. The following soft constraints are minimized: two in a row: minimize the number of students that are scheduled to write two examinations one after the other; two in a day: minimize the number of students that are scheduled to write two examinations in a day; period spread: examinations must be well-spread for each student, if the examinations for a student is within a specified period, a penalty will be allocated; mixed durations: if examinations of different durations are scheduled in the same period in the same venue a penalty is allocated; largest examinations at the beginning of the examination period: minimizes the number of large examinations not scheduled at the beginning of the examination period; room penalty: minimize the usage of rooms that have a penalty associated with using it; period penalty: minimize the usage of periods that have a penalty associated with using it. The problem instances for the ITC2007 benchmark set are depicted in Table 2.

Table 2. ITC2007 examination timetabling benchmark set

Instance	Exams	Students	Periods	Rooms	Conflict density (%)
Exam1	607	7891	54	7	5.05
Exam2	870	12743	40	49	1.17
Exam3	934	16439	36	48	2.62
Exam4	873	5045	21	1	15.00
Exam5	1018	9253	42	3	0.87
Exam6	242	7909	16	8	6.16
Exam7	1096	14676	80	15	1.93
Exam8	598	7718	80	8	0.08
Exam9	169	655	25	3	4.55
Exam10	214	1577	32	48	0.05
Exam11	934	16439	26	40	0.03
Exam12	78	1653	12	50	0.18

The Yeditepe benchmark set is comprised of eight real world problem instances from Yeditepe University taken over eight semesters over three years. There are two hard constraints, namely, there must be no clashes, i.e. no student must be scheduled to write more than one examination in a period, and the overall examination capacity must not be exceeded. The soft constraint is that the number of examinations scheduled in consecutive periods for students must be minimized. The Yedetipe problem instances are listed in Table 3.

Table 3. Yedetipe examination timetabling benchmark set

Instance	Exams	Students	Capacity	Conflict density (%)
20011	559	126	450	0.18
20012	591	141	450	0.18
20013	234	26	150	0.25
20021	826	162	550	0.18
20022	869	182	550	0.17
20023	420	38	150	0.2
20031	1125	174	550	0.15
20032	1185	210	550	0.14

Previous work was used to set the initial parameter values for SBPSS and SBPSS-I. Each parameter value was then in-turn tuned by keeping the other parameter values constant and performing trial runs. The tuning process produced the following parameter values: $simThresh$-5; $unsche$-5;$exploreIts$-50;$exploitIts$-10; $noAttempts$-50.

Hypothesis tests using the Z statistic were performed to test the statistical significance of the difference in performance in SBPSS and SBPSS-I in solving each of the three examination timetabling problems. The tests were conducted for three levels of significance, namely, 1%, 5% and 10%.

The approach was developed on a computer running Windows 7 professional (64-bit Operating System). The approach was developed using Java version 1.6 and the open source Eclipse IDE from IBM. The technical specifications of the desktop are Intel® Core^{RM} i7-6700 CPU @ 3.40 Ghz with 7.88 GB usable RAM. Simulations were run on a national multicore cluster.

5   Results and Discussion

This section discusses the performance of the improved version of the structured-based partial solution space SBPSS-I on each of the benchmark sets. The performance of SBPSS-I is compared to SBPSS and the state-of-the-art approaches described in Sect. 2. Thirty runs of SBPSS-I and SBPSS, each with a different random number generator seed, was performed for each problem instance in each of the benchmark sets. In all results tables the best results obtained are indicated in bold.

Both SBPSS and SBPSS-I have found feasible solutions for all the Carter benchmark problem instances. Table 4 lists the best soft constraint cost, average soft constraint cost, variance and average runtime in seconds over the thirty runs for SBPSS-I and SBPSS. It is evident from Table 4 that SBPSS-I has outperformed

SBPSS on all problem instances for the Carter benchmark set. These results were found to be statistically significant at the 5% level of significance for hec-92 and at the 1% level of significance for all other problem instances. Table 5 compares the performance of the SBPSS-I with the state-of-the-art approaches described in Sect. 2. SBPSS-I produced best known results for two of the problem instances.

Table 4. Carter Benchmark: Comparison of the Performance of SBPSS-I and SBPSS

Instance	SBPSS-I				SBPSS			
	Best	Average	Variance	Runtime	Best	Average	Variance	Runtime
car-f-92 I	**3.701**	3.880.51	0.0116	172800	3.892	3.954	0.001	110922
car-f-91 I	**4.390**	4.632	0.0177	172800	4.551	4.703	0.002	172800
ear-f-83 I	**32.588**	33.120	0.0584	86400	33.012	33.223	0.012	12521
hec-s-92 I	**10.032**	10.165	0.0048	86400	10.122	10.183	0.003	1568
kfu-s-93	**12.810**	13.003	0.0087	172800	13.078	13.185	0.007	56606
lse-f-91	**9.804**	9.993	0.0079	86400	9.936	10.044	0.004	79551
rye-s-93	**7.849**	8.079	0.0081	172800	8.265	8.411	0.002	172800
sta-f-83 I	**157.032**	157.048	0.0001	86400	157.034	157.078	0.001	7395
tre-s-92	**7.619**	7.891	0.0125	86400	7.897	7.984	0.002	14985
uta-s-92 I	**3.035**	3.182	0.0061	172800	3.171	3.212	0.0007	129895
ute-s-92	**24.759**	24.871	0.0041	86400	24.761	24.906	0.005	12779
yor-f-83 I	**34.413**	35.240	0.09589	86400	35.205	35.746	0.062	5978

Table 5. Carter Benchmark Comparison of with State-of-the-Art

Instance	SBPSS-I	Caramia et al. [4]	Burke et al. [2]	Mandal et al. [7]	Leite et al. [6]	Bellio et al. [1]
car-f-91 I	4.39	6.6	4.6	4.58	4.31	**4.25**
car-f-92 I	3.701	6.0	3.9	3.82	3.68	**3.66**
ear-f-83 I	32.588	**29.3**	32.8	33.23	32.48	32.42
hec-s-92 I	10.032	**9.2**	10.0	10.32	10.03	10.03
kfu-s-93	12.81	13.8	13.0	13.34	12.81	**12.80**
lse-f-91	9.804	**9.6**	10.0	10.24	9.78	9.77
rye-s-93	7.849	**6.8**	-	9.79	7.89	7.9
sta-f-83 I	157.032	158.2	**156.9**	157.12	157.03	157.03
tre-s-92	**7.619**	9.4	7.9	7.84	7.66	7.68
uta-s-92 I	3.035	3.5	3.2	3.13	3.01	**2.97**
ute-s-92	24.759	**24.4**	24.8	25.28	24.80	24.79
yor-f-83 I	**34.413**	36.2	34.9	35.46	34.45	34.48

Table 6 compares the performance of SBPSS-I and SBPSS over thirty runs for each problem instance for the ITC2007 benchmark set. As SBPSS-I and SBPSS were able to find feasible solutions on all runs for all problem instances, the minimum, average and variance values listed are for the software constraint cost and average runtimes in seconds over the thirty generations.

Table 6. ITC2007 Benchmark: Comparison of the Performance of SBPSS-I and SBPSS

Instance	SBPSS-I				SBPSS			
	Best	Average	Variance	Runtime	Best	Average	Variance	Runtime
Exam1	**3925**	4208.80	19501.06	43200	5192	5303	8850	27702
Exam2	**376**	478.43	2115.56	43200	546	565.1	163.69	24241
Exam3	**8245**	8924.40	133525.77	43200	9109	9454.5	24836.25	31765
Exam4	**12493**	13050.67	100582.92	43200	14073	14181.6	215108.24	5997
Exam5	**2659**	2761.80	2908.51	86400	2825	2994.33	15896.27	6226
Exam6	**25205**	25903.50	115897.91	43200	26030	26253	14771	9613
Exam7	**3901**	4429.23	91177.70	86400	4697	4921	37658	64024
Exam8	**6756**	8146.10	556412.16	86400	8570	8910	36101.33	40511
Exam9	**910**	1008.83	3568.01	43200	1044	1075.5	193.64	2612
Exam10	**12939**	13091.43	5136.19	86400	13352	13534.4	6244.63	6907
Exam11	**24809**	25684.50	292857.84	86400	28181	28751.8	94696.96	45031
Exam12	**5095**	5192.83	2248.14	43200	5138	5305.4	14787.04	396

SBPSS-I has outperformed SBPSS on all problem instances. This result was found to be statistically significant at the 1% level of significance for all problem instances. Table 7 compares the performance of SBPSS-I with state-of-the-art approaches. SBPSS-I has outperformed the other approaches on four of the problem instances, producing the best known result for these instances.

Table 7. ITC2007 Benchmark: Comparison of with State-of-the-Art

Instance	SBPSS-I	Muller et al. [9]	Gogos et al. [5]	Bykov et al. [3]	Leite et al. [6]
Exam1	3925	4370	4128	**3647**	5050
Exam2	**376**	400	380	385	395
Exam3	8245	10049	7769	**7487**	19574
Exam4	12493	18141	13103	**11779**	12299
Exam5	2659	2988	2513	**2447**	3115
Exam6	**25205**	26585	25330	25210	25750
Exam7	3901	4213	**3537**	3563	4308
Exam8	6756	7742	7087	**6614**	7506
Exam9	**910**	1030	913	924	977
Exam10	12939	16682	13053	**12931**	13449
Exam11	24809	34129	24369	**23784**	30112
Exam12	**5095**	5535	**5095**	5097	5148

This section examines the performance of the SBPSS-I on the Yedetipe benchmark set. SBPSS-I and SBPSS were able to find feasible solutions for all problem instances in the Yedetipe Benchmark Set on all thirty runs performed. Table 8 lists minimum, average and variance of the software constraint cost and the average runtimes in seconds over the thirty runs.

Table 8. Yedetipe Benchmark: Comparison of the Performance of SBPSS-I and SBPSS

Instance	SBPSS-I				SBPSS			
	Best	Average	Variance	Runtime	Best	Average	Variance	Runtime
20011	**47**	49.8	1.2	21600	50	51.7	2.61	21600
20012	**102**	108.68	14.14	21600	105	110.1	20.29	21600
20013	**29**	29	0	3600	29	29	0	3600
20021	**47**	55.16	15.01	21600	55	60.1	13.89	21600
20022	**129**	151.04	146.52	21600	151	164.5	49.25	21600
20023	**56**	56	0	3600	56	56	0	3600
20031	**99**	129.76	147.94	21600	117	132.2	93.75	21600
20032	**359**	385.28	162.92	21600	375	391.4	103.64	21600

As for the previous two benchmark sets, SBPSS-I has produced better results than SBPSS for all problem instances. For problem instances 20013 and 20031 this result was not found to be statistically significant. For the rest of the problem instances it was found to be significant at the 5% level of significance. Table 9 compares the performance of SBPSS-I to the two state-of-the-art approaches previously applied to the Yedetipe benchmark set. SBPSS-I has produced the best result for all problem instances.

Table 9. Yedetipe Benchmark: Comparison of with State-of-the-Art

Instance	SBPSS-I	Muller [9]	Muklason et al. [8]
20011	**47**	62	56
20012	**102**	125	122
20013	**29**	29	29
20021	**47**	70	76
20022	**129**	170	162
20023	**56**	70	56
20031	**99**	223	143
20032	**359**	440	434

6   Conclusion

The research presented in this paper investigated improvements to the SBPSS approach identified in previous work. These included the use of additional exploitation operators, a mechanism to control exploration and exploitation during the deconstruction phase of SBPSS and further evaluation of SBPSS on additional examination timetabling benchmark sets. The extended version of SBPSS with the improvements, namely, SBPSS-I, was found to outperform SBPSS, producing best results for 2 problem instances in the Carter benchmark set, 4 in the ITC2007 benchmark set and all problem instances in the Yedetipe benchmark set. However, the runtimes of SBPSS-I are higher than that of SBPSS and future work will investigates methods, such as high performance computing, to reduce the runtimes.

Future work will also examine the use of transfer learning, involving transferring and reusing knowledge in solving the problem instances of one benchmark set to other benchmark sets. Future work will also apply the improved SBPSS to other educational timetabling problems, e.g. university course timetabling and school timetabling, and other discrete optimization as well as continuous optimization problems.

References

1. Bellio, R., Ceschia, S., Di Gaspero, L., Schaerf, A.: Two-stage multi-neighourhood simulated annealing for uncapacitated examination timetabling. Comput. Oper. Res. **132**, 105300 (2021)
2. Burke, E.K., Eckersley, A.J., McCollum, B., Petrovic, S., Qu, R.: Hybrid variable neighbourhood approaches to university exam timetabling. Eur. J. Oper. Res. **206**(1), 46–53 (2010)
3. Bykov, Y., Petrovic, S.: A step counting hill climbing algorithm applied to university examination timetabling. J. Sched. **19**, 479–492 (2016)
4. Caramia, M., Dell'Olmo, P., Italiano, G.F.: Novel local-search-based approaches to university examination timetabling. INFORMS J. Comput. **20**(1), 86–99 (2008)
5. Gogos, C., Alefragis, P., Housos, E.: An improved multi-staged algorithmic process for the solution of the examination timetabling problem. Ann. Oper. Res. **194**(1), 203–221 (2012)
6. Leite, N., Melício, F., Rosa, A.C.: A fast simulated annealing algorithm for the examination timetabling problem. Expert Syst. Appl. **122**, 137–151 (2019)
7. Mandal, A.K., Kahar, M.: Solving examination timetabling problem using partial exam assignment with great deluge algorithm. In: Proceedings of the 2015 International Conference on Computer, Communications, and Control Technology (I4CT), pp. 530–534 (2015)
8. Muklason, A., Parkes, A.J., Özcan, E., McCollum, B., McMullan, P.: Fairness in examination timetabling: Student preferences and extended formulations. Appl. Soft Comput. **55**, 302–318 (2017)
9. Muller, T.: ITC 2007 solver description: a hybrid approach. Ann. Oper. Res. **172**(1), 429–446 (2009)

10. Qu, R., Burke, E., McCollum, B., Merlot, L., Lee, S.: A survey of search methodologies and automated approaches for examination timetabling. J. Sched. **12**, 55–89 (2009)
11. Rajah, C., Pillay, N.: A structure-based partial solution search for the examination timetabling problem. In: Proceedings of the IEEE 2019 Congress on Evolutionary Computation, pp. 81–86 (2019)

Human Activity Recognition for Online Examination Environment Using CNN

S. Ramu$^{(\boxtimes)}$, Ram Mohana Reddy Guddeti, and Biju R. Mohan

Department of Information Technology, National Institute of Technology Karnataka,
Surathkal, Mangalore, India
{ramu.197it006,profgrmreddy,biju}@nitk.edu.in

Abstract. Human Activity Recognition (HAR) is an intelligent system that recognizes activities based on a sequence of observations about human behavior. Human activity recognition is essential in human-to-human interactions to identify interesting patterns. It is not easy to extract patterns since it contains information about a person's identity, personality, and state of mind. Many studies have been conducted on recognizing human behavior using machine learning techniques. However, HAR in an online examination environment has not yet been explored. As a result, the primary focus of this work is on the recognition of human activity in the context of an online examination. This work aims to classify normal and abnormal behavior during an online examination employing the Convolutional Neural Network (CNN) technique. In this work, we considered two, three and four layered CNN architectures and we fine-tuned the hyper-parameters of CNN architectures for obtaining better results. The three layered CNN architecture performed better than other CNN architectures in terms of accuracy.

Keywords: Convolutional neural network · Human activity recognition · Online exam environment

1 Introduction

Due to its convenience, accessibility, and user-friendliness, e-learning has gained popularity in recent years. As e-Learning courses gain popularity, students are more likely to cheat on tests by using multiple windows, asking classmates for answers, and even bringing malpractice materials to online examinations. The identification of activities is accomplished by using an intelligent system known as Human Activity Recognition (HAR), which analyses a sequence of observations relating to human behavior. Pattern recognition (i.e., classification) has been used in the literature to handle Human Activity Recognition (HAR) [12]. In order to predict future behavior or activities, the researchers collect data from smart environments about individuals' behavior and actions. Machine learning algorithms are used to process the collected data from the smart environment to predict future action. Human activity recognition has created interest among

© The Author(s), under exclusive license to Springer Nature Switzerland AG 2023
L. Rutkowski et al. (Eds.): ICAISC 2022, LNAI 13589, pp. 327–335, 2023.
https://doi.org/10.1007/978-3-031-23480-4_27

researchers and academicians due to various smart application environments such as intelligent homes, healthcare, etc.

Recognizing human activity is a difficult task. Sensor data is used to forecast a person's movement, and traditional methods from signal processing are used to accurately extract the features from the raw data for a machine learning model. Machine learning is an intelligent system in which a computer "learns" about data without being explicitly programmed [1]. Machine learning approaches have recently demonstrated capability and even achieved state-of-the-art outcomes by learning characteristics from raw sensor data [10]. Deep learning approaches like convolutional neural networks and recurrent neural networks are the most widely used models for HAR [4,7]. There are several works on HAR using machine learning methods. However, HAR for the online examination environment has not been carried out. Therefore in this work, the main focus is on human activity recognition for the online examination environment. The work contribution of this paper is to classify normal and abnormal activity during the online examination using the CNN method.

The remainder of this paper is organized as follows. Section 2 summarizes the Related Work on HAR, and Sect. 3 describes the Proposed System, Sect. 4 discusses the Experimental Results and Analysis. Finally, Sect. 5 concludes the paper with future directions.

2   Related Work

Hassan et al. [9] proposed reliable system for recognising human activity using data from smartphones' sensors. Multiple robust features were recovered from the sensor signals and then kernel principal component analysis (KPCA) was used to reduce the dimensions. Deep Belief Network (DBN), a deep learning approach, has been integrated with robust features for activity training and recognition. However, their proposed method outperformed the traditional multiclass SVM method. GCHAR (Group-based Context-aware human activity recognition), a hierarchical group-based classification method with context awareness, was proposed [6] to considerably improve classification performance. Furthermore, GCHAR has been thoroughly tested on a publicly available HAR repository, which includes six common human actions such as walking, going upstairs, going downstairs, standing, sitting, and lying.

For the recognition of physical activities on cellphones, the Transition-Aware Human Activity Recognition (TAHAR) system architecture was presented [14]. A set of inertial sensors is used to do real-time categorization while addressing the concerns such as transitions between activities and unknown activities that may arise throughout the learning process. Further, the probabilistic output of successive SVM activity forecasts is combined with a heuristic filtering strategy to achieve the said real-time task.

An online examination proctoring multimedia analytics system is presented in order to keep academic integrity in e-learning at a high level [3]. Since only two low-cost cameras and a microphone are needed, the system is both affordable

Table 1. Related works

Authors	Methodology	Merit	Demerit
Hassan et al. [9]	Human activity recognition system using Deep Learning Method	The features are further processed by a kernel principal component analysis (KPCA) and linear discriminant analysis (LDA) to make them more robust	Not considered other feature selection method for improvement of accuracy
Cao et al. [6]	Human activity recognition system based on Group clustering methods	Improve the classification efficiency, and reduces the classification error	algorithm does not work when the datasets contain heteroskedasticity and nonlinearity
Reyes et al. [14]	Human activity recognition system using Support Vector Machine (SVM) approach	Improves the classification efficiency, and reduces the classification error	Standard feature selection methods are not robust for nonlinear data
San et al. [16]	Human activity recognition system using Deep Convolutional Neural Networks (CNN)	Improve the classification efficiency, and reduces the classification error	Standard feature selection methods are not robust for nonlinear data
Ronao et al. [15]	Human activity recognition system using deep neural networks	Improves the classification efficiency, and reduces the classification error	Standard feature selection methods are not robust for nonlinear data
Atoum et al. [3]	Automated online proctoring using SVM classifier with radial basis function	Efficient cheat detection	Accuracy is low (61%)
Ashwinkumar et al. [2]	Online exam proctoring using deep learning and transfer learning	Efficient face detection	Audio detection is not considered
Ganidisastra et al. [8]	Online exam proctoring using deep learning	efficient face detection	Audio detection is not considered

and simple to use from the perspective of the test taker. Six key components are extracted from the video and audio captured: identification of the user, detection of text and speech in the window being viewed or spoken in (also known as "activate window detection"); gaze estimate; and phone detection.

An examination proctoring algorithm that uses transfer learning to realise deep learning was developed for detecting whether a pupil is exhibiting any anomalies such as impersonation, use of gadgets (or fraudulent objects for engaging in malpractice), and abnormal gaze detection [2]. Because of transfer learning, their system was capable of easy implementation on a broad scale when compared to other conventional models. Deep learning based face recognition training was proposed by applying an incremental training to the training process [8]. The summary of related work in terms of merits and demerits is shown in Table 1.

3   Proposed Work

The overall framework of the proposed system is depicted in Fig. 1. We have captured the different images of the online examination and stored them in the database. The primary goal of this work is to identify the normal and abnormal activities of online examination using the CNN. We have collected samples from the database and preprocessed the images in this work. Normalization of images into equal size is done in the preprocessing task. Later these images are given into CNN to classify the normal and abnormal activities of online examinations. The details of CNN are briefed in the next section.

3.1   CNN Model to Classify Normal and Abnormal Activities of Online Examination

Researchers in aritificial intelligence and machine learning (AIML), computer vision (CV), and language processing have been attracted with significant attention in the CNN [5,9,11,13]. CNN models have seen tremendous success in the practical application of image classification. Therefore in this proposed system, we have considered the CNN model to classify normal and abnormal activities of online examination as shown in Fig. 2. Here, abnormal activity is related to an unusual behavior of human during the examination. The input images are

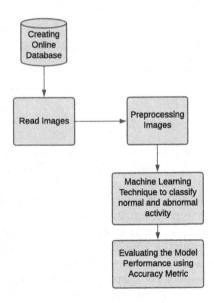

Fig. 1. Overall flow of proposed system

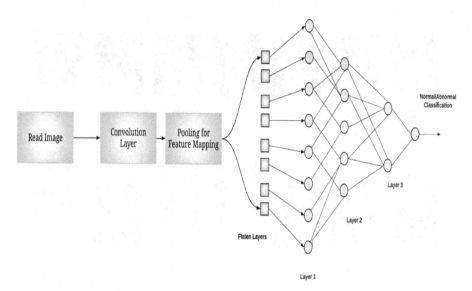

Fig. 2. Human activity recognition based on image data to identify normal activity or abnormal activity using machine learning classification method.

given into the CNN model. Each image has a size of 150 pixels wide by 150 pixels high (22500 pixels). The input images are scaled to 150×150 to reduce training time; as a result, the input shape is 64. We considered kernel size 3×3, pooling size 2×2 and dropout layer 0.25 in CNN model. We considered the fully connected layers with 256 neurons. Sigmoid and softmax functions are used to express probability distributions over a binary variable, i.e., normal and abnormal activities. z denotes the values of activation function, if z is very negative, this activation function saturates to 0 and saturates to 1 when z is very positive, respectively. The information in a neural network propagates forward from input nodes to output nodes via hidden layers. W_i denotes the weights that have been assigned to input data and B is the bias. The summation of weighted input data is output of each neurons. In Eq. 1, The weighted summation of input is defined by Eq. 1 [9].

$$\alpha = \left(\sum_{i=1}^{n} W_i Input + B \right) \tag{1}$$

CNN is trained using gradient descent iterative optimizers. Here accuracy metric is used as a cost function for the correct classification of normal and abnormal activities during online examination.

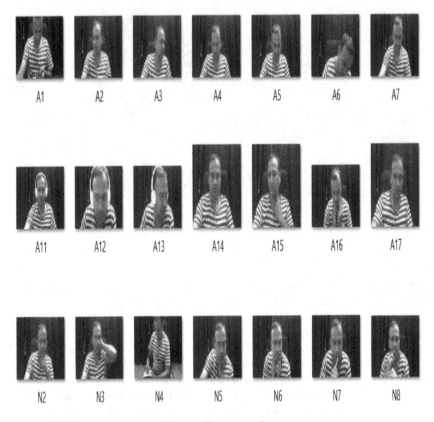

Fig. 3. Data sample

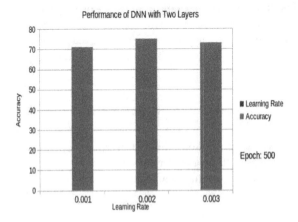

Fig. 4. Performance of CNN with two layers

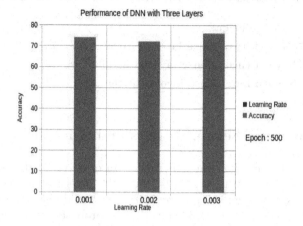

Fig. 5. Performance of CNN with three layers

Table 2. Hyper-parameters for experiment

Hyper-parameters	Range
Number of Layers	2 to 4
Number of Neurons	256 to 500
Learning Rate	0.01 to 0.06
Epochs	300 to 1000

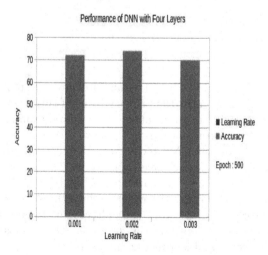

Fig. 6. Performance of CNN with four layers

Table 3. Comparision of CNN architectures with different layers.

Prediction technique	Accuracy (%)
CNN with two dense layered Architecture	75
CNN with three dense layered Architecture	76
CNN with four dense layered Architecture	74

4   Experimental Setup, Results and Analysis

R Studio is used to conduct the experiment. In order to shorten training time, the input images have been scaled to 150×150 pixels; as a result, the input shape is 150 pixels in size. Each image is 150 pixels wide by 150 pixels high and has a resolution of 150 pixels (22500 pixels). The proposed system considered the 250 samples, out of these 70% samples are used for training and the remaining 30% samples are used for testing. Few samples of data are depicted in Fig. 3. The class labels in the datasets are normal and abnormal. CNN model is used to classify normal and abnormal activities of online examination. The performance of the CNN model was evaluated using the Accuracy metric, and it is defined by Eq. 2. We have fine tuned the hyper parameters of CNN to minimize the error rate in the model as shown in Table 2. The performance of the model with different learning rates is shown in Figs. 4, 5 and 6. In experimental setup, we considered CNN architecture with two dense layer, three dense layer and four dense layers. We found that CNN with three dense layer architecture performed better than other architectures as shown in Table 3.

In the experiment, we performed tenfold cross-validation.

$$Accuracy = \frac{T_P + T_N}{T_P + T_N + F_P + F_N} \qquad (2)$$

5   Conclusion

Human Activity Recognition (HAR) is an intelligent system that recognizes human activities. Consequently, the primary focus of this proposed system is on classifying human activity in the context of an online examination. This work aims to distinguish between normal and abnormal behavior during an online examination by utilizing the CNN technique. CNN architectures with two layers, three layers, and four layers have been studied. In order to achieve better outcomes, we have fine-tuned the hyper-parameters of the CNN. The three layered CNN model outperformed the other architectures in terms of accuracy, scoring 76% on average. Future work can enhance the hybrid optimizer for the CNN model for parameter optimization. Further, the propose system can be implemented using federated learning architecture.

References

1. Akour, I., Alshurideh, M., Al Kurdi, B., Al Ali, A., Salloum, S.: Using machine learning algorithms to predict people's intention to use mobile learning platforms during the COVID-19 pandemic: machine learning approach. JMIR Med. Educ. **7**(1), e24032 (2021)

2. Ashwinkumar, J.S., Kumaran, H.S., Sivakarthikeyan, U., Rajesh, K.P.B.V., Lavanya, R.: Deep learning based approach for facilitating online proctoring using transfer learning. In: 2021 5th International Conference on Computer, Communication and Signal Processing (ICCCSP), pp. 306–312. IEEE (2021)

3. Atoum, Y., Chen, L., Liu, A.X., Hsu, S.D.H., Liu, X.: Automated online exam proctoring. IEEE Trans. Multimed. **19**(7), 1609–1624 (2017)

4. Bock, M., Hölzemann, A., Moeller, M., Van Laerhoven, K.: Improving deep learning for Har with shallow LSTMS. In: 2021 International Symposium on Wearable Computers, pp. 7–12 (2021)

5. Bougoudis, I.,Iliadis, L., Papaleonidas, A.: Fuzzy inference ANN ensembles for air pollutants modeling in a major urban area: the case of Athens, pp. 1–14 (2014)

6. Cao, L., Wang, Y., Zhang, B., Jin, Q., Vasilakos, A.V.: GCHAR: an efficient group-based context-aware human activity recognition on smartphone. J. Parallel Distrib. Comput. **118**, 67–80 (2018)

7. Challa, S.K., Kumar, A., Semwal, V.B.: A multibranch CNN-BiLSTM model for human activity recognition using wearable sensor data. Vis. Comput. **38**, 1–15 (2021)

8. Ganidisastra, A.H.S., Bandung, Y.: An incremental training on deep learning face recognition for m-learning online exam proctoring. In: 2021 IEEE Asia Pacific Conference on Wireless and Mobile (APWiMob), pp. 213–219. IEEE (2021)

9. Hassan, M.M., Uddin, M.Z., Mohamed, A., Almogren, A.: A robust human activity recognition system using smartphone sensors and deep learning. Fut. Gener. Comput. Syst. **81**, 307–313 (2018)

10. Janiesch, C., Zschech, P., Heinrich, K.: Machine learning and deep learning. Electron. Mark. **31**, 1–11 (2021)

11. Lazaridis, P.C., et al.: Structural damage prediction under seismic sequence using neural networks (2021)

12. Li, X., He, Y., Fioranelli, F., Jing, X.: Semisupervised human activity recognition with radar micro-doppler signatures. IEEE Trans. Geosci. Remote Sens. **60**, 1–2 (2021)

13. Papaleonidas, A., Iliadis, L.: Hybrid and reinforcement multi agent technology for real time air pollution monitoring. In: Iliadis, L., Maglogiannis, I., Papadopoulos, H. (eds.) AIAI 2012. IAICT, vol. 381, pp. 274–284. Springer, Heidelberg (2012). https://doi.org/10.1007/978-3-642-33409-2_29

14. Reyes-Ortiz, J.-L., Oneto, L., Sama, A., Parra, X., Anguita, D.: Transition-aware human activity recognition using smartphones. Neurocomputing **171**, 754–767 (2016)

15. Ronao, C.A., Cho, S.-B.: Human activity recognition with smartphone sensors using deep learning neural networks. Expert Syst. Appl. **59**, 235–244 (2016)

16. San-Segundo, R., Blunck, H., Moreno-Pimentel, J., Stisen, A., Gil-Martín, M.: Robust human activity recognition using smartwatches and smartphones. Eng. Appl. Artif. Intell. **72**, 190–202 (2018)

BIM-Based Approach to House Renovation Projects Assessment

Barbara Strug[ID] and Grażyna Ślusarczyk[✉][ID]

Institute of Applied Computer Science, Jagiellonian University, Łojasiewicza 11,
30-059 Kraków, Poland
{barbara.strug,grazyna.slusarczyk}@uj.edu.pl

Abstract. This paper deals with the problem of house renovation, which is often necessary to improve the value of human existence. The new space-use requirements demanded by new life-styles and changes in family living needs require creating renovation projects, the feasibility of which should be properly assessed. The information about house construction elements required for verifying a project is derived from an IFC file of a building. In order to select and process multiple building elements a dedicated IFC browser application has been developed. This application improves the workflow of the renovation projects and makes accessing relevant information fast and easy. The presented approach is illustrated by an example of changing the spatial arrangement in a single-family house.

Keywords: House renovation · BIM technology · IFC files · Design assessment

1 Introduction

This paper deals with the problem of feasibility assessment of house renovation projects. In the long-term use of buildings usually either usage behaviour or the family composition change, making some renovations necessary. To meet family living needs in the various stages of their life, indoor spaces should be adjusted from time to time. The house models in a BIM form can then be used to predict the renovation options to suit future occupancy needs.

The objective of this study is to develop a design support tool for building renovation. The proposed prototype application assists the designer in creating renovation projects for single and multi-family houses by taking advantage of using information about their construction elements, which is included in the BIM models of houses.

Nowadays architectural building designs are often created with the use of CAD tools, where BIM (Building Information Modelling) technology is applied. This technology enables to represent syntactic and semantic building information with respect to the entire life cycle of designed objects. Therefore the project

L. Rutkowski et al. (Eds.): ICAISC 2022, LNAI 13589, pp. 336–345, 2023.
https://doi.org/10.1007/978-3-031-23480-4_28

created in BIM technology can be treated as a database that allows to record both technical information about building elements, and its purpose and history.

Many studies in architecture, engineering, construction (AEC) and facility management (FM) industries have indicated that the adoption of BIM technology is beneficial to building projects [2,8,25]. However, although BIM is suitable mainly for larger and more complex buildings, like these of commercial, residential, educational and healthcare types [17,29], it also begins to be used for small residential houses. In [3] the embodied energy of the materials involved in construction of a typical dwelling house in Scotland and their environmental impact has been estimated.

In recent years many approaches to building assessment and renovation have been developed. The need for tools which would provide the easy way of changing and improving spaces to the different needs and conditions in order to achieve better comfort and quality of life standards is expressed in [13]. In [9] shape grammars are used for the adaptation of existing houses to new requirements. Three design proposals of building modifications that target different expected service life periods and use BIM technology to simulate design performance based on a building renovation scenario analysis are described in [17]. Several assessment tools have been adapted for renovation purposes [27], while others specifically developed for renovation [20]. Reviews of tools which can support the decision makers in the renovation process can be found in [1,10,18,21]. The authors underline the need to connect these tools to BIM models. A methodology for digitalization and assessment of existing buildings in order to provide cost-related analyses of potential interventions is presented in [7]. Most of these approaches evaluate building conditions in order to establish renovation needs and priorities and to explore alternatives for building retrofitting including energy performance evaluation [23].

Access to design data contained in the BIM database is possible using the IFC (Industry Foundation Classes) format [5]. However, despite the fact that the IFC is an open standard, its complex nature makes finding useful information difficult. Therefore efficient algorithms that allow for an easy access to building elements and their attributes essential for solving user-defined tasks are needed [16,19]. Most of IFC-based BIM systems lack efficient semantic query and reasoning capabilities. To tackle that problem, researchers began to combine ontologies described in Web Ontology Language (OWL) with BIM to enhance semantic interoperability and provide a better representation of the knowledge and collaborative management capabilities of BIM [11,28]. Due to the considerable size and complexity of the proposed IfcOWL-based ontologies [4,22] current research focuses on developing small, modular vocabularies that are easier to use and maintain [24].

The proposed prototype application supports the designer in evaluating the feasibility of a planed renovation. This evaluation is based on the tool which enables the designer to browse, search and display information about data stored in the IFC of a given building. It is assumed that to save IFC files the format IFC STEP Physical File (SPF) is used [15]. The designer has the possibility to filter records of the IFC by a given category. The objects useful for the task under consideration, which can be extracted from the IFC file include supporting elements, like pillars, beams and floor slabs, and infill elements, like separating walls, floorboards, ceilings, doors and windows, along with attributes describing their characteristics.

In order to propose a house renovation project the designer creates a new floor layout on the basis of the existing one. Creating such a new layout usually requires removing and adding some walls. For each wall that is to be removed, the designer should check that it is not a load-bearing wall. In such a case the design has to be modified. For each wall to be added, it must be checked that it does not overload the building structure. If the location of the bathroom or kitchen is to be changed, the pipeline system should be checked. Information obtained from the IFC file allow the designer to modify the renovation project in order to make it feasible. After completing the required changes and creating the final project, the dedicated software can be used to simulate the ventilation performance and daylighting of a renovated house.

2   IFC Files of Dwelling Houses

Nowadays projects of small residential houses are often created with the use of BIM technology. Such a project can be treated as a database that allows to record all technical information about building elements. Access to design data contained in the BIM database is possible using the IFC (Industry Foundation Classes) format. An IFC model is composed of IFC entities arranged in a hierarchical manner [14]. Each entity includes a fixed number of attributes and any number of properties encoding its detailed characteristics.

In the proposed prototype application the evaluation of the feasibility of a planned renovation is based on information about house supporting elements like pillars, beams and floor slabs, and infill elements, like separating walls, floorboards, ceilings, doors and windows. Attributes describing the characteristics of these elements are extracted from the IFC file of a given house [26].

In this paper it is assumed that IFC files are stored in the format IFC STEP Physical File (SPF). The designer has the possibility to indicate elements on the floor layout and the system returns characteristics of these elements. There is also a possibility to filter records of the IFC by a given category (Fig. 1) and then browse all useful items. In this way the designer can obtain information about elements of the electrical, hvac and plumbing systems.

The visualization of the building from data stored in the IFC file is obtained using the BIMVision viewer [6], which allows the user to manipulate the visual representations of the file as well as to select elements and generate 2D plans. In Fig. 2 a visualization of an example IFC file is depicted (an example from [12] is used here). In this figure a wall is selected for the further analysis. The identifier of this wall is used to locate it in our IFC browser application, where its detailed characteristics can be searched for.

Fig. 1. A filter dialog window.

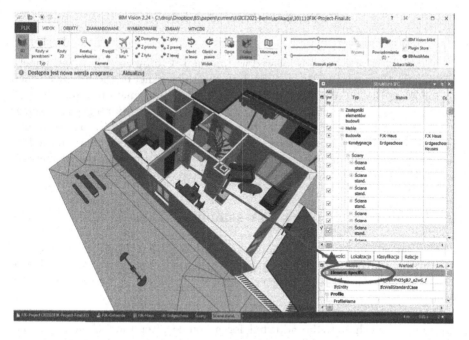

Fig. 2. A visualization of a single floor in BIMVision.

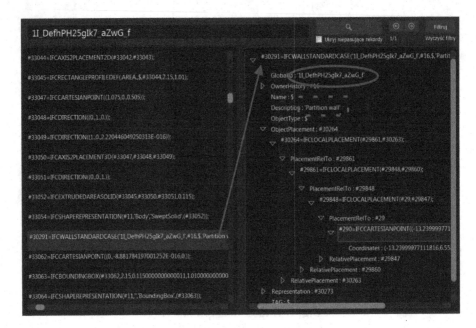

Fig. 3. A IFC browser window.

When the selected wall is located within the IFC browser (Fig. 3, left panel), clicking on the found wall transfers it to the right panel of the browser. Then by following the IFCWall element hierarchy all the information about the wall can be accessed. This information also includes the description denoting that it is a partition wall (denoted by a green dashed rectangle) and thus not a load bearing one. The application allows for selecting multiple elements from the IFC file and provides an intuitive and user friendly way of accessing the geometrical and design (semantic) information about selected elements.

3   Renovation Project Assessment

The designer usually starts a house renovation project from considering the existing floor layout and indicating walls which are to be removed. For each wall that is to be removed, the designer should verify that it is not a load-bearing wall.

In the running example let us consider a renovation project of the first floor apartment, which is to be adjusted for a couple after their child left home. According to user demand a new layout should contain one bedroom, a study and a kitchen open to the living room. A layout of the renovated floor and the same floor layout with the depicted walls to be removed are presented in Fig. 4.

In the proposed application for each wall depicted by the designer as the one to be removed the IFC browser checks its type. In the next step load-bearing

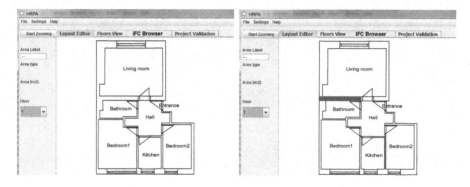

Fig. 4. A layout of the renovated first floor and the same floor layout with the depicted walls to be removed.

walls are marked on the layout. While the remaining walls can be removed, the marked ones force the planned renovation to be changed.

In Fig. 5 it is shown that the IFC browser identifies the wall between the living room and the bathroom as a load-bearing one. In the layout on the left-hand side of Fig. 6 this wall is marked in red. On the right-hand side of this figure the floor layout with removed walls is shown. It should be noted that the load-bearing wall has not been removed completely. Only the middle part of it

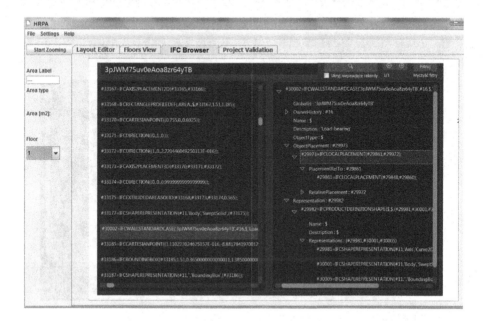

Fig. 5. Identification of a load-bearing wall.

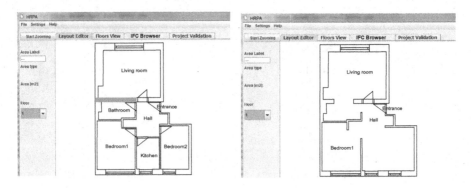

Fig. 6. A load-bearing wall marked in red and the layout with removed walls. (Color figure online)

has been removed creating the wide passage between the living room and the adjacent space.

After removing selected walls, in the successive design step the new walls are planned. For each wall to be added, it must be checked that it does not overload the building structure. Therefore for each new wall a new element is added to the IFC file together with its characteristics such as width, length and material. Then, the wall supporting elements, like pillars, beams and floor slabs, can be extracted from the IFC file and serve as parameters of one of

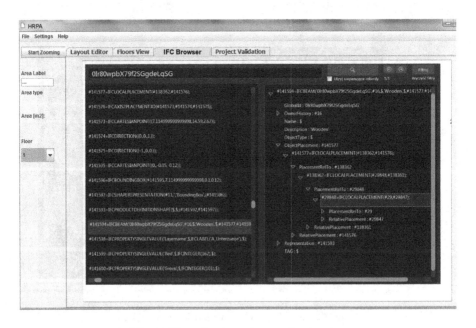

Fig. 7. Identification of a wooden beam.

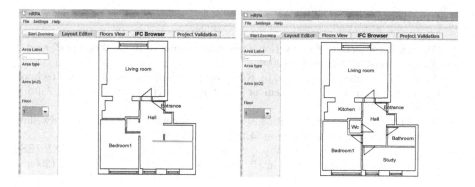

Fig. 8. A wrongly situated wall marked in red and the new floor layout. (Color figure online)

the available structural analysis software determining the effect of loads on the physical structures and their components.

In Fig. 7 one of the wooden beams supporting the floor under the rooms labelled Bedroom 2, Kitchen and Hall is found in the IFC browser. On the left-hand side of Fig. 8 new walls, which are to be added to the floor layout, are drawn. The wall depicted in red overloads the beneath structure, so it must be moved to the south. The redesigned floor layout with new spaces for the study, bathroom and wc is shown on the right-hand side of Fig. 8.

If the location of the bathroom or kitchen is to be changed in the renovation project, the pipeline system should be checked, as not always such changes are possible. The required information can be obtained by browsing the given IFC file. The proposed application supports the designer in checking and modifying the created project in order to make it feasible. For the final project the dedicated software can be used to simulate the ventilation performance and daylighting of a renovated house.

4   Conclusion

The renovation of existing housing stock is an important problem which must meet the new space-use requirements demanded by new life-styles. In this paper an application that can be used as a part of the renovation projects work-flow is proposed. It makes accessing information about house elements easy and straightforward. As the renovation projects are highly prone to unexpected situations a possibility to access and analyse all the available information can significantly improve the quality and shorten time of creating such a project.

In future an extension to the workflow, which would make possible adding annotations to the IFC file to reflect different renovation options, is planned. The other development will allow for the assessment of the renovation costs. Further plans include also an extension of the application that would allow for the 3D visualization of the building without the need for an external viewer.

References

1. Alanne, K.: Selection of renovation actions using multi-criteria "knapsack" model. Autom. Constr. **13**, 377–391 (2004)
2. Arayici, Y., Coates, P., Koskela, L., Kagioglou, M., Usher, C., O'reilly, K.: Technology adoption in the BIM implementation for lean architectural practice. Autom. Constr. **20**, 189–195 (2011)
3. Asif, M., Muneer, T., Kelley, R.: Life cycle assessment: a case study of a dwelling home in Scotland. Build. Environ. **42**, 1391–1394 (2007)
4. Beetz, J., Van Leeuwen, J., De Vries, B.: IfcOWL: a case of transforming EXPRESS schemas into ontologies. Artif. Intell. Eng. Des. Anal. Manuf. **23**, 89–101 (2009)
5. buildingSMART. IFC2x3. Retrieved from buildingSMART (2013). http://www.buildingsmart-tech.org/ifc/IFC2x3/TC1/html/index.htm
6. Datacomp: Technical documentation BIMVision, bimvision.eu. Accessed 11 May 2022
7. Di Giuda, G., Seghezzi, E., Schievano, M., Paleari, F.: A digital workflow for building assessment and renovation. In: Conference: EUBIM 2020 - BIM International Conference, Valéncia (2020)
8. Eastman, C., Lee, J.M., Jeong, Y.S., Lee, J.K.: Automatic rule-based checking of building designs. Autom. Constr. **18**, 1011–1033 (2009)
9. Eloy, S., Duarte, J.: A transformation grammar for housing rehabilitation. Nexus Netw. J. **13**, 49–71 (2011)
10. Ferreira, J.E., Pinheiro, M.D., Brito, J.D.: Refurbishment decision support tools review-Energy and life cycle as key aspects to sustainable refurbishment projects. Energy Policy **62**, 1453–1460 (2013)
11. Guitao, C.: A BIM and ontology-based approach for the building operation and maintenance management. J. Inf. Technol. Civ. Eng. Archit. **9**, 67–73 (2017)
12. IFCRepository: Open IFC repository. http://openifcmodel.cs.auckland.ac.nz/. Accessed 11 May 2022
13. Iker, G.I., Alfonso, A.G.: Indoor flexibility by industrialized methods: a way to improve use of dwellings. In: Proceedings of the 16th Annual Conference of the CIB W104 Open Building Implementation, Bilbao, Spain (2010)
14. Ismail, A., Nahar, A., Scherer, R.: Application of graph databases and graph theory concepts for advanced analysing of BIM models based on IFC standard. In: Proceedings of the 24th EG-ICE International Workshop, Nottingham, UK (2017)
15. ISO 10303–21. Industrial automation systems and integration. Product data representation and exchange. Part 21: Implementation methods: Clear text encoding of the exchange structure (2015)
16. Jin, C., Xu, M., Lin, L., Zhou, X.: Exploring BIM data by graph-based unsupervised learning. In: Proceedings of the ICPRAM (2018)
17. Juan, Y.K., Hsing, N.P.: BIM-based approach to simulate building adaptive performance and life cycle costs for an open building design. Appl. Sci. **7**, 837 (2017)
18. Kolokotsa, D., Diakaki, C., Grigoroudis, E., Stavrakakis, G.S., Kalaitzakis, K.: Decision support methodologies on the energy efficiency and energy management in buildings. Adv. Build. Energy Res. **3**, 121–146 (2009)
19. Langenhan, C., Weber, M., Liwicki, M., Petzold, F., Dengel, A.: Graph-based retrieval of building information models for supporting the early design stages. Adv. Eng. Inform. **27**, 413–426 (2013)
20. Mancik, S.: Růžička multicriterion assessment of existing buildings in reSBToolCZ. J. Int. J. Sustain. Constr. **1**, 73–81 (2012)

21. Nielsen, A.N., Jensen, R.L., Larsen, T.S., Nissen, S.B.: Early stage decision support for sustainable building renovation - a review. Build. Environ. **103**, 165–181 (2016)
22. Pauwels, P., Terkaj, W.: EXPRESS to OWL for construction industry: towards a recommendable and usable ifcOWL ontology. Autom. Constr. **63**, 100–133 (2016)
23. Sanhudo, L., et al.: Building information modeling for energy retrofitting - a review. Renew. Sustain. Energy Rev. **89**, 249–260 (2018)
24. Schneider, G.F., Rasmussen, M.H., Bonsma, P., Oraskari, J., Pauwels P.: Linked building data for modular building information modelling of a smart home. In: Proceedings of the 12th European Conference on Product and Process Modelling (ECPPM), CRC Press, Copenhagen, Denmark, pp. 407–414 (2018)
25. Singh, V., Gu, N., Wang, X.: A theoretical framework of a BIM-based multi-disciplinary collaboration platform. Autom. Constr. **20**, 134–144 (2011)
26. Szczepańczyk, A.: IFCdocuments processing (in polish). BS thesis. Jagiellonian University, Kraków (2020)
27. Thuvander, L., Femenías, P., Mjörnell, K., Meiling, P.: Unveiling the process of sustainable renovation. Sustainability **4**, 1188–1213 (2012)
28. Werbrouck, J.M., Pauwels, P., Bonduel, M., Beetz, J., Bekers, W.: Scan-to-graph: semantic enrichment of existing building geometry. Autom. Constr. **119**, 103286 (2020)
29. Volk, R., Stengel, J., Schultmann, F.: Building information modeling (BIM) for existing buildings - literature review and future needs. Autom. Constr. **38**, 109–127 (2014)

Machine Learning-Based Conditioning and Drying of Sewage Sludge as Part of the Management of Co-fermentation Processes

Anna Tuchołka and Magdalena Scherer[✉]

Czestochowa University of Technology, Czestochowa, Poland
magdalena.scherer@pcz.pl

Abstract. The article describes the issues related to the need for energy management of sewage sludge. The constant increase in the number of people and the expansion of sewage networks together with sewage treatment plants translates into a constant increase in the amount of sewage sludge. These sediments can be used for further development in various economic branches. The article focuses on the preparation of sewage sludge for use as fuel and raw material for the production of gaseous fuels. Sewage sludge is characterized by a high water content, which worsens its calorific value, makes further use difficult and increases the costs of transporting the sludge to the place of its processing. An important aspect is its proper preparation, where the key is to remove water and sterilize the sediment. The text also provides guidelines for the treatment and preparation of sewage sludge in force in Polish legislation. We focused on neural network-based modeling conditioning and drying of sewage sludge from a selected sewage treatment plant in terms of determining its further suitability for industrial use.

1 Introduction

The development of urban planning entails the constant expansion of sewage systems along with the necessary infrastructure of the treatment plant. Obviously, it is necessary to maintain hygienic standards, but it is associated with the problem of increasing the amount of generated wastewater. Thus, the amount of sewage sludge also increases. This sludge, depending on the management possibilities, can be treated both as waste and valuable raw material for further processing. Sludge management is in part forced by the search for alternative energy sources and the need to reduce the consumption of natural resources. Another argument for their further use is the fact that there is a constant increase in "resources" of sewage sludge. Integrated management of water resources and wastewater management is an integral part of socio-economic development and thus takes into account the interdependencies throughout the entire system, which includes: society, economy and environment. Achieving sustainable development of the economy is possible only with the use of appropriate management methods,

L. Rutkowski et al. (Eds.): ICAISC 2022, LNAI 13589, pp. 346–355, 2023.
https://doi.org/10.1007/978-3-031-23480-4_29

both in all areas of production and consumer activities, including appropriate management of sewage waste. The sludge itself is formed at various stages of wastewater treatment and is a suspension contained in the mixture of domestic and domestic wastewater. The physico-chemical characteristics of the sediments depend on the type and technical condition of the sewage system, the amount of water used for sewage purposes, the treatment technology and the standard of living of the inhabitants of a given area. At the same time, there are no uniform characteristics of sewage sludge and their properties are different for each locality. Additionally, they can still change in annual, monthly and even daily cycles. The high efficiency of modern wastewater treatment technologies allows the assumption that practically the entire amount of pollutants carried in the wastewater is accumulated in the biomass in the form of sludge. Due to this fact, each sewage treatment plant is forced to include in its balance the mass of pollutant load at the input and the sewage sludge at the output. This necessitates proper management of the generated sewage sludge. The possibilities of sludge management are determined mainly by the technologies used in a given facility. For buildings under construction or modernization, it is necessary to define the direction of sludge management prior to the commencement of works. Incorrect assumptions at this stage may in the future translate into the generation of sludge by sewage treatment plants that cannot be processed or managed [6].

2 Sewage Sludge Management

2.1 Properties of Sewage Sludge

The high degree of hydration translates into technical problems related to their further transformation. The problem is mostly related to their thermal processing. Increasing the share of latent heat in the energy balance of exhaust gases translates into their lower temperature, which may hinder the combustion process and the proper discharge of exhaust gases, e.g. to a cleaning device. An additional problem may be the high humidity of the exhaust gases, favoring the condensation of moisture and accelerating the corrosion of devices or exhaust pipes.

Thus, in order to be able to incinerate or process sewage sludge, it must be stripped of as much water as possible before further treatment. Water in sediments occurs in four "states":

- free water, not associated with solid particles of sewage sludge;
- colloidal water collected inside the fluff structure;
- capillary water, mainly bound by adhesive forces
- hydration water, i.e. water biologically bound by organisms living in sewage sludge [6].

2.2 Treatment and Management Methods of Sewage Sludge

The management of sewage sludge already covered the period before the implementation of sewage treatment technology. Previously, sewage sludge was discharged directly into international waters. Unfortunately, many low-developed countries still practice this type of practice. In the process of wastewater treatment, apart from the aforementioned sewage sludge, also other by-products such as solids or sand are produced, which are separated during straining and mechanical treatment. These substances can be dehydrated and used in further processing or as an additive to energy fuels. Sand is most often cleaned and used as a structural material. On the other hand, sewage sludge requires a much more technologically advanced treatment path. In their cases, the issue of further use is related to the problems arising from:

– Reduction of the water content and the volume of sewage sludge;
– Energy use;
– sterilization or at least reduction of the amount and harmfulness of microorganisms;
– Recovery of phosphorus. [1]

The Baltic countries use various technologies for the processing of sewage sludge. At the same time, many standard methods are used in each of them, albeit to a different extent. The technologies themselves differ in terms of diversity and prevalence in these countries. It should be noted here that some methods are mutually exclusive, e.g. different sludge dewatering technologies.

2.3 Methods and Technology Used in Selected Sewage Treatment Plants of the Baltic Countries

Gdańsk, East Sewage Treatment Plant. The East sewage treatment plant in Gdańsk is equipped with equipment for the biological removal of phosphorus and nitrogen and has the ability to chemically precipitate phosphorus. Annually, it processes about 18 thousand tons of sewage sludge, counting their dry mass. The technological line includes a large preliminary sedimentation tank with a long retention time of sewage, which translates into a significant amount of preliminary sediments. They are then concentrated in a gravity thickener and then sent to fermentation. On the other hand, excess sludge is thickened by two devices: a gravity thickener and a screw press. This is due to the need for a higher degree of sludge thickening (6% instead of 4.8%). The fermentation process itself is assisted by ultrasound. The fermented sludge is dewatered using centrifuges. Biogas produced in the sludge fermentation processes is produced with an efficiency of 40.5% and used for the production of electricity and heat. The amount of energy generated by the facility allows for 100% coverage of the needs of sewage sludge treatment plants and incinerators. Heat from biogas is used in technological processes. Due to the low demand for phosphorus fertilizers in this region, sewage sludge from the East sewage treatment plant is not used in agriculture. The second reason for the low demand is the high content of heavy metals in these sediments. Part of the sewage sludge is stored.

Brest, Brestwodokanal Sewage Treatment Plant (Belarus). The wastewater treatment plant in Brest uses the traditional BOD removal process with the use of activated sludge. As part of its modernization, changes were made to enable chemical phosphorus precipitation. Since 2010, sewage sludge from this treatment plant is sent to the waste processing plant in Brest, which processes excess and primary sludge. The processing plant mixes processed sewage sludge with municipal solid waste and sends it to landfills. The effluents from the processing plant are pre-treated and then directed to the sewage treatment plant for complete treatment. In the past, sewage sludge generated in the Brest sewage treatment plant was stored in sediment lagoons with a volume of $100,000\,m^2$ located along the Bug River.

Chełm, Bielawin Sewage Treatment Plant. The Bielawin sewage treatment plant cleans municipal sewage from Chełm and the surrounding area. Mainly domestic and industrial sewage and industrial sewage from dairy production and fruit concentrates are sent here. In 2011–2015, the facility was modernized. The scope of modernization covered practically the entire technological system. At that time, the biological system of the sewage treatment plant was changed, and instead of SBR technology, a flow system equipped with secondary settling tanks was used and a line of closed fermentation chambers was built. Since 2016, the facility has started trials with co-fermentation with waste whey. Initially, it covered a laboratory scale, but in the same year full-scale tests were undertaken. The whey was introduced, after gravity concentration, to the initial sludge in the amount of 5–15 $m^3/24h$, and only the mixture was directed to the mixed sludge tank. The excess sludge was additionally subjected to ultrasonic disassembly. Then the whole was directed to the methane fermentation process. The addition of whey had a positive effect on the amount of biogas produced, but resulted in a significant increase in sulfur in the produced biogas. Due to the lack of stability in the supply of whey and problems with maintaining the quality of biogas, the trials were completed at the end of 2016.

3 Preparation and Processing of Sewage Sludge

Due to the limitations in the storage of certain types of sludge, it is necessary to look for alternative methods of their management. One of the methods is methane fermentation with high biogas production capacity. Biogas is a gas mixture with a high proportion of methane (up to 70%). Due to its high calorific value and flammability, it can be converted into electricity and heat. The biogas itself can be produced in agricultural biogas plants, landfill and sewage treatment plants. Depending on the place of its generation, biogas may differ in its methane content. The highest levels are usually found in agricultural biogas, where substrates rich in proteins and fats were used for fermentation. Biogas from sewage sludge is generally characterized by a slightly lower methane volume fraction— up to approx. 58% [5] From a technological point of view, an important aspect is to supply the digestion chamber with the right amount of ingredients, but at the same time not to overload it with organic matter. Each material sent to the chamber should be previously checked for chemical and microbiological, which is to protect the fermentation process against the influence of any inhibitors contained in the substrates [9, 10]. In order to increase the biogas yield and shorten the decomposition time, fermentation under thermophilic conditions (52–56 °C) is also used. It is a less energy-efficient solution due to the need to reheat the fermentation chambers. In the context of sludge management, waste heat or heat from the incineration of sludge from a fermentation or co-fermentation plant is often transferred to the digestion chambers [8]. Sewage sludge is a very good substrate for biogas production due to the high content of nutrients necessary for microorganisms. This also includes micro and macro elements. In addition, sewage sludge itself has the necessary microflora necessary for the production of biogas by methane fermentation and is a substrate for the development of microorganisms. At the same time, a technological solution consisting in modification of the fermentation process is used more and more often as part of sewage sludge management. A new solution is co-fermentation. This process is based on the controlled decomposition of at least two different substrates in one fermentation chamber. However, it requires prior determination of the proportion of nutrients for bacteria carrying out fermentation processes. The substrates for co-fermentation must be properly selected due to their distribution during the process. Determining the correct ratio of carbon and nitrogen is particularly important here. The lack or deficiency of any of these elements may stop the fermentation processes in the chamber. The most frequently used substrates added to sewage sludge are fatty waste, agri-food waste and stillage [7]. Sewage sludge as a substrate in the co-fermentation process allows to provide a large amount of nutrients for microorganisms. This is especially important in the case of co-fermentation with a high carbon component, e.g. dry vegetable waste. This is due to the fact that the dominant component in the sewage sludge is nitrogen, which is necessary in the methane fermentation process. At the same time, too high a nitrogen content can lead to ammoniacal nitrogen build-up in the fermentation liquid. This may stop biogas production in the fermentation chamber due to the high pH, which disrupts the ion exchange with various forms of nitrogen (NH_4 and NH_3). For technological reasons, the optimal pH for methane fermentation and co-fermentation is considered to be 7.2 to 8.2. The advantage of co-fermentation is also the possibility of reducing the mass and volume of the substrates used, and the fermentation itself stabilizes and hygienises sewage sludge and other used substrates. Long fermentation time and anaerobic conditions allow for complete elimination of pathogens from sewage sludge. This is despite the fact that the process takes place at a temperature of approx. 37 °C, which is not a high level for the conditions of sewage sludge hygienisation. An additional effect of methane fermentation, apart from biogas, is post-fermentation, i.e. post-fermentation pulp. Due to the high abundance of nutrients necessary by plants, the digestate is often used for fertilizing crops. The use of co-fermentation in biogas plants allows for the management of several wastes at the same time. It is beneficial in the context of the biogas plant becoming independent from possible interruptions in the supply of raw materials or a sudden deterioration of the quality of any of the substrates used. In the case of fermentation of one component, the occurrence of the above-mentioned events translates into a suspension of biogas production or a clear deterioration of the product. The use of co-fermentation in production processes can therefore be crucial for the stable production of biogas. The use of sludge in the methane fermentation process is also supported by its hydration (enabling pumping), a high content of organic matter and a pH close to neutral. At the same time, it should be remembered that the sewage sludge itself is a specific product with high physicochemical variability. Therefore, their use in fermentation processes requires constant monitoring in terms of their quantitative and qualitative composition. Additionally, sewage sludge is often chemically and biologically contaminated. This is especially true of heavy metals and pathogenic microorganisms [12].

The purpose of the stabilization process is to minimize the occurrence of biological and chemical reactions. The oldest and so far the most frequently used method of sludge stabilization is anaerobic digestion. The first anaerobic digestion chambers were commissioned in the United States over a century ago. Concentrated organic and inorganic sludges undergo anaerobic microbiological decomposition into methane and inorganic substances. The main benefits of the fermentation process are the stabilization of sewage sludge, reduction of its volume (reduction of the organic matter content of the sludge) and the production of biogas. Sewage sludge from municipal treatment plants is characterized by a relatively high degree of hydration. This creates technological and economic problems related to the transport, storage and storage of sewage sludge. This means that proper preparation of the sludge is necessary for further activities. This applies to their stabilization, hygienization and, above all, the removal of water by means of dehydration or thickening and drying. Thickening is of particular importance due to the reduction of the volume of the sludge with simultaneous removal of excess water from them. It is assumed in the literature that an increase in the proportion of dry matter by a factor of two translates into a reduction in the volume of the sludge by 50%. The second and equally important process is sludge dewatering. It allows to reduce the volume and increase the dry mass, which facilitates further management of sewage sludge. With proper drainage, it is possible to achieve up to 80% dry matter content in the sludge [2].

4   Drying of Sediments in Laboratory Conditions

For digestate sludge collected from the Warta S.A. sewage treatment plant pressure filtration tests were performed at the filtration pressure equal to 0.4 MPa and 0.8 MPa. At the beginning, filtration was performed for the raw sludge sample and then for the sludge samples conditioned with polyelectrolytes, then polyelectrolyte 1 in doses d1 and d2, polyelectrolyte 2 in doses d1 and d2. Polyelectrolyte doses were selected on the basis of the CSK test. Then, based on the obtained filtration results, the following graphs were developed comparing the filtration efficiency, final hydration and filtration resistance of raw sludge and conditioned sludge (Figs. 1 and 3).

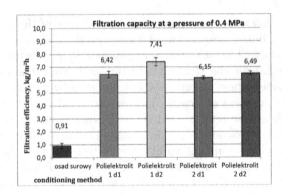

Fig. 1. Filtration efficiency at 0.4 MPa pressure of sludge conditioned with polyelectrolytes.

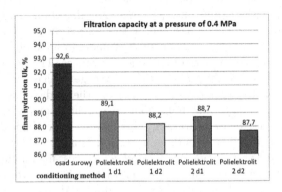

Fig. 2. Final hydration at a pressure of 0.4 MPa of sludge conditioned with polyelectrolytes.

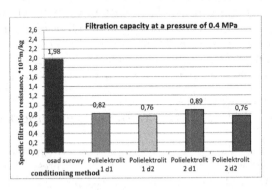

Fig. 3. Specific filtration resistance at a pressure of 0.4 MPa of sludge conditioned with polyelectrolytes.

It can be noticed that after adding polyelectrolytes, then polyelectrolyte 1 in the dose d1 and d2 and polyelectrolyte 2 in the dose d1 and d2, the filtration efficiency increased. The highest value obtained was with the use of polyelectrolyte 1 and the dose d2. In relation to the filtration efficiency of raw sludge, it is over eight times higher. On the other hand, the lowest efficiency in relation to raw sludge was obtained for the test with polyelectrolyte 2 in the d1 dose. The final hydration of raw sludge after pressure filtration at a pressure of 0.4 MPa was 92.6%. After the use of polyelectrolytes, its decrease by about 3–5% can be noticed (Fig. 2). The lowest final hydration was obtained for sludge conditioned with polyelectrolyte 2. Compared to the final hydration of raw sludge, after using this polyelectrolyte we obtain 3.9% lower hydration for the dose of 6 mg/g d.m.o. In the case of using polyelectrolyte 1 for the same dose, the final hydration was 3.4% lower than the final hydration of the raw sludge, i.e. slightly higher than for polyelectrolyte 2. A decrease in the filtration resistance can be noticed after adding polyelectrolytes, which has a positive effect on the filtration efficiency. With the increase in the dose of polyelectrolyte, the resistance also decreased. In the course of laboratory tests, studies of capillary suction time were also carried out for excess deposits for the weakly cationic electrolyte C-494, 2 medium cationic electrolytes C-498 and PRESTOL855 BS and 2 highly cationic electrolytes 655 BC-S and PROESTOL 863 BC. Measurements for excess sludge showed fluctuations in capillary suction. This could be due to the differences in the composition of the individual sediment samples. The average process time was around 65 s. In subsequent measurements, the sludge was enriched with the addition of polyelectrolytes in various doses. This was to determine their influence on the dynamics of the capillary suction process. The research showed that the addition of polyelectrolytes translates into an improvement in the filtration properties of sewage sludge compared to raw sludge. In the case of C-494 polyelectrolyte, a decrease in capillary suction time was observed for each dose. At the same time, increasing the sample was characterized by large fluctuations in values, but they were still clearly better than for the sludge not enriched with polyelectrolyte. In the case of Prestol 855 BS, an improvement in filtration properties and an increase in susceptibility to dewatering were also observed. In the case of this polyelectrolyte, such fluctuations in the values of individual samples were not demonstrated (Figs. 4, 5 and 6).

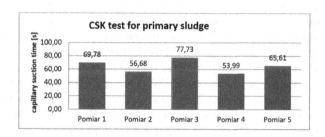

Fig. 4. Capillary suction time of excess sludge without the addition of polyelectrolyte.

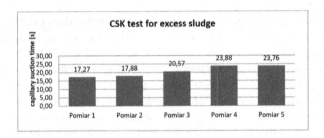

Fig. 5. Capillary suction time of excess sediments with the addition of polyelectrolyte C-494 in a dose of 5 ml/dm^3 excess sediment water solution.

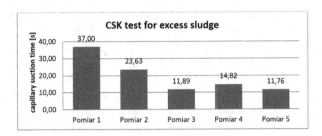

Fig. 6. Capillary suction time of excess sediments with the addition of polyelectrolyte Prestolu 855 BS in a dose of 5 ml/dm^3 excess sediment water solution.

We used a multilayer neural network [3,4,11] with fifteen neurons in each of two hidden layers and one output neuron to predict the effects of conditioning and drying of sewage sludge. After fifty epochs we achieved 98.7% accuracy.

5 Conclusions

The management of sewage sludge should be planned strictly with regard to its final management. Many activities and solutions should be determined and planned at the stage of construction or modernization of the treatment plant. Poor design assumption for the management of sewage sludge can entail huge problems and costs during the operation of the sewage treatment plant. Methane fermentation is quite an effective solution for sewage sludge management, but it depends on the supply and quality of the raw material. Methane co-fermentation allows to eliminate the disadvantages of monofermentation by independence from supplies and quality of the raw material. Co-fermentation allows better control of the amount of nutrients in the fermentation chamber. One of the main problems of wastewater treatment is and will always be high hydration, which makes transport and storage difficult. Both mono- and co-fermentation require prior preparation of the sludge by dehydration. Dehydration can be implemented in many ways using various technological solutions. The addition of appropriate polyelectrolytes may translate into an improvement in the drainage of sewage sludge. We used nonlinear multilayer perceptron neural network to predict the effects of conditioning and drying of sewage sludge. The results were almost perfect. Not achieving 100% accuracy was possibly caused by measurement errors when creating the dataset.

References

1. ATV-DVWK-M 368E: Biological Stabilisation of Sewage Sludge, German water and waste water association. http://www.dwa.de (2011). Accessed 14 Aug 2021
2. Babel, S., del Mundo Dacera, D.: Heavy metal removal from contaminated sludge for land application: a review. Waste Manage. **26**(9), 988–1004 (2006)
3. Bilski, J., Kowalczyk, B., Marjański, A., Gandor, M., Zurada, J.: A novel fast feedforward neural networks training algorithm. J. Artif. Intell. Soft Comput. Res. **11**(4), 287–306 (2021). https://doi.org/10.2478/jaiscr-2021-0017
4. Bishop, C.M., et al.: Neural Networks for Pattern Recognition. Oxford University Press (1995)
5. Chandra, R., Takeuchi, H., Hasegawa, T.: Methane production from lignocellulosic agricultural crop wastes: a review in context to second generation of biofuel production. Renew. Sustain. Energy Rev. **16**(3), 1462–1476 (2012)
6. Czekala, J.: Dewatering of sewage sludge - selected issues. Wodociagi-Kanalizacja (12) (2003)
7. Gazda, M., Rak, A., Sudak, M.: Research on cofermentation of sewage sludge with waste fats for the wastewater treatement plant in brzeg. Infrastruktura i Ekologia Terenów Wiejskich III(III) (2012)
8. Kozłowski, K., Dach, J., Lewicki, A., Cieślik, M., Czekała, W., Janczak, D.: Environmental and process parameters of methane fermentation in continuosly stirred tank reactor (CSTR). Inżynieria Ekologiczna **50**, 153–160 (2016)
9. Kumaran, P., Hephzibah, D., Sivasankari, R., Saifuddin, N., Shamsuddin, A.H.: A review on industrial scale anaerobic digestion systems deployment in Malaysia: opportunities and challenges. Renew. Sustain. Energy Rev. **56**, 929–940 (2016)
10. Mao, C., Feng, Y., Wang, X., Ren, G.: Review on research achievements of biogas from anaerobic digestion. Renew. Sustain. Energy Rev. **45**, 540–555 (2015)
11. Rumelhart, D.E., Hinton, G.E., Williams, R.J.: Learning internal representations by error propagation. California Univ San Diego La Jolla Inst for Cognitive Science, Technical report (1985)
12. Smith, S.: Organic contaminants in sewage sludge (biosolids) and their significance for agricultural recycling. Philos. Trans. R. Soc. A Math. Phys. Eng. Sci. **367**(1904), 4005–4041 (2009)

An Efficient Algorithm to Find a Maximum Weakly Stable Matching for SPA-ST Problem

Nguyen Thi Uyen[1(✉)] and Tran Xuan Sang[2]

[1] School of Engineering and Technology, Vinh University, Vinh, Vietnam
uyennt@vinhuni.edu.vn
[2] Cyber School, Vinh University, Vinh, Vietnam
sangtx@vinhuni.edu.vn

Abstract. This paper presents a heuristic algorithm to seek a maximum *weakly stable* matching for the *Student-Project Allocation with lecturer preferences over Students containing Ties* (SPA-ST) problem. We extend Gale-Shapley's idea to find a stable matching and propose two new heuristic search strategies to improve the found stable matching in terms of maximum size. The experimental results show that our algorithm is more effective than AP in terms of solution quality and execution time for solving the MAX-SPA-ST problem of large sizes.

Keywords: SPA · Heuristic Search · Weakly Stable Matching · MAX-SPA-ST · Undominated Blocking Pairs

1 Introduction

The *Student-Project Allocation problem with lecturer preferences over Students containing Ties* (SPA-ST) is an extension of the Student-Project Allocation problem (SPA) [6,7,9,17,18,20]. This extension makes the original SPA problem more practical because lecturers have preference lists over students, and students also have preference lists over projects with allowing ties in order. The goal of SPA-ST is to seek a *stable matching* like SPA, which includes pairs of students and projects based on their preference lists. Note that each student is eligible for only one project, and capacity constraints of both projects and lecturers meet requirements and satisfactions. According to ties given in the SPA-ST problem, there are three stability criteria of matching consists of *weakly stable*, *strongly stable*, and *super-stable* matching [21,22].

Recently, several researchers have focused on solving the SPA-ST problem because of its applications to large-scale matching schemes in university departments around the world, such as Glasgow University [14], Southern Denmark University [20], York University [15], and elsewhere [2–4,8,10]. Several algorithms have been proposed to solve the SPA-ST problem. Cooper et al. [6] presented a 3/2- approximation algorithm, called AP, to find a *weakly stable*

© The Author(s), under exclusive license to Springer Nature Switzerland AG 2023
L. Rutkowski et al. (Eds.): ICAISC 2022, LNAI 13589, pp. 356–366, 2023.
https://doi.org/10.1007/978-3-031-23480-4_30

matching based on Király's idea for the HRT problem [13,16]. Besides, they also modeled the SPA-ST problem as an Integer Programming (IP) problem. Olaosebikan et al. [21] described the polynomial-time algorithm to find a *strongly stable* matching, and they proved that it might not exist for SPA-ST problem. Their algorithm runs in $O(m^2)$ time, where m is the total length of the students' preference lists. In addition, Olaosebikan et al. [22] proposed an approximation algorithm for solving SPA-ST problem in terms of finding a *super-stable* matching.

Practically, the problem of finding *weakly stable* matching is the most suitable for real-life applications. Irving et al. [11] showed that *weakly stable* matchings always exist and have different sizes [19]. This research aims to find a *weakly stable* matching with maximum size, called a MAX-SPA-ST problem, meaning that as many students as possible are assigned to projects. However, the MAX-SPA-ST problem is known as NP-hard, and therefore, finding an efficient algorithm to solve the MAX-SPA-ST of large sizes is a challenge for the research community.

Our contribution. This paper presents an effective heuristic algorithm to solve the MAX-SPA-ST problem of large sizes. Our main idea is to start from a stable matching, then define two heuristic strategies promoting *unmatched students* and *under-subscribed* lecturers to improve the matching size by breaking stable pairs. Our algorithm terminates when it finds a *perfect matching* or reaches a maximum number of iterations. The experimental results show that our proposed algorithm is more efficient than the AP algorithm [6] in terms of solution quality and execution time.

The rest of this paper is organized as follows: Sect. 2 presents preliminaries of SPA-ST, Sect. 3 describes our proposed algorithm, Sect. 4 discusses our experimental results, and Sect. 5 concludes our work.

2   Preliminaries

An SPA-ST instance consists of a set of students, denoted by $\mathcal{S} = \{s_1, s_2, \cdots, s_n\}$, a set of projects, denoted by $\mathcal{P} = \{p_1, p_2, \cdots, p_q\}$, and a set of lecturers, denoted by $\mathcal{L} = \{l_1, l_2, \cdots, l_m\}$. Each lecturer l_k offers a set of projects and ranks a set of students in her/his preference list. Each student s_i ranks a set of projects in her/his preference list. Both lecturers' and students' preference lists allow ties in order. Each lecturer has a capacity $d_k \in \mathbb{Z}^+$ indicating the maximum number of students that can be matched to l_k. Each project is offered by one lecturer and has a capacity $c_j \in \mathbb{Z}^+$ indicating the maximum number of students that can be matched to p_j. For any pair $(s_i, p_j) \in \mathcal{S} \times \mathcal{P}$ where p_j is offered by l_k, we consider (s_i, p_j) as an *acceptable pair* if s_i and p_j both find each other acceptable, i.e. p_j is ranked by a student s_i and s_i is ranked by a lecturer l_k who offers p_j. We denote the rank of p_j in s_i's preference list by $R_{s_i}(p_j)$ and the rank of s_i in l_k's preference list by $R_{l_k}(s_i)$. Note that we will use the term rank list instead of the preference list in the implementation process.

A matching M of a SPA-ST instance is a set of acceptable pairs (s_i, p_j) or $(s_i, \varnothing)$ such that $|M(s_i)| \leq 1$ for all $s_i \in \mathcal{S}$, $|M(p_j)| \leq c_j$ for all $p_j \in \mathcal{P}$, and $|M(l_k)| \leq d_k$ for all $l_k \in \mathcal{L}$. A project p_j is *under-subscribed, full* or *over-subscribed* according as $|M(p_j)| < c_j$, $|M(p_j)| = c_j$, or $|M(p_j)| > c_j$, respectively. Similarly, lecturer l_k is *under-subscribed, full* or *over-subscribed* according as $|M(l_k)| < d_k$, $|M(l_k)| = d_k$, or $|M(l_k)| > d_k$, respectively. If $(s_i, p_j) \in M$, then s_i is matched to p_j, denoted by $M(s_i) = p_j$. If $M(s_i) = \varnothing$, then s_i is *unmatched* in M.

Let $(s_i, p_j) \in (\mathcal{S} \times \mathcal{P}) \setminus M$ be a blocking pair for a weakly stable matching M if the following conditions are satisfied:

1. s_i and p_j find accept each other;
2. s_i prefers p_j to $M(s_i)$ or $M(s_i) = \varnothing$;
3. either (a), (b) or (c) holds as follows:
   (a) $|M(p_j)| < c_j$ and $|M(l_k)| < d_k$;
   (b) $|M(p_j)| < c_j$, $|M(l_k)| = d_k$ and;
      i. either $s_i \in M(l_k)$ or;
      ii. l_k prefers s_i to the worst student in $M(l_k)$;
   (c) $|M(p_j)| = c_j$ and l_k prefers s_i to the worst student in $M(p_j)$.

Suppose that we have two blocking pairs (s_i, p_j) and (s_i, p_k), we say that (s_i, p_j) *dominates* (s_i, p_k) from the student's point of view if s_i prefers p_j to p_k. A pair (s_i, p_j) is *undominated* if there are no blocking pairs that dominate it from the student's point of view.

A matching M is called *weakly stable* if it admits no blocking pair, otherwise it is called *unstable*. In this paper, we consider a *weakly stable* matching as a stable matching. The size of a stable matching M, denoted by $|M|$, is the number of matched students in M. If $|M| = n$, then M is a *perfect* matching, otherwise, M is a *non-perfect* matching.

3   Proposed Algorithm

3.1   HA Algorithm

This section describes our heuristic algorithm for the MAX-SPA-ST problem, called HA, in Algorithm 1. Our main idea is to start stable matching, which is adapted from Gale-Shapely's idea [7]. Then, if a stable matching is *non-perfect*, we improve its size by proposing two heuristic search strategies with two tasks as follows:

Task 1: Algorithm 1 selects a random *unmatched* student s_i from a stable matching M. Then, the algorithm considers one by one project $p_j \in \mathcal{P}$ in order of s_i's rank list in which p_j is offered by l_k. The algorithm finds a student s_t which is the same ties with s_i in l_k's rank list, i.e. $R_{l_k}(s_t) = R_{l_k}(s_i)$. Then, the algorithm satisfies either condition in case (1) or (2) as follows: *Case* (1): p_j is *under-subscribed* and $v(s_i) \geq v(s_t)$. *Case* (2): p_j is full, $s_t \in M(p_j)$ and $v(s_i) \geq v(s_t)$, then the algorithm replaces (s_t, p_z) where $p_z = M(s_t)$ by (s_i, p_j) in M and

---

Algorithm 1: HA Algorithm for MAX-SPA-ST problem

Input: - An SPA-ST instance I.
    - *max_iter* is the maximum number of iterations.
Output: A maximum stable matching M.

1. **function** HA(I)
2. $M := $ EGS(I);   ▷ **generate a stable matching**
3. $v(s_i) := 0, (1 \le i \le n)$;   ▷ **mark the replacing time of** s_i
4. $v(p_i) := 0, (1 \le i \le q)$;   ▷ **mark the replacing time of** p_i
5. $iter := 0$;
6.   **while** *(iter $\le$ max_iter)* **do**
7. $iter := iter + 1$;
8.   **if** $|M| = n$ **then** break ;
9. $s_i :=$ a random unmatched student in M;   ▷ Task 1
10. $R'_{s_i} := s_i$'s ranks list;
11.   **while** R'_{s_i} *is non-empty* **do**
12. $p_j := \text{argmin}(R'_{s_i} > 0), \forall p_j \in \mathcal{P}$;
13. $l_k :=$ a lecturer who offers p_j;
14.   **for** *(each $s_t \in M(l_k) | R_{l_k}(s_t) = R_{l_k}(s_i)$)* **do**
15.   **if** $(|M(p_j)| < c_j)$ *or* $(s_t \in M(p_j)$ *and* $|M(p_j)| = c_j)$ **then**
16.   **if** $v(s_i) \ge v(s_t)$ *or a small probability* **then**
17. $M := M \setminus \{(s_t, p_z)\} \cup \{(s_i, p_j)\} | p_z = M(s_t)$;
18. $v(s_i) := v(s_i) + 1$;
19. $Repair(p_z, l_k)$;
20. $M := $ Break_Student(M, s_t);
21.   break;
22.   **if** $M(s_i) \ne \varnothing$ **then** break;
23.   **else** $R'_{s_i}(p_j) := 0$;
24. $p_i :=$ a random *under-subscribed* project in M;   ▷ Task 2
25. $l_k :=$ a lecturer who offers p_i;
26.   **for** *each $s_j \in \mathcal{S} | R_{s_j}(p_t) = R_{s_j}(p_i) | p_t = M(s_j)$* **do**
27.   **if** $(|M(l_k)| < d_k)$ *or* $(|M(l_k)| = d_k$ *and* $s_j \in M(l_k))$ **then**
28.   **if** $(v(p_i) \ge v(p_t)$ *or a small probability* **then**
29. $M := M \setminus \{(s_j, p_t)\} \cup \{(s_j, p_i)\}$;
30. $v(p_i) := v(p_i) + 1$;
31. $M := $ Break_Lecturer(M, p_t);
32.   break;
33.   **return** M;
34. **end function**

---

increases the value of $v(s_i)$. It should be noted that the condition $v(s_i) \ge v(s_t)$ means that the number of replacements of s_i is higher than s_t, meaning that s_i is prioritized to match with p_j. If p_z is removed and became *under-subscribed*, we call the function $Repair(p_z, l_k)$ to break blocking pairs for M.

Algorithm 2: Breaking blocking pair from the student of M

    Input: A matching M.

    Output: A stable matching M.

1. **function** Break_Student(M, s_t)
2.   **while** *(there exists blocking pairs)* **do**
3. $(s_t, p_u) :=$ an undominated blocking pair from s_t;
4. $l_k :=$ a lecturer who offers p_u;
5. $M := M \cup \{(s_t, p_u)\}$;
6.   **if** p_u *is over-subscribed* **then**
7. $s_w :=$ a worst student of p_u;
8. $M := M \setminus \{(s_w, p_u)\}$;
9. $s_t := s_w$;
10.   **else if** l_k *is over-subscribed* **then**
11. $s_r :=$ a worst student of l_k;
12. $M := M \setminus \{(s_r, p_z)\}$, where $p_z = M(s_r)$;
13. $Repair(p_z, l_k)$; ▷ repair blocking pair of type *(3bi)*
14. $s_t := s_r$;
15.   **return** M;
16. **end function**

To avoid a local minimum with a small probability, we prioritize s_i without considering the value of $v(s_i)$. As a result, s_t is now *unmatched*, thus the algorithm calls the Algorithm 2 to break blocking pairs for M. Finally, Algorithm 1 returns a stable matching that is equal to or greater in size than the current matching M.

Task 2: Algorithm 1 selects a random *under-subscribed* project p_i from a stable matching M. Then, the algorithm finds a student s_j which ranks p_i at the same rank as $M(s_j)$ in s_j's rank list, i.e. $R_{s_j}(p_i) = R_{s_j}(p_t)$ where $M(s_j) = p_t$. Then, the algorithm satisfies either condition in case (1) or (2) as follows: *Case* (1): l_k is *under-subscribed* and $v(p_i) \geq v(p_t)$. *Case* (2): l_k is full, $s_j \in M(l_k)$, and $v(p_i) \geq v(p_t)$, then the algorithm replaces (s_j, p_t) by (s_j, p_i) in M and increases the value of $v(p_i)$. Note that $v(p_i) \geq v(p_t)$ means that the number of replacements of p_i is higher than p_t, i.e. p_i is prioritized to match with s_j. To avoid a local minimum, with a small probability, we always prioritize p_i without considering the value of $v(p_i)$. As a result, p_t and l_w who offers p_t are *under-subscribed*, thus the algorithm calls the Algorithm 3 to break blocking pairs for M. Finally, the Algorithm 1 returns a stable matching that is equal to or greater in size than the current matching M. Our HA algorithm stops if a perfect matching is found or it is reached to the maximum number of iterations.

Algorithm 3: Breaking blocking pair from the lecturer of M

Input: A matching M.
Output: A stable matching M.

1. **function** Break_Lecturer(M, p_t)
2.   **while** *(there exists blocking pairs)* **do**
3. $l_w :=$ a lecturer who offers p_t;
4.   **if** p_t *is under-subscribed* **then**
5. $Repair(p_t, l_w)$; ▷ **repair blocking pair of type** *(3bi)*
6. $(s_z, p_u) :=$ an undominated blocking pair from l_w;
7.   **if** *there exists pair* (s_z, p_u) **then**
8. $M := M \setminus \{(s_z, p_h)\} \cup \{(s_z, p_u)\}$, where $p_h = M(s_z)$;
9.   **if** p_u *is over-subscribed* **then**
10. $s_w :=$ a worst student of p_u;
11. $M := M \setminus \{(s_w, p_u)\}$;
12. $M := $ Break_Student(M, s_w);
13.   **else if** l_w *is over-subscribed* **then**
14. $s_r :=$ a worst student of l_w;
15. $M := M \setminus \{(s_r, p_z)\}$, where $p_z = M(s_r)$;
16. $Repair(p_z, l_w)$;
17. $M := $ Break_Student(M, s_r);
18. $p_t := p_h$;
19.   **else**
20.   break;
21.   **return** M;
22. **end function**

We use Algorithm 2 to break blocking pairs when a student s_t is removed and becomes an *unmatched* student. The algorithm finds an undominated blocking pair (s_t, p_u) from s_t's point of view. If there exists, then we add (s_t, p_u) into M, where p_u is offered by l_k. This process repeats until there are no existing blocking pairs for only *unmatched* students who have just been removed.

We use Algorithm 3 to break blocking pairs when a project p_t is replaced and becomes *under-subscribed*. The algorithm uses the function $Repair(p_t, l_w)$ to remove blocking pairs type of *(3bi)*, then we find an undominated blocking pair (s_z, p_u) from l_w's point of view. If there exists, then we remove (s_z, p_h) where $p_h = M(s_z)$ and add (s_z, p_u) into M. This process repeats until there are no existing blocking pairs for only p_h which have just been removed.

3.2 Example

This section presents an example execution of our HA algorithm for the SPA-ST instance consisting of seven students, eight projects, and three lecturers in Table 1. Starting from a stable matching $M = \{(s_1, p_1), (s_2, \varnothing), (s_3, p_4), (s_4, p_2),$ $(s_5, \varnothing), (s_6, p_5), (s_7, p_3)\}$ of size $|M| = 5$, HA runs as follows: HA algorithm takes a random *unmatched* student s_2 and finds s_6 such that $R_{l_2}(s_2) = R_{l_2}(s_6)$ from l_2's rank list. Then, HA removes (s_6, p_5) and adds (s_2, p_5) into M, thus

s_6 becomes *unmatched*. Next, the algorithm calls Algorithm 2 to break blocking pairs. The algorithm finds an undominated blocking pair (s_6, p_8) from s_6's point of view and adds (s_6, p_8) into M to generate a new stable matching $M = \{(s_1, p_1),$ $(s_2, p_5), (s_3, p_4), (s_4, p_2), (s_5, \varnothing), (s_6, p_8), (s_7, p_3)\}$ of size $|M| = 6$. Next, HA considers a random *under-subscribed* project p_7 and seeks a project p_1 which has $R_{s_1}(p_7) = R_{s_1}(p_1)$ from s_1's rank list. Then, the algorithm removes (s_1, p_1) and adds (s_1, p_7) into M. Next, the algorithm calls the Algorithm 3 to break blocking pairs for M. Finally, HA returns a perfect matching $M = \{(s_1, p_7),$ $(s_2, p_5), (s_3, p_1), (s_4, p_2), (s_5, p_4), (s_6, p_8), (s_7, p_3)\}$ of size $|M| = 7$.

Table 1. An instance of SPA-ST

Student's preferences	Lecturer's preferences	
s_1: $(p_1\ p_7)$	l_1: $(s_7\ s_4)\ s_1\ s_3\ (s_2\ s_5)\ s_6$	l_1 offers p_1, p_2, p_3
s_2: $p_1\ p_2\ (p_3\ p_4)\ p_5\ p_6$	l_2: $s_3\ (s_2\ s_6)\ s_7\ s_5$	l_2 offers p_4, p_5, p_6
s_3: $(p_2\ p_1)\ p_4$	l_3: $(s_1\ s_7)$	l_3 offers p_7, p_8
s_4: p_2		
s_5: $(p_1\ p_2)\ p_3\ p_4$		
s_6: $(p_2\ p_3)\ p_4\ p_5\ p_6$	Project capacities	$c_1 = 2,\ c_i = 1,\ (2 \le i \le 8)$
s_7: $(p_5\ p_3)\ p_8$	Lecturer capacities	$d_1 = 3,\ d_2 = 2,\ d_3 = 2$

4   Experimental Results

In this section, we compared the solution quality and execution time of HA with those of AP which is an approximation algorithm [6] for the MAX-SPA-ST problem. We implemented these algorithms by Matlab R2019a software on a Xeon-R Gold 6130 CPU 2.1 GHz computer with 16 GB RAM. To perform the experiments, we generated randomly SPA-ST instances with five parameters (n, m, q, p_1, p_2), where n is the number of students, m is the number of lecturers, q is the number of projects, p_1 is the probability of incompleteness, and p_2 is the probability of ties. By this setting, on average, each student ranks about $q \times (1 - p_1)$ projects. In our experiments, we set the total capacity of projects and lecturers as $C = 1.2n$ and $D = 1.1n$, respectively.

4.1   Comparison of Solution Quality

This section presents two experiments to compare the solution quality found by HA with that found by AP [6].

Experiment 1. Firstly, we randomly generated 100 instances of SPA-ST for parameters (n, m, q, p_1, p_2) with $n \in \{100, 200\}$, $m = 0.05n$, $q = 0.1n$, $p_1 \in [0.1, 0.8]$ with step 0.1, and $p_2 \in [0.0, 1.0]$ with step 0.1. Then, we ran HA and AP, averaged results, and compared the percentage of perfect matchings and the average number of *unmatched* students found by these two algorithms. Our experimental results show that when $p_1 \in [0.1, 0.6]$ with every the value of p_2, both our HA and AP obtain approximately 100% of perfect matchings, so we do not show the experiment results here. Figures 1(a) and 1(c) show the percentage of perfect matchings found by HA and AP. When $p_1 = 0.7$ or $p_1 = 0.8$, HA finds a much higher percentage of perfect matchings than AP does. Figures 1(b) and 1(d) show the average number of *unmatched* students found by HA and AP. When $p_1 = 0.7$ or $p_1 = 0.8$, HA finds a fewer number of *unmatched* students in stable matchings than AP does.

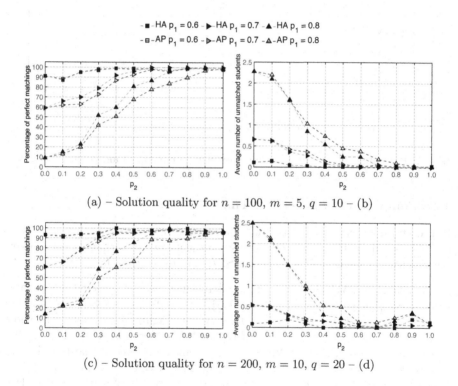

(a) – Solution quality for $n = 100$, $m = 5$, $q = 10$ – (b)

(c) – Solution quality for $n = 200$, $m = 10$, $q = 20$ – (d)

Fig. 1. Percentage of perfect matching and average number of unmatched students

Experiment 2. As we saw in Experiment 1, when p_1 increases, both HA and AP are hard to find perfect matchings since the number of projects ranked in students' preference lists decreases. In this experiment, we changed $n \in \{300, 400\}$, $p_1 \in \{0.82, 0.84, 0.86\}$ and kept the values of m, q, and p_2 as in Experiment 1. Figure 2 shows the percentage of perfect matchings found by HA and AP. Again, we see that HA finds a much higher percentage of perfect matchings than AP does.

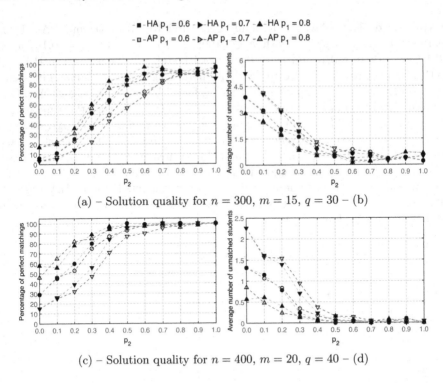

(a) – Solution quality for $n = 300$, $m = 15$, $q = 30$ – (b)

(c) – Solution quality for $n = 400$, $m = 20$, $q = 40$ – (d)

Fig. 2. Percentage of perfect matching and average number of unmatched students

4.2 Comparison of Execution Time

In the above experiments, n is small, and therefore, the execution time of HA and AP is almost the same. This section presents two experiments to compare the execution time of HA and AP for SPA-ST instances of large sizes.

Experiment 3. We randomly generated 100 instances of SPA-ST for parameters (n, m, q, p_1, p_2) with $n \in \{1000, 2000\}$, $m = 0.05n$, $q = 0.4n$, $p_1 \in [0.1, 0.8]$ with step 0.1, and $p_2 \in [0.0, 1.0]$ with step 0.1. Figures 3 (a) and 3(b) show the average execution time over p_1 of HA and AP. When p_2 increases from 0.0 to 1.0, the execution time of AP almost remains unchanged, while HA slightly decreases, except for $p_2 = 1.0$, the execution time of HA significantly increases. When $n = 1000$, HA runs about 9 times faster than AP. When $n = 2000$, HA runs about 12.5 times faster than AP.

Experiment 4. Finally, we kept the values of n, m, q, and p_2 as in Experiment 3, increased the values of $p_1 \in [0.81, 0.89]$ with step 0.01, and randomly generated 100 instances of SPA-ST for each combination of values (p_1, p_2). By increasing the values of p_1, we aim to reduce the number of projects ranked by each student compared to Experiment 3. Figures 3(c) and 3(d) show the average execution time over p_1 of HA and AP. As in Experiment 3, we saw that

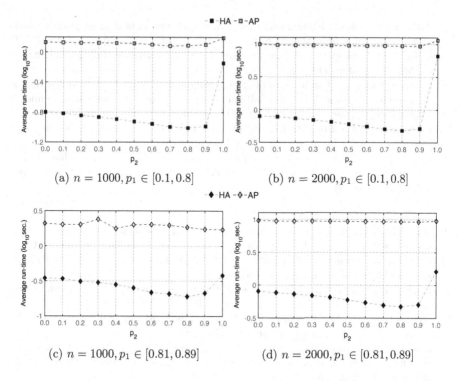

(a) $n = 1000, p_1 \in [0.1, 0.8]$ (b) $n = 2000, p_1 \in [0.1, 0.8]$

(c) $n = 1000, p_1 \in [0.81, 0.89]$ (d) $n = 2000, p_1 \in [0.81, 0.89]$

Fig. 3. Average of execution time for $n = 1000, 2000$

when p_2 increases from 0.0 to 1.0, the execution time of AP almost remains unchanged, while HA slightly decreases, except for $p_2 = 1.0$, the execution time of HA increases. When $n = 1000$, HA ran faster about 5 times than AP. When $n = 2000$, HA runs faster about 12.5 times than AP.

5   Conclusions

In this study, we presented a heuristic algorithm for solving the MAX-SPA-ST problem. We started with a stable matching and improved the matching size by defining two heuristic strategies to pair the *unmatched* students and *under-subscribed* projects. The experimental results showed that our proposed algorithm is efficient in terms of solution quality and execution time for the MAX-SPA-ST problem of large sizes. In the future, we will extend this proposed approach to solve the other variants of the SPA problem [1,5,12,20].

References

1. Abraham, D., Irving, R., Manlove, D.: Two algorithms for the student-project allocation problem. J. Discret. Algorithms **5**(1), 73–90 (2007)

2. Aderant, F., Amosa, R., Oluwatobiloba, A.: Development of student project allocation system using matching algorithm. In: ICONSEET, vol. 1, pp. 153–160 (2016)
3. Binong, J.: Solving student project allocation with preference through weights. In: COMSYS, pp. 423–430 (2021)
4. Calvo-Serrano, R., Guillén-Gosálbez, C., Simon, K., Andrew, M.: Mathematical programming approach for optimally allocating students' projects to academics in large cohorts. Educ. Chem. Eng. **20**, 11–21 (2017)
5. El-Atta, A.A., Ibrahim, M.M.: Student project allocation with preference lists over (student, project) pairs. In: ICCEE, vol. 1, pp. 375–379 (2009)
6. Frances, C., Manlove, D.: A 3/2-approximation algorithm for the student-project allocation problem. In: SEA, vol. 103, pp. 8:1–8:13 (2018). https://doi.org/10.4230/LIPIcs.SEA.2018.8
7. Gale, D., Shapley, L.S.: College admissions and the stability of marriage. Am. Math. Mon. **9**(1), 9–15 (1962)
8. Gani, M., AHamid, R., et al.: Optimum allocation of graduation projects: survey and proposed solution. J. Al-Qadisiyah Comput. Sci. Math. **13**(1), 58–66 (2021)
9. Hamada, K., Iwama, K., Miyazaki, S.: The hospitals-residents problem with lower quotas. Algorithmica **74**(1), 440–465 (2016)
10. Harper, R., Senna, V., Vieira, I., Shahani, A.: A genetic algorithm for the project assignment problem. Comput. Oper. Res. **32**(5), 1255–1265 (2005)
11. Irving, R., Manlove, D., Sandy, S.: The hospitals/residents problem with ties. In: SWAT, pp. 259–271. Bergen, Norway, July 2000
12. Ismaili, A., Yahiro, K., Yamaguchi, T., Yokoo, M.: Student-project-resource matching-allocation problems: two-sided matching meets resource allocation. In: AAMAS, pp. 2033–2035 (2019)
13. Iwama, K., Miyazaki, S., Yanagisawa, H.: Improved approximation bounds for the student-project allocation problem with preferences over projects. J. Discret. Algorithms **13**, 59–66 (2012)
14. K. Augustine, Irving, R., Manlove, D., S.Colin: profile-based optimal matchings in the student/project allocation problem. In: IWOCA, pp. 213–225 (2014)
15. Kazakov, D.: Co-ordination of student-project allocation. Manuscript (2002)
16. Király, Z.: Linear time local approximation algorithm for maximum stable marriage. Algorithms **6**(1), 471–484 (2013)
17. Kwanashie, K., Manlove, D.: An integer programming approach to the hospitals/residents problem with ties. In: GOR, pp. 263–269 (2013)
18. Manlove, D., Gregg, O.: Student-project allocation with preferences over projects. J. Discret. Algorithms **6**(4), 553–560 (2008)
19. Manlove, D., Irving, R., Iwama, K., Miyazaki, S., Morita, Y.: Hardvariants of stable marriage. Theoret. Comput. Sci. **276**(1), 261–279 (2002)
20. Marco, C., Rolf, F., Stefano, G.: Handling preferences in student-project allocation. Ann. Oper. Res. **275**(1), 39–78 (2019)
21. Olaosebikan, S., Manlove, D.: An algorithm for strong stability in the student-project allocation problem with ties. In: CALDAM, pp. 384–399 (2020)
22. Olaosebikan, S., Manlove, D.: Super-stability in the student-project allocation problem with ties. J. Comb. Optim. **43**, 1–37 (2020)

Author Attribution of Literary Texts in Polish by the Sequence Averaging

Tomasz Walkowiak[(✉)](iD)

Faculty of Information and Communication Technology,
Wroclaw University of Science and Technology, Wroclaw, Poland
`tomasz.walkowiak@pwr.edu.pl`

Abstract. The paper concerns the authorship recognition in a collection of Polish literary texts from the late 19th and early 20th centuries, consisting of 99 novels from 33 authors. The authors divide the books into smaller parts and analyze the classification based on a book part. To mimic the real task of testing an unknown book, the data set has been divided in such a way that parts of the same book do not appear simultaneously in the training and test set. The authors compare the approaches by working with raw texts, and, to avoid the semantic features of the text, they represent texts in the form of a sequence of grammatical class bigrams. In the case of raw text analysis, classical TF-IDF, supervised fastText, and contemporary transformer-based BERT are analyzed. In the case of grammatical classes, only TF-IDF and fastText are applied. In addition, the authors propose a sequence averaging method that works by dividing the text into smaller parts, classifying each part separately, and making the final classification based on averaging results from each part of the text. The study suggests that the TF-IDF on the raw text outperforms other methods and the sequence averaging improves the classification results for most of the analyzed schemas. Surprisingly, the BERT based method is the worst. This phenomenon is carefully analyzed and explained.

Keywords: Stylometry · Natural language processing · Polish · BERT · fastText · TF-IDF

1 Introduction

Authorship attribution [15] is part of stylometry [7], an interdisciplinary area of science that aims to study the associations between the statistical properties of texts and their meta-properties (such as authorship). Most of the stylometric analysis concerns texts in English and the authorship attribution [4]. The author attribution is mainly oriented in two types of texts: messages (e-mails, posts) [3,9,11] and literary texts [5].

This study aims to extend the research analysis presented in [24] where the authors compared three groups of methods (bag-of-words [12] with TF-IDF [21], grammatical bigrams [7] and supervised fastText [10]) in the authorship attribution of Polish literary texts from the late 19th and early 20th centuries. The

© The Author(s), under exclusive license to Springer Nature Switzerland AG 2023
L. Rutkowski et al. (Eds.): ICAISC 2022, LNAI 13589, pp. 367–376, 2023.
https://doi.org/10.1007/978-3-031-23480-4_31

results showed that the grammatical bigrams representing the style information included in the text outperformed other methods analyzed. The authors of [24] classify the whole book and divide it into parts (c.a. 20,000 bytes). The size of the chunk (part of the book) was chosen arbitrarily, raising the question of what the size of the chunk should be and how the classification of the chunks should be organized. This paper attempts to answer three research questions.

1. What is the influence of the size of tested text (i.e., chunk size) on the authorship attribution?
2. Does averaging of classifiers results [1] give better results than building a classifier on the whole chunk?
3. Does the BERT [6] deep neural network, the SOTA [23] for text classification, is suitable for authorship analysis [2,9] of literary texts?

The paper is structured as follows. Section 2 describes the classification methods (TF-IDF, fastText, and BERT) that work on raw texts. Next, a description of the methods for working on a sequence of grammatical classes is presented. It is followed by the introduction of the sequence averaging method. The next section introduces the analyzed corpus and shows the numerical results. They are followed by a discussion. Finally, conclusions are given.

2   Methods

2.1   Text Classification

Authorship attribution can be viewed as a classical text classification task, that is, assigning labels to text using a statistical classifier trained on labeled texts.

The most basic method of text classification is a bag-of-words based TF-IDF [21]. The method is based on counting the occurrences of words in a text and weighting these frequencies by the maximum word frequency in a document and by IDF (inverse document frequency). The weighted frequencies are standardized, that is, we remove the mean and scale the feature vectors to the unit variance. Standardization is a widely used transformation of data in machine learning and is also known in statistics as the z-score. As a classifier, we have used the multilayer perceptron (MLP) [13] trained using the stochastic gradient descent method (SGD) [13]. The list of words, corresponding IDFs, mean, and variance of each feature is set up on the train set and used for the calculation of features on the test set.

The big step in the area of text classification was the introduction of the word2vec method [16]. In these approaches, individual words are represented by high-dimensional feature vectors (word embeddings) trained on large text corpora, that is, why such methods are called language models (LM). The next method used, fastText [14] is similar to word2vec in building the embedding of words. However, word embeddings are learned in a supervised way on the analyzed data set, not on the external large corpora. The main idea behind fastText (a supervised version) is to perform word embedding and classifier learning simultaneously. Since classification is done for the whole text (sequence of words), fastText needs a method of document representation; it is done by averaging the embedding of words in text.

The newest approaches to language modeling are inspired by deep learning algorithms and are context-aware methods. The state of the art is BERT [6] based on the transformers [22] architecture. We have extended the BERT network by the classification layer (fully connected) and tuned the whole model (BERT part and classifier) on the training set. In this study, we used the HerBert (base)[1] [20] language model.

2.2 Classification Based on Grammatical Classes

Classical (non-deep) machine learning is based on representing objects by features and applying a statistical classification. The TF-IDF method, described in the previous section, is an example of such an approach, where the features are weighted vectors of word occurrences. In author attribution, the features should reveal characteristics of the text that are specific to the author and his or her style. They should not be correlated with the content of the text. Because books by the same author may cover different subject areas. Therefore, the use of word-based features raises the doubt that a classifier will pick some characteristic semantic words, like the names of heroes, and will base a classification on them. Therefore, we follow the ideas proposed in [17] and [24] of representing texts by bigrams (a sequence of two elements) of grammatical classes. For sure, a sequence of grammatical classes does not contain semantic information and contains some information of the text style.

Text in Polish is automatically converted (using tagger [19]) in the sequence of tokens representing pairs of part-of-speech tags following the National Corpus of Polish tagset [18] with an additional tag at the end of a sentence. An example of such a representation is shown in Fig. 1. It resulted in a vocabulary of size equal to 1195. We can treat such sequences as texts in some artificial language and use the same classification methods as described in Sect. 2.1. However, in the case of BERT, we are lacking the grammatical bigram language model to be tuned on our dataset. Theoretically, it could be possible to train such models, but the results could be doubtful, due to a small vocabulary of bigram sequences (compared to natural language texts). Therefore, only TF-IDF and fastText could be used in performed experiments.

2.3 Sequence Classification by Averaging

The text classification algorithms assign a class label to a sequence of words. Therefore, they are inherently dealing with a sequence classification problem. It is solved differently depending on the algorithm. TF-IDF builds the feature vector for the whole text. fastText averages the embeddings of each word in the sequence. In the case of BERT, it is a much more sophisticated mechanism. It

---

[1] https://huggingface.co/allegro/herbert-base-cased.

prep_subst subst_subst subst_qub qub_adv adv_praet praet_interp interp_conj
conj_subst subst_interp interp_interp interp_pcon pcon_adj adj_subst subst_interp
interp_adv adv_qub qub_praet praet_adj adj_subst subst_interp interp_prep prep_adj
adj_subst subst_interp interp_praet praet_ppron3 ppron3_subst subst_interp
interp_prep prep_subst subst_ign ign_ppas ppas_interp interp_conj conj_praet
praet_adv adv_prep prep_subst subst_conj conj_adj adj_subst subst_prep

Fig. 1. Example of a sequence of grammatical bigrams. The first sentence from "Pan burmistrz z Pipidówki" by Michał Bałucki.

builds a sequence representation by dynamically weighted averages (attention mechanism) of subword (token) embeddings and next repeats the same mechanism in each layer, building a new representation (sequence of vectors) based on results from the previous layer.

We propose to test the other mechanism: to divide each text into smaller chunks, build a classifier on chunked texts, and during the inference build the final decision based on the sequence of decisions for each chunk. It is similar to the ensemble approach in machine learning, however, here we have one classifier and different inputs. As the classifier of the sequence, we propose to use a simple averaging rule [1]. Let $p_c(x_i)$ be the a posteriori probability of the chunk x_i and the label c. The sum of c is equal to 1. Technically, it is a softmax layer output (the last one) presented in each of the analyzed text classifiers. Let n be the length of the text in chunks. The classification principle is to find the maximum value of the average probability

$$\arg\max_c \frac{1}{n} \sum_i p_c(x_i). \tag{1}$$

3   Experiments and Results

3.1   Data Set

The study was carried out on 99 Polish novels written by 33 authors in the late 19th and early 20th centuries (the corpus from [8], without one novel by Magdalena Samozwaniec). We have randomly divided the books into training and testing so that two books of each author were placed in the training set and one in the test set. The books were then divided into chunks that count more or less the same number of words. We have made nine sets of data, depending on the length of the chunk, i.e. 100, 300, 500, 1000, 3000, 5000, 10000, 50000 words, and whole books. The order of chunks in the books was preserved to allow sequential averaging. It is important to state that the chunks in the test set are from different books than chunks in the training set. It allows us to mimic the real case when analyzed, the unknown author book (or a chunk of it) was not seen during learning.

3.2   Results for Text Classifiers

We have performed a set of experiments. Each of the three analyzed text classifiers (described in Sect. 2.1) was trained on each dateset. Data sets differ in the size of the chunks used in training. The results are presented in Tables 1, 2, 3. The values on the diagonals show the accuracy of the text classifiers (without any averaging). The values above the diagonals show the results of the sequence averaging method presented in Sect. 2.3. For example, the cell in row 300 and column 3,000 shows the results for the classifier trained on chunks of 300 words size. Texts with a length of 3000 words were classified based on an average of the ten subsequent classifier outputs. Tables 4, 5 show the results of the classification based on the grammatical class bigrams. The accuracies marked in red indicate the best results for the given text length.

Table 1. Accuracy of the text-based TF-IDF method, trained on chunks of different sizes (rows) and tested on texts of different sizes (columns). The label book means that a given classifier was tested or/and trained on the whole book. Values marked in bold show the best accuracy for a given size of chunks in the inference procedure. The values in bold italics are the best for a given text size among all methods, that is, Tables 1, 2, 3, 4 and 5.

Train chunk size	Test text size in words								
	100	300	500	1,000	3,000	5,000	10,000	50,000	Book
100	**33.07**	48.40	58.09	69.29	80.16	83.22	83.93	81.08	78.79
300	——	**59.46**	59.46	75.49	*87.83*	89.80	*91.98*	93.42	*96.97*
500	——	——	**67.23**	*76.55*	87.11	*90.53*	91.30	*96.00*	93.94
1,000	——	——	——	74.81	86.81	89.34	91.22	94.67	93.94
3,000	——	——	——	——	82.18	82.18	88.66	93.42	93.94
5,000	——	——	——	——	——	78.80	85.53	93.33	90.91
10,000	——	——	——	——	——	——	79.14	86.49	84.85
50,000	——	——	——	——	——	——	——	56.92	45.45
Book	——	——	——	——	——	——	——	——	69.70

The best results for texts of equal length or more than 500 words were achieved for TF-IDF based on text. Shorter texts (100, 300 words) are best recognized by fastText based on text. Surprisingly, BERT gave the worst results. The problem will be discussed in the next section, and the results of Table 3 are not analyzed more in this section.

Text methods overcome methods based on grammatical bigrams. However, bigram-based fastText should be considered an interesting method, as it is positioned mainly as a second method and guarantees that it does not take into account the semantic features of the text.

The answer to the first question raised in the Introduction could be found by looking at the diagonals of the tables. For each method, there is an optimal

Table 2. Accuracy of the text-based fastText method trained on chunks of different sizes (rows) and tested on texts of different sizes (columns).

Train chunk size	Test text size in words								
	100	300	500	1,000	3,000	5,000	10,000	50,000	Book
100	*49.03*	*61.81*	**66.06**	**70.90**	76.91	**80.54**	**83.61**	**87.84**	84.85
300	——	59.47	58.16	70.02	77.80	78.52	81.48	81.58	84.85
500	——	——	62.58	70.10	**78.52**	80.10	81.37	86.67	**90.91**
1,000	——	——	——	67.15	76.38	79.64	81.50	85.33	84.85
3,000	——	——	——	——	69.57	65.67	72.09	78.95	75.76
5,000	——	——	——	——	——	67.11	70.39	74.67	78.79
10,000	——	——	——	——	——	——	63.91	67.57	66.67
50,000	——	——	——	——	——	——	——	69.23	60.61
Book	——	——	——	——	——	——	——	——	54.55

Table 3. Accuracy of BERT trained on chunks of different sizes (rows) and tested on texts of different sizes (columns).

Train chunk size	Test text size in words								
	100	300	500	1,000	3,000	5,000	10,000	50,000	Book
100	**42.29**	**49.42**	**52.23**	**54.65**	**57.27**	**58.72**	**58.03**	**59.46**	**54.55**
300	——	42.22	42.22	47.69	53.65	55.80	58.33	57.89	54.55
500	——	——	40.11	43.26	48.05	48.80	52.17	53.33	51.52
1,000	——	——	——	39.84	44.00	45.88	46.39	42.67	42.42
3,000	——	——	——	——	32.23	32.23	37.21	40.79	30.30
5,000	——	——	——	——	——	22.87	25.66	25.33	15.15
10,000	——	——	——	——	——	——	10.26	9.46	6.06
50,000	——	——	——	——	——	——	——	1.54	0.00
Book	——	——	——	——	——	——	——	——	0.00

Table 4. Accuracy of grammatical classes based TF-IDF trained on chunks of different sizes (rows) and tested on texts of different sizes (columns).

Train chunk size	Test text size in words								
	100	300	500	1,000	3,000	5,000	10,000	50,000	Book
100	**31.08**	44.66	52.06	60.08	70.50	73.32	75.41	72.97	66.67
300	——	**46.45**	46.45	62.18	75.85	79.75	82.41	81.58	75.76
500	——	——	**53.72**	**62.40**	**76.56**	**79.94**	**82.61**	82.67	**81.82**
1000	——	——	——	60.92	73.52	78.51	81.19	82.67	**81.82**
3000	——	——	——	——	64.26	64.26	76.45	**82.89**	**81.82**
5000	——	——	——	——	——	67.11	71.05	77.33	75.76
10000	——	——	——	——	——	——	55.96	66.22	60.61
50000	——	——	——	——	——	——	——	50.77	42.42
Book	——	——	——	——	——	——	——	——	33.33

Table 5. Accuracy of grammatical classes fastText trained on chunks of different sizes (rows) and tested on texts of different sizes (columns).

Train chunk size	Test text size in words								
	100	300	500	1,000	3,000	5,000	10,000	50,000	Book
100	**25.64**	39.61	48.56	60.05	74.06	77.35	80.33	79.73	75.76
300	——	**51.00**	50.05	66.62	80.23	83.00	86.11	**90.79**	87.88
500	——	——	**60.34**	68.98	80.47	**85.87**	**87.89**	90.67	**90.91**
1,000	——	——	——	67.82	77.26	80.78	85.27	89.33	87.88
3,000	——	——	——	——	74.17	72.27	81.98	86.84	87.88
5,000	——	——	——	——	——	73.29	75.99	76.00	81.82
10,000	——	——	——	——	——	——	71.85	78.38	75.76
50,000	——	——	——	——	——	——	——	64.62	63.64
Book	——	——	——	——	——	——	——	——	54.55

value of the size of text when the accuracy for a given classifier (built on the same-size chunks) is the best. It is mostly 3,000 to 5,000 words long. For other sizes, it constantly decreases.

One could also notice that averaging of classifier outputs for texts longer than 500 words gives better results than building a classifier on the whole text. It is manifested in Tables 1, 2, 3, 4 and 5 by no values in bold (best accuracy for a given method) occurring on the diagonal (the case where there is no classifier average) for values over 500 words in the train chunk. The optimal chunk size used in the training depends on the method used. It is relatively low, i.e. 100 for fastText on raw texts, goes to 300 for TF-IDF on raw texts, and increases to 500 for bigram-based methods.

3.3 Analysis of BERT Results

Reported results show that the BERT-based classifier fails in the attribution of literary texts. It was not expected since BERT is the SOTA for many text classification tasks [6,20,23]. There are also reports of good performance of BERT in the area of author attribution of short texts (such as emails, reviews, chats and discussions) [2,9]. The diagonal of Table 3 shows that BERT gives the best results for short texts (less than 500 words), which was expected due to the limited length of the input of BERT, which is 512 tokens (subwords). However, for texts of 300 words (ca. input limit of BERT), TF-IDF and fastText outperform BERT by 17% points. The explanation of this situation could be found in the construction of the test-train split and the fact that the train set contains chunks of other books than included in the test set. Probably, BERT aligns its model so well with the books in the training set that it cannot generalize to another book by the same author (from the test set). To make this thesis credible, we conducted an experiment when the train-test split is done randomly on all book chunks, i.e., chunks from the same book could exist in the train and test set. We used chunks of 300 words size. For BERT, we got an accuracy equal to 99.15%, and only 92.34% for TF-IDF. This shows that BERT fits the data much better than TF-IDF, but it is a drawback in the case of attributing the authorship of literary texts.

Comparing the rows in Table 3 with other rows in the table, we can see that the accuracy of BERT increases much slower with increasing length of the analyzed text than in other methods. To analyze the reasons, we have carefully investigated the accuracies for each class for BERT and TF-IDF. We have noticed that in the case of BERT there are more than 7 labels (among 33 labels) with results less than 5%, while it never happens in the case of TF-IDF (where the smallest value of the accuracy is 11%). This shows that the class accuracy for BERT has a large variance (some classes are not recognized at all), and that is probably why the sequence accuracy increases poorly as a function of its length.

4   Conclusion

The authors presented and tested a set of methods aimed at finding the author of a book or a part of the book based on a set of labeled books. Achieved results, accuracy ca. 97%, show that even in a rigorous task, when no part of the tested book was seen during learning, the machine learning methods are capable of almost perfect (an error in case of only one book among 33) authorship attribution. The results presented for the non-sequence analysis show that the input text size for the authorship attribution has some optimal value (3,000-5,000 words), and the accuracy for longer texts starts to decrease. This effect could be overcome by sequence averaging, i.e., by a proposed by the authors classification method that trains a classifier on smaller-size chunks and makes the final decision based on an average of classifier outputs for a sequence of chunks. Sequence averaging significantly improves the results.

The surprising conclusion of the work is that the SOTA in most of the text classification tasks, BERT with a classification layer trained on the authorship attribution of literary texts fails. Performed analyses suggest that it is caused by a small diversity in the training set (only two books of each author, chunked in smaller pieces) and the ability of BERT to match the data very accurately.

Future plans include the improvement of the performance of the BERT based methods. We plan to construct more robust data sets and test some modifications of the training goals, i.e., to apply the metric learning approach. Moreover, we plan to test the proposed methods on the corpora of literary texts in other languages.

References

1. Alexandre, L.A., Campilho, A.C., Kamel, M.: On combining classifiers using sum and product rules. Pattern Recogn. Lett. **22**(12), 1283–1289 (2001). https://doi.org/10.1016/S0167-8655(01)00073-3

2. Barlas, G., Stamatatos, E.: Cross-domain authorship attribution using pre-trained language models. In: Maglogiannis, I., Iliadis, L., Pimenidis, E. (eds.) AIAI 2020. IAICT, vol. 583, pp. 255–266. Springer, Cham (2020). https://doi.org/10.1007/978-3-030-49161-1_22

3. Calix, K., Connors, M., Levy, D., Manzar, H., McCabe, G., Westcott, S.: Stylometry for e-mail author identification and authentication (2008)

4. Can, M.: Authorship attribution using principal component analysis and competitive neural networks. Math. Comput. Appl. **19**(1), 21–36 (2014)

5. Craig, H., Kinney, A.: Shakespeare, Computers, and the Mystery of Authorship. Cambridge University Press, Cambridge (2009)

6. Devlin, J., Chang, M.W., Lee, K., Toutanova, K.: Bert: pre-training of deep bidirectional transformers for language understanding. arXiv preprint arXiv:1810.04805 (2018)

7. Eder, M., Piasecki, M., Walkowiak, T.: Open stylometric system based on multi-level text analysis. Cogn. Stud. — Études cognitives **17** (2017). https://doi.org/10.11649/cs.1430

8. Eder, M., Rybicki, J.: Late 19th- and early 20th-century polish novels (2015). http://hdl.handle.net/11321/57. CLARIN-PL digital repository

9. Fabien, M., Villatoro-Tello, E., Motlicek, P., Parida, S.: BertAA: BERT fine-tuning for authorship attribution. In: Proceedings of the 17th International Conference on Natural Language Processing (ICON), pp. 127–137. NLP Association of India (NLPAI), Indian Institute of Technology Patna, Patna, India (2020). https://aclanthology.org/2020.icon-main.16

10. Grave, E., Bojanowski, P., Gupta, P., Joulin, A., Mikolov, T.: Learning word vectors for 157 languages. In: Proceedings of the International Conference on Language Resources and Evaluation (LREC 2018), pp. 3483–3487 (2018)

11. Grivas, A., Krithara, A., Giannakopoulos, G.: Author profiling using stylometric and structural feature groupings. In: Working Notes of CLEF 2015 - Conference and Labs of the Evaluation forum, Toulouse, France, 8–11 September 2015. CEUR Workshop Proceedings, vol. 1391. CEUR-WS.org (2015). http://ceur-ws.org/Vol-1391/68-CR.pdf

12. Harris, Z.S.: Distributional structure. Word **10**(2–3), 146–162 (1954)

13. Hastie, T.J., Tibshirani, R.J., Friedman, J.H.: The Elements of Statistical Learning: Data Mining, Inference, and Prediction. Springer Series in Statistics, Springer, New York (2009). https://doi.org/10.1007/978-0-387-84858-7

14. Joulin, A., Grave, E., Bojanowski, P., Mikolov, T.: Bag of tricks for efficient text classification. In: Proceedings of the 15th Conference of the European Chapter of the Association for Computational Linguistics: Volume 2, Short Papers, pp. 427–431. Association for Computational Linguistics (2017). http://aclweb.org/anthology/E17-2068

15. Juola, P.: Authorship attribution. Found. Trends Inf. Retr. **1**(3), 233–334 (2006). https://doi.org/10.1561/1500000005

16. Le, Q., Mikolov, T.: Distributed representations of sentences and documents. In: International Conference on Machine Learning, pp. 1188–1196 (2014)

17. Piasecki, M., Walkowiak, T., Eder, M.: Open stylometric system WebSty: Integrated language processing, analysis and visualisation. Comput. Methods Sci. Technol. **24**(1), 43–58 (2018)

18. Przepiórkowski, A., Bańko, M., Górski, R.L., Lewandowska-Tomaszczyk, B. (eds.): Narodowy Korpus Jezyka Polskiego [Eng.: National Corpus of Polish]. Wydawnictwo Naukowe PWN (2012). http://nkjp.pl/settings/papers/NKJP_ksiazka.pdf

19. Radziszewski, A.: A tiered CRF tagger for Polish. In: Bembenik, R., Skonieczny, L., Rybinski, H., Kryszkiewicz, M., Niezgodka, M. (eds.) Intelligent Tools for Building a Scientific Information Platform. Studies in Computational Intelligence, vol. 467, pp. 215–230. Springer, Heidelberg (2013). https://doi.org/10.1007/978-3-642-35647-6_16

20. Rybak, P., Mroczkowski, R., Tracz, J., Gawlik, I.: KLEJ: comprehensive benchmark for polish language understanding. In: Proceedings of the 58th Annual Meeting of the Association for Computational Linguistics, pp. 1191–1201. Association for Computational Linguistics (2020). https://www.aclweb.org/anthology/2020.acl-main.111

21. Salton, G., Buckley, C.: Term-weighting approaches in automatic text retrieval. Inf. Process. Manag. **24**(5), 513–523 (1988)

22. Vaswani, A., et al.: Attention is all you need. In: Guyon, I., et al. (eds.) Advances in Neural Information Processing Systems, vol. 30. Curran Associates, Inc. (2017). https://proceedings.neurips.cc/paper/2017/file/3f5ee243547dee91fbd053c1c4a845aa-Paper.pdf

23. Walkowiak, T.: Subject classification of texts in polish - from TF-IDF to transformers. In: Zamojski, W., Mazurkiewicz, J., Sugier, J., Walkowiak, T., Kacprzyk, J. (eds.) DepCoS-RELCOMEX 2021. AISC, vol. 1389, pp. 457–465. Springer, Cham (2021). https://doi.org/10.1007/978-3-030-76773-0_44

24. Walkowiak, T., Piasecki, M.: Stylometry analysis of literary texts in polish. In: Rutkowski, L., Scherer, R., Korytkowski, M., Pedrycz, W., Tadeusiewicz, R., Zurada, J.M. (eds.) ICAISC 2018. LNCS (LNAI), vol. 10842, pp. 777–787. Springer, Cham (2018). https://doi.org/10.1007/978-3-319-91262-2_68

Bioinformatics, Biometrics and Medical Applications

Prediction of Protein Molecular Functions Using Transformers

Felipe Lopes de Mello(ID), Gabriel Bianchin de Oliveira$^{(\boxtimes)}$(ID), Helio Pedrini(ID), and Zanoni Dias(ID)

Institute of Computing, University of Campinas, Campinas, SP, Brazil
f171119@dac.unicamp.br, {gabriel.oliveira,helio,zanoni}@ic.unicamp.br

Abstract. At the end of 2021, there were more than 200 million proteins in which their molecular functions were still unknown. As the empirical determination of these functions is slow and expensive, several research groups around the world have applied machine learning to perform the prediction of protein functions. In this work, we evaluate the use of Transformer architectures to classify protein molecular functions. Our classifier uses the embeddings resulting from two Transformer-based architectures as input to a Multi-Layer Perceptron classifier. This model got $F_{\max}$ of 0.562 in our database and, when we applied this model to the same database used by DeepGOPlus, we reached the value of 0.617, surpassing the best result available in the literature.

Keywords: Protein function prediction · Transformers · BERT

1 Introduction

Proteins have an important role in biology, being responsible from structural functions to the catalysis of vital reactions for the metabolism of living beings [17]. In December 2021, according to data from the protein database UniProtKB [15], out of a total of more than 225 million proteins, less than 0.3% had their functions discovered. More than that, in many samples present in this database, the only information available is the amino acid sequence. There are biochemical methods capable of empirically determining the molecular functions of the proteins, but their high cost makes their use in practice on such a large basis unfeasible.

The Genetic Ontology (GO), proposed by Ashburner *et al.* [2], is the main method used in the literature to classify the functions performed by proteins, in which protein functions are structured hierarchically. In GO, there are already more than 40,000 terms structured hierarchically and distributed in three domains: Molecular Function Ontology, Biological Process Ontology, and Cellular Component Ontology. In all three ontologies, each protein can be assigned to multiple functions, making it a multi-label task.

Several works have already been proposed with the objective of finding the best solution for this problem. Kulmanov and Hoehndorf [9] combined the predictions of a Convolutional Neural Network with predictions based on sequence

L. Rutkowski et al. (Eds.): ICAISC 2022, LNAI 13589, pp. 379–387, 2023.
https://doi.org/10.1007/978-3-031-23480-4_32

similarity, improving the results obtained with their previous method [10]. You *et al.* [20] created GOLabeler, a method that combines five different classifiers to make the protein function prediction, surpassing DeepText2GO [19], a model that analyzes the text of abstracts that cite the function of the protein and its sequence through BLAST [1] and predicted through logistic regression and k-NN, on the 2016 Swiss-Prot database.

Among the three domains, the Molecular Function Ontology is the most dependent on the amino acid sequence [3]. Considering that proteins are phrases and amino acids are words, classifiers based on Natural Language Processing (NLP) techniques can achieve good results. Strodthoff *et al.* [14] proposed a method based on LSTM neurons [7] that learned general representations from amino acids sequences using self-supervised learning and then fine-tuned for different tasks using proteins, which includes protein function prediction, achieving competitive results with the methods in the literature. Ranjan *et al.* [13] applied the tf-idf technique after employing k-mers from the amino acid sequence, followed by an ensemble of neural networks, reaching competitive results with other methods in the literature mainly in Molecular Function Ontology (followed by Biological Process Ontology) in the UniProtKB dataset[1].

In this work, we present and discuss a classifier based on the ensemble of embeddings extracted from two Transformer-based [16] architectures to predict Molecular Function Ontology terms. We opted for this classifier because of their recent success on NLP tasks [4,5,11]. Our main contributions include: (i) as far as we know, this is the first time that Transformer-based architectures on the protein function prediction task, (ii) our method surpassed the results achieved with DeepGOPlus [9] on the database used in that work and (iii) we detected some inconsistencies on DeepGOPlus dataset and created a new dataset without these issues.

The remaining of the text is organized as follows. In Sect. 2, we describe the datasets used to evaluate our classifier. In Sect. 3, we present the pipeline applied to classify protein molecular function terms. In Sect. 4, we discuss the results achieved with our method. In Sect. 5, we present our conclusions and highlight possible research lines for future work.

2   Datasets

In this work, we used the same database described by Kulmanov and Hoehndorf [9], considering only the Molecular Function Ontology. The work describes a dataset derived from the CAFA3 challenge (Critical Assessment of Functional Annotation) [22], published in September 2016. The authors filtered molecular function terms that are presented in at least 50 proteins in the database, leaving 677 terms at different levels on the ontology. The number of proteins used in the training, validation and testing sets can be seen in Table 1, indicated as the original database.

---

[1] https://www.ebi.ac.uk/uniprot.

Table 1. Number of proteins in the training, validation and test sets of the original and the preprocessed dataset.

Database	Training	Validation	Test	Total
Original	32,468	3,642	1,137	37,247
Preprocessed	32,421	3,587	1,137	37,145

After checking the proteins contained in the training, validation and testing sets of the original database, we noticed that there were duplicated proteins in different sets, that is, there were cases in which proteins were both present in the training set and in the validation set, proteins in the training set and the test set, and also proteins that were in both the validation set and the test set. Therefore, in order to obtain more reliable results with the reality of the problem, we chose to also create another database - called preprocessed - in which the duplicated proteins are removed. To do that, we first removed the duplicates from validation (considering training and test sets) and then in the training set (considering the test set). The final treatment result can also be seen in Table 1.

In Fig. 1, we present a real example of a protein present in the training set to illustrate how each protein was represented in the database. In this image, we can see that each amino acid is represented by a letter, while the protein is a sequence of amino acids. Each function is indicated by a label that has been assigned by the GO. We indicate with 0 if the protein does not have that term and 1 otherwise.

```
Sequence of amino acids   GO:0004497   GO:0003674   GO:0003824
...GGPGRSYTADAGYA...           0            1            1
```

Fig. 1. Example of a protein in the database with only 3 functions. Values 0 and 1 indicate whether or not the protein has the term, respectively.

3 Methodology

In this section, we describe our classifier and a baseline method, as well as the evaluation metric.

3.1 Baseline

In this work, we implement an algorithm that predicts that a protein has a label based on the frequency of its presence in the training set. Thus, if a label is present in 10% of the proteins in the training set, the algorithm predicts, for each protein in the validation and testing set, the presence of this label with a 10% chance. This type of classifier is commonly used in the literature as a baseline.

3.2   Proposed Classifier

In this subsection, we present our classifier for the protein function prediction task. We divided this subsection into Transformers and Ensemble topics. The codes are available in our GitHub[2].

Transformers. To make the prediction of protein functions using Molecular Function Ontology terms, we applied three architectures based on BERT [5], namely ProtBERT [6], ProtBERT-BFD [6] and UncasedBERT [5]. ProtBERT and ProtBERT-BFD were pre-trained considering proteins databases, while UncasedBERT was pre-trained with texts in English.

As the proteins in our datasets have different sizes and the architectures have a limit of input size, we split them into slices, considering the size of 100 amino acids. For instance, if a protein has 1,568 amino acids, we split it into 15 slices with 100 amino acids and 1 slice with 68 amino acids. As the input treatment adds new artificial samples (the slices of each protein), we also need to perform a post-processing step to aggregate the slices and thus making the predictions refer to the original input. For this purpose, we apply the mean operation of the output values for each term to all slices of a protein. Figure 2 illustrates this pipeline.

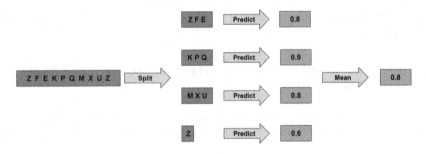

Fig. 2. Pipeline for the prediction of a molecular function of a protein, considering slices with size 3.

After the predictions of Transformers models, we performed the true path rule of the GO [21]. Therefore, we propagated the prediction from the children nodes to their parents, considering that the value of the parent node is the largest prediction value of all the children nodes.

ProtBERT, ProtBERT-BFD, and UncasedBERT were fine-tuned to our task using `ktrain` [12] and `huggingface` [18], during 100 epochs with early stopping technique with patience equal to 1, batch size equal to 16, and learning rate equal to 10^{-5}.

---

[2] https://github.com/gabrielbianchin/ProteinFunctionTransformers.

Ensemble. Based on the two best Transformer-based architectures applied to our task (ProtBERT and ProtBERT-BFD), we made the ensemble between them considering two approaches, soft voting and embeddings. In the soft voting method, we performed the ensemble using the mean operation between the predicted values for each term considering the ProtBERT and ProtBERT-BFD architectures.

In addition to the soft voting approach, we also explored the use of embeddings after training (fine tuning) for our datasets. In this method, we extracted the embeddings from the "CLS" token, which is a special token from BERT-based architectures that can capture the semantic meaning of the entire sentence [5] (in our case, the slice of the protein). After that, we considered two different approaches.

In the first one (called here approach #1), we got the unique representation for each protein using the mean operation between its slices, so we had different representations from ProtBERT and ProtBERT-BFD for the same protein. After that, we employed operations (concatenation, mean, minimum and maximum) between the ProtBERT and ProtBERT-BFD representations, and used them as input for a classifier.

In the second approach (called here approach #2), we applied operations (concatenation, mean, minimum and maximum) between the same slice from ProtBERT and ProtBERT-BFD embeddings, and used them as input to a classifier. After that, we employed the mean operation between all the slices of the same protein to have the protein prediction.

As an ensemble classifier, we used a Multi-Layer Perceptron (MLP) for both approaches (#1 and #2). In all of the cases, we made a grid search considering 1 up to 3 layers with 1,000 neurons each layer. To train our ensemble classifier, we used 100 epochs with early stopping technique with patience equal to 10, batch size equal to 32, learning rate equal to 10^{-3}, and Adam [8] optimizer.

3.3 Evaluation Metric

The evaluation metric used in this work was the $F_{\max}$, which is the same metric used in the CAFA3 challenge. This metric uses the concepts of precision and recall. Precision measures the number of correctly predicted labels by the number of predicted labels. Recall, on the other hand, measures the number of labels correctly predicted by the number of labels the protein has. Precision and recall are defined in Eqs. 1 and 2, respectively.

$$pr(\tau) = \frac{1}{m(\tau)} \sum_{i=1}^{m(\tau)} \frac{\sum_f I(f \in P_i(\tau) \wedge f \in T_i)}{\sum_f I(f \in P_i(\tau))} \tag{1}$$

$$rc(\tau) = \frac{1}{n_e} \sum_{i=1}^{n} \frac{\sum_f I(f \in P_i(\tau) \wedge f \in T_i)}{\sum_f I(f \in T_i)} \tag{2}$$

In the previous equations, $P_i(\tau)$ denotes the set of terms that have predictions greater than or equal to τ for the protein i, T_i denotes the labels that the protein actually has, $m(\tau)$ is the number of sequences with at least one prediction greater than or equal to τ, $I(\cdot)$ is an identity function, n_e is the number of proteins used in the evaluation and f is an ontology label.

With these definitions, we can describe the CAFA3 evaluation metric as the harmonic mean between precision and recall considering several values of the threshold τ, represented in Eq. 3. In this work, we vary the threshold $\tau \in [0, 1]$ with 0.01 step.

$$F_{\max} = max_\tau \{\frac{2 \times pr(\tau) \times rc(\tau)}{pr(\tau) + rc(\tau)}\} \tag{3}$$

4   Results and Discussion

In this section, we present and discuss the results achieved by our classifier on the preprocessed and original datasets.

4.1   Preprocessed Dataset

In our first experiment, we applied the Transformer-based architectures pre-trained to the protein databases (ProtBERT and ProtBERT-BFD), Uncased-BERT (which was trained on English text corpora), and the baseline model on the validation set. The results reached by each classifier are presented in Table 2, showing that ProtBERT and ProtBERT-BFD, which were pre-trained with amino acid sequences, are more adequate to our problem and, therefore, their obtained results reached better values when compared to UncasedBERT.

Table 2. $F_{\max}$ achieved with each classifier on the validation set of preprocessed dataset.

Classifier	$F_{\max}$
ProtBERT-BFD	0.620
ProtBERT	0.609
UncasedBERT	0.515
Baseline	0.417

Then, we made the ensemble between the two best classifiers, ProtBERT and ProtBERT-BFD. To do this, we considered the soft voting and embeddings approaches, as explained in Sect. 3. Table 3 presents the $F_{\max}$ achieved with each method through the MLP configuration that reached the best result. From the results, we can see that, although we have tested the MLP with 1, 2, and 3 layers, the $F_{\max}$ with configurations with less than 3 layers obtained the best values, which justifies the reason for not having tested it with more layers. Furthermore, all tests done with approach #1 were superior to approach #2. This fact suggests that, by breaking the protein into several slices (and then predicting the labels of each slice), we are missing an important piece of information, while training because of the amino acids neighbors in the original data are separated into different samples.

Table 3. $F_{\max}$ achieved with each ensemble method on the validation set of preprocessed dataset considering the MLP configuration that reached the best result.

Ensemble method	Approach	Number of layers	$F_{\max}$
Soft voting	—	—	0.620
Embeddings			
Mean	1	1	0.650
Concatenate	1	2	0.649
Minimum	1	1	0.648
Maximum	1	2	0.647
Concatenate	2	1	0.534
Mean	2	2	0.553
Maximum	2	2	0.520
Minimum	2	3	0.519

Finally, we apply the best model, the MLP with one layer that uses as input the mean of the embeddings of ProtBERT and ProtBERT-BFD architectures for each protein, that is, training the MLP with the embeddings of the protein as a whole, in the test set, and compared to the baseline classifier. The results are shown in Table 4. With these final results, we can observe an increase of 0.116 in the $F_{\max}$ obtained with our model in relation to the baseline classifier.

Table 4. $F_{\max}$ achieved with our model on the test set of preprocessed dataset.

Classifier	$F_{\max}$
Our model	0.562
Baseline	0.446

4.2 Original Dataset

To compare our classifier with DeepGOPlus in a fair way, we trained and evaluated our model on the original dataset. The result obtained and its comparison with the result of DeepGOPlus are shown in Table 5. With an increase of 0.060 in $F_{\max}$, we can conclude that the model proposed in this work is superior to that of DeepGOPlus (which, in turn, is the best result in the literature for this database).

Table 5. F_{max} achieved with DeepGOPlus and our model on the test set of original dataset.

Classifier	F_{max}
Our model	0.617
DeepGOPlus	0.557
Baseline	0.446

A final observation can be made regarding the difference between the results achieved in the original database and in the preprocessed dataset. Since the original database has duplicate proteins in the training and validation, training and testing, and validation and testing sets, it is to be expected that the predictions made in the original database have better F_{max} score, as the model was optimized for some proteins in the test. As our model reached 0.562 of F_{max} in the preprocessed dataset and 0.616 in the original dataset, we conclude that this hypothesis was confirmed empirically. This fact shows that the result obtained with DeepGOPlus was influenced by the duplicated proteins and, therefore, on a properly treated basis, its F_{max} would probably be even smaller.

5   Conclusions and Future Work

The problem of predicting molecular functions in proteins is one of the greatest current challenges in the field of biology. At the end of 2021, the number of proteins that have not yet been labeled with the genetic ontology is greater than 200 million. As laboratory tests that empirically determine the functions of proteins are slow and generally costly, a new method that performs this procedure automatically with high precision would be of great value.

In this work, we present and discuss a classifier that makes the ensemble of the embeddings generated by two Transformer-based architectures, ProtBERT and ProtBERT-BFD, for predicting Molecular Function Ontology terms. Our method achieved 0.562 of F_{max} in our dataset. Furthermore, applying our model to the database used by DeepGOPlus, we obtained 0.617 of F_{max}, surpassing the results reached by DeepGOPlus.

As directions for future work, we suggest the investigation of different techniques to make the ensemble of the embeddings, as well as the application on Biological Process Ontology and Cellular Component Ontology.

Acknowledgements. This research was supported by São Paulo Research Foundation (FAPESP) [grant numbers 2015/11937-9, 2017/12646-3, 2017/16246-0, 2017/12646-3 and 2019/20875-8], the National Council for Scientific and Technological Development (CNPq) [grant numbers 161015/2021-2, 304380/2018-0 and 309330/2018-1], and Coordination for the Improvement of Higher Education Personnel (CAPES).

References

1. Altschul, S.F., et al.: Gapped BLAST and PSI-BLAST: a new generation of protein database search programs. Nucleic Acids Res. **25**(17), 3389–3402 (1997)
2. Ashburner, M., et al.: Gene Ontology: tool for the unification of biology. Nat. Genet. **25**(1), 25–29 (2000)
3. Bonetta, R., Valentino, G.: Machine learning techniques for protein function prediction. Proteins: Struct. Funct. Bioinform. **88**(3), 397–413 (2020)
4. Brown, T.B., et al.: Language models are few-shot learners. arXiv:2005.14165 (2020)
5. Devlin, J., Chang, M.W., Lee, K., Toutanova, K.: BERT: pre-training of deep bidirectional transformers for language understanding. arXiv:1810.04805 (2018)
6. Elnaggar, A., et al.: ProtTrans: Towards Cracking the Language of Life's Code Through Self-Supervised Deep Learning and High Performance Computing. arXiv:2007.06225 (2021)
7. Hochreiter, S., Schmidhuber, J.: Long short-term memory. Neural Comput. **9**(8), 1735–1780 (1997)
8. Kingma, D.P., Ba, J.: Adam: A Method for Stochastic Optimization. arXiv:1412.6980 (2014)
9. Kulmanov, M., Hoehndorf, R.: DeepGOPlus: improved protein function prediction from sequence. Bioinformatics **36**(2), 422–429 (2019)
10. Kulmanov, M., Khan, M.A., Hoehndorf, R.: DeepGO: predicting protein functions from sequence and interactions using a deep ontology-aware classifier. Bioinformatics **34**(4), 660–668 (2018)
11. Liu, Y., et al.: RoBERTa: a robustly optimized bert pretraining approach. arXiv:1907.11692 (2019)
12. Maiya, A.S.: ktrain: A Low-Code Library for Augmented Machine Learning. arXiv:2004.10703 (2020)
13. Ranjan, A., Fernandez-Baca, D., Tripathi, S., Deepak, A.: An ensemble Tf-Idf based approach to protein function prediction via sequence segmentation. IEEE/ACM Trans. Comput. Biol. Bioinf. **14**(8), 1–12 (2021)
14. Strodthoff, N., Wagner, P., Wenzel, M., Samek, W.: UDSMProt: universal deep sequence models for protein classification. Bioinformatics **36**(8), 2401–2409 (2020)
15. UniProt: UniProt Database (2021). https://www.uniprot.org/
16. Vaswani, A., et al.: Attention is all you need. In: Advances in Neural Information Processing Systems (NIPS), pp. 5998–6008 (2017)
17. Weaver, R.F.: Molecular Biology, 5th edn. McGraw-Hill, New York (2012)
18. Wolf, T., et al.: Huggingface's transformers: state-of-the-art natural language processing. arXiv:1910.03771 (2019)
19. You, R., Huang, X., Zhu, S.: DeepText2GO: improving large-scale protein function prediction with deep semantic text representation. Methods **145**(1), 82–90 (2018)
20. You, R., Zhang, Z., Xiong, Y., Sun, F., Mamitsuka, H., Zhu, S.: GOLabeler: improving sequence-based large-scale protein function prediction by learning to rank. Bioinformatics **34**(14), 2465–2473 (2018)
21. Zhao, Y., Wang, J., Chen, J., Zhang, X., Guo, M., Yu, G.: A literature review of gene function prediction by modeling gene ontology. Front. Genet. **11**, 400 (2020)
22. Zhou, N., et al.: The CAFA challenge reports improved protein function prediction and new functional annotations for hundreds of genes through experimental screens. Genome Biol. **20**(1), 244 (2019)

Dynamic Signature Verification Using Selected Regions

Marcin Zalasiński[2(✉)] [ID], Piotr Duda[2] [ID], Stanisław Lota[1] [ID],
and Krzysztof Cpałka[2] [ID]

[1] University of Lower Silesia, Wrocław, Poland
stanislaw.lota@dsw.edu.pl
[2] Department of Computational Intelligence, Częstochowa University of Technology,
Częstochowa, Poland
{marcin.zalasinski,piotr.duda,krzysztof.cpalka}@pcz.pl

Abstract. Identity verification takes into account biometric attributes. Behavioral ones are particularly important. One of them is a dynamic signature. An analysis of such type of a signature uses signals describing the signing process. In this paper, we consider the velocity signal and propose a new method for dividing the dynamic signature into groups. In this case, a group is a subset of consecutive discretization points corresponding to similar velocity values. We have also assumed here that the signature fragments characterized by the highest pen velocity are the most characteristic of each user, therefore we reject partitions related to medium and low velocity values. As a result, we individually create a unique set of partitions of different sizes for each user. We do not use skilled forgeries, which is an additional advantage of our approach. The proposed method has been tested using the BioSecure dynamic signature database. The obtained results have confirmed the effectiveness of the proposed approach.

Keywords: Behavioral biometrics · Identity verification · Dynamic signature · Signature partitioning · Signature groups

1 Introduction

Identity verification takes into account biometric attributes, in which behavioral ones are particularly important, and one of them is the dynamic signature. Usually, a graphics tablet is used to create such a signature. The dynamic signature is represented by discrete waveforms describing how the pen is guided. They are, e.g., pen trajectory signals which can be used to determine pen velocity and acceleration. There are different approaches to analyzing dynamic signatures and they often use population-based algorithms [19, 20, 26, 29, 54, 56], fuzzy systems [28, 35], neural networks [7, 8, 15, 31], and other methods of artificial intelligence and machine learning [21, 32, 47]. The most common approaches:

L. Rutkowski et al. (Eds.): ICAISC 2022, LNAI 13589, pp. 388–397, 2023.
https://doi.org/10.1007/978-3-031-23480-4_33

- extract features from the waveforms describing the signing process [22,52]. Values of these features depend on the specificity of the user's signatures and are used in the verification of test signatures. The set of features can be additionally selected individually for each user and various metaheuristic methods can be used for the selection [53,55].
- divide signatures into partitions that may have different interpretations [16, 17]. Partitions are usually created by points similar to each other in the sense of the adopted similarity criterion. For example, they are related to a similar velocity value or the same time moment of signing. Moreover, partitions can be selectable, like features [51].
- select characteristic fragments of the signature and analyze them [34]. This selection can be made individually for each user.
- analyze the shape of the signature and its dynamics [24]. Thus, such approaches are hybrid in nature-they can use many different methods of shape and dynamics analysis, aggregating the results of the component methods.
- transform the waveforms describing the signature dynamics in order to select unique properties that increase the effectiveness of signature verification [1, 14].
- generate common properties of the real signatures of all available users. They most often use skilled forgeries [49], which reduces the verification of signatures to two-class classification and eliminates the problem of designing one-class classifiers. The disadvantage of such solutions is the use of skilled forgeries, which in practice are usually not available.

In this paper, we consider the discrete waveform of the signature velocity and propose a new method for dividing the signature into groups. By a group we mean a subset of successive discretization points corresponding to similar velocity values. In this paper we have assumed that the most characteristic of the user are the fragments of the signature written at the highest pen velocity, hence we discard the partitions related to the medium and low velocity. Therefore, for each user, we create an individual set of partitions with different sizes. We do not use skilled forgeries, which is an additional advantage of the proposed approach.

1.1 Motivation

The motivation for the preparation of the method considered in this paper can be summarized as follows:

- In our previous research on dynamic signature verification, we also created partitions and determined their importance. Partitions related to higher pen velocity were most often more important than others (see e.g. [17]). Therefore, in this work, we only focus on the partitions related to the high pen velocity and the signature areas related to them.
- In the proposed algorithm, the signature partitions have a different interpretation than in our previous work. Then, the partition associated with e.g. high velocity was a set of discretization points that did not have to be adjacent to

each other (see e.g. [17]). In this paper, a partition has a group of adjacent discretization points. Such groups of points are easier to interpret and relate to signature areas.

1.2 Contribution of the Paper

The elements of the novelty presented in this paper can be summarized as follows:

- We propose a new interpretation of the dynamic signature partitions. In our algorithm, a partition is a group of contiguous discretization points that are not distributed as they were in our previous methods.
- We propose a new method of dynamic signatures verification that focuses on the areas with the highest pen velocity. In our previous papers, we considered all the discretization points regardless of what velocity they were related to.

1.3 Structure of the Paper

In Sect. 2 we have described the proposed approach to the dynamic signature verification using selected regions. Section 3 contains sample simulation results. Section 4 presents a summary of the most important conclusions and plans for future research.

2 Description of the Proposed Method

The algorithm for the dynamic signature verification using selected regions implements partitioning in the domain of velocity. It searches for partitions for each user independently, so the number of partitions may consequently be different for each user. The proposed method requires a training phase which can be followed by a testing phase (practical use). A scheme of the training phase is shown in Fig. 1. The steps - for a single user - of this phase are as follows:

- Rejection of random discretization points. It consists in selecting and rejecting $Ndisprej$ (in %) discretization points associated with the highest and lowest pen velocity values. As a result, points corresponding to random pen movements are rejected and do not determine the signature verification procedure. Each user usually creates several reference signatures in the training phase, so this step is performed independently for each of the signatures.
- Averaging values of the corresponding discretization points from different reference signatures. Before this process, the points should be matched using the Dynamic Time Warping [25] algorithm in the context of reference signatures. As a result of this step, a template of reference signatures is created for each user. The template is processed in the subsequent steps of the training phase.
- Normalization of the signature template discretization points. It facilitates the partitioning of signatures and their classification.

- Rejection of the signature template discretization points corresponding to the pen velocity lower than V_{LimHi}. It is a parameter of the algorithm that is common to all users. The reduction of discretization points results in the creation of discretization points groups of the signature template.
- Rejection of discretization points groups containing less than $Ndisppar$ points. This step allows you to eliminate the signature groups associated with short (random) moments of increased pen velocity. This is especially important in the context of recognizing skilled forgeries.
- Determination of the weights corresponding to the selected points groups. These weights are created by product-type aggregation of two components. The first is the average velocity in the group of points (its greater value means a greater weight value). The second component is the size of the group (its greater value means a greater weight value).
- Determination of the remaining parameters of the one-class signature classifier (in addition to partition groups weights). The operation of the classifier and the procedure for determining its parameters were described in our previous work [17].

The discrete test signature verification phase begins with matching its discretization points to the template points determined in the learning phase. Then, the test signature is normalized and its similarity to the template is determined. The determination of this similarity is based on the determination of absolute errors for each partition independently. The values of these errors are then given to the inputs of the classifier, which determines the fuzzy similarity of signatures. It is the basis for the test signature verification [50].

3   Simulations

The proposed method has been tested for the dynamic signature database BioSecure DS2 distributed by the BioSecure Association [23]. It contains the signatures of 210 users. In the training phase, we used 5 randomly selected genuine signatures of each signer and in the test phase, we used 10 genuine signatures and 10 skilled forgeries of each signer. Value of $Ndisprej$ was set to 10%, value of V_{LimHi} was set to average velocity signal value of the base signature multiplied by 1.15, and value of $Ndispar$ was set to 3% of the total number of the signature discretization points. The obtained results are presented in Table 1.

The simulation conclusions can be summarized as follows:

- The method proposed in this paper works correctly (see Table 1). The adopted interpretation of the discretization points group and the concept of focusing on the areas of the signature corresponding to the high velocity of the pen are correct.
- The proposed method works with an accuracy comparable to other partitioning methods (see Table 1). However, the accuracy of the method proposed in this paper was obtained for a reduced subset of partitions (selected fragments of the signature), which reduces its computational complexity.

Table 1. Comparison of the results for selected dynamic signature verification methods using partitioning.

Id.	Method	Average FAR	Average FRR	Average error
1.	Method using vertical partitioning presented in [16]	3.13 %	4.15 %	3.64 %
2.	Method using horizontal partitioning presented in [17]	2.94 %	4.45 %	3.70 %
3.	Method using hybrid partitioning presented in [18]	3.36 %	3.30 %	3.33 %
4	Our method	**3.78 %**	**5.26 %**	**4.52 %**

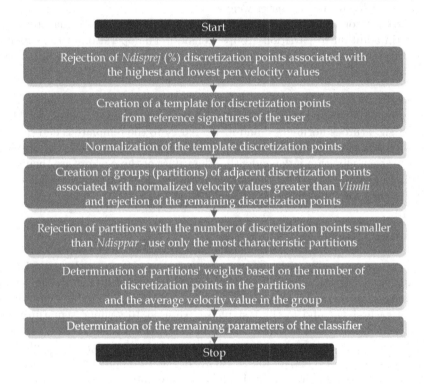

Fig. 1. The training phase scheme for a single user.

- Looking at the simulation results, one can get the impression that focusing only on the areas related to high pen speed is not the optimal solution for typical use cases of the method. Probably, this solution can only work for tests involving a large number of skilled forgeries. Therefore, no significant increase in accuracy was observed for the signature database used in the simulations.

4   Conclusions

In this paper, we have proposed an original method for dynamic signature verification. It uses signature partitioning and a new partition formula. The method was tested with the use of BioSecure DS2-an authentic signature database. The obtained results confirm that focusing on the areas of performance related to the highest pen velocity is the correct approach. This was confirmed by the simulation results. The resulting accuracy is similar to that obtained by other partitioning methods but was achieved with a reduced set of partitions.

Our plans for the dynamic signature verification include the use of population based algorithms [33, 40–46], fuzzy systems [36–39, 48], neural networks and deep learning methods [2–13, 27, 30] to determine the impact of skilled forgeries on the effectiveness of the signature partitioning procedure.

Acknowledgment. This paper was financed under the program of the Minister of Science and Higher Education under the name 'Regional Initiative of Excellence' in the years 2019–2022, project number 020/RID/2018/19 with the amount of financing PLN 12 000 000.

References

1. Alpar, O.: Signature barcodes for online verification. Pattern Recogn. **124**, 108426 (2022)
2. Bilski, J., Wilamowski, B.M.: Parallel learning of feedforward neural networks without error backpropagation. In: Rutkowski, L., Korytkowski, M., Scherer, R., Tadeusiewicz, R., Zadeh, L.A., Zurada, J.M. (eds.) ICAISC 2016. LNCS (LNAI), vol. 9692, pp. 57–69. Springer, Cham (2016). https://doi.org/10.1007/978-3-319-39378-0_6
3. Bilski, J., Wilamowski, B.M.: Parallel Levenberg-Marquardt algorithm without error backpropagation. In: Rutkowski, L., Korytkowski, M., Scherer, R., Tadeusiewicz, R., Zadeh, L.A., Zurada, J.M. (eds.) ICAISC 2017. LNCS (LNAI), vol. 10245, pp. 25–39. Springer, Cham (2017). https://doi.org/10.1007/978-3-319-59063-9_3
4. Bilski, J., Kowalczyk, B.: A new variant of the GQR algorithm for feedforward neural networks training. In: Rutkowski, L., Scherer, R., Korytkowski, M., Pedrycz, W., Tadeusiewicz, R., Zurada, J.M. (eds.) ICAISC 2021. LNCS (LNAI), vol. 12854, pp. 41–53. Springer, Cham (2021). https://doi.org/10.1007/978-3-030-87986-0_4
5. Bilski, J., Kowalczyk, B., Cader, A.: Modifications of the givens training algorithm for artificial neural networks. In: Rutkowski, L., Scherer, R., Korytkowski, M., Pedrycz, W., Tadeusiewicz, R., Zurada, J.M. (eds.) ICAISC 2019. LNCS (LNAI), vol. 11508, pp. 14–28. Springer, Cham (2019). https://doi.org/10.1007/978-3-030-20912-4_2
6. Bilski, J., Kowalczyk, B., Grzanek, K.: The parallel modification to the Levenberg-Marquardt algorithm. In: Rutkowski, L., Scherer, R., Korytkowski, M., Pedrycz, W., Tadeusiewicz, R., Zurada, J.M. (eds.) ICAISC 2018. LNCS (LNAI), vol. 10841, pp. 15–24. Springer, Cham (2018). https://doi.org/10.1007/978-3-319-91253-0_2

7. Bilski, J., Kowalczyk, B., Marchlewska, A., Żurada, J.: Local Levenberg-Marquardt algorithm for learning feedforward neural networks. J. Artif. Intell. Soft Comput. Res. **10**(4), 299–316 (2020). https://doi.org/10.2478/jaiscr-2020-0020

8. Bilski, J., Kowalczyk, B., Marjański, A., Gandor, M., Żurada, J.: A novel fast feedforward neural networks training algorithm. J. Artif. Intell. Soft Comput. Res. **11**(4), 287–306 (2021). https://doi.org/10.2478/jaiscr-2021-0017

9. Bilski, J., Kowalczyk, B., Zurada, J.M.: Application of the givens rotations in the neural network learning algorithm. In: Rutkowski, L., Korytkowski, M., Scherer, R., Tadeusiewicz, R., Zadeh, L.A., Zurada, J.M. (eds.) ICAISC 2016. LNCS (LNAI), vol. 9692, pp. 46–56. Springer, Cham (2016). https://doi.org/10.1007/978-3-319-39378-0_5

10. Bilski, J., Kowalczyk, B., Żurada, J.M.: Parallel implementation of the givens rotations in the neural network learning algorithm. In: Rutkowski, L., Korytkowski, M., Scherer, R., Tadeusiewicz, R., Zadeh, L.A., Zurada, J.M. (eds.) ICAISC 2017. LNCS (LNAI), vol. 10245, pp. 14–24. Springer, Cham (2017). https://doi.org/10.1007/978-3-319-59063-9_2

11. Bilski, J., Rutkowski, L., Smolag, J., Tao, D.: A novel method for speed training acceleration of recurrent neural networks. Inf. Sci. **553**, 266–279 (2021). https://doi.org/10.1016/j.ins.2020.10.025

12. Bilski, J., Smolag, J.: Fast conjugate gradient algorithm for feedforward neural networks. In: Rutkowski, L., Scherer, R., Korytkowski, M., Pedrycz, W., Tadeusiewicz, R., Zurada, J.M. (eds.) ICAISC 2020. LNCS (LNAI), vol. 12415, pp. 27–38. Springer, Cham (2020). https://doi.org/10.1007/978-3-030-61401-0_3

13. Bilski, J., Smolag, J., Najgebauer, P.: Modification of learning feedforward neural networks with the BP method. In: Rutkowski, L., Scherer, R., Korytkowski, M., Pedrycz, W., Tadeusiewicz, R., Zurada, J.M. (eds.) ICAISC 2021. LNCS (LNAI), vol. 12854, pp. 54–65. Springer, Cham (2021). https://doi.org/10.1007/978-3-030-87986-0_5

14. Chavan, M., Singh, R.R., Bharadi, V.A.: Online signature verification using hybrid wavelet transform with hidden Markov model. In: 2017 International Conference on Computing, Communication, Control and Automation (ICCUBEA), pp. 1–6 (2017). https://doi.org/10.1109/iccubea.2017.8463660

15. Duda, P., Jaworski, M., Cader, A., Wang, L.: On training deep neural networks using a streaming approach. J. Artif. Intell. Soft Comput. Res. **10**(1), 15–26 (2020). https://doi.org/10.2478/jaiscr-2020-0002

16. Cpałka, K., Zalasiński, M.: On-line signature verification using vertical signature partitioning. Expert Syst. Appl. **41**, 4170–4180 (2014)

17. Cpałka, K., Zalasiński, M., Rutkowski, L.: New method for the on-line signature verification based on horizontal partitioning. Pattern Recogn. **47**, 2652–2661 (2014)

18. Cpałka, K., Zalasiński, M., Rutkowski, L.: A new algorithm for identity verification based on the analysis of a handwritten dynamic signature. Appl. Soft Comput. **43**, 47–56 (2016)

19. Dziwiński, P., Bartczuk, Ł, Paszkowski, J.: A new auto adaptive fuzzy hybrid particle swarm optimization and genetic algorithm. J. Artif. Intell. Soft Comput. Res. **10**(2), 95–111 (2020). https://doi.org/10.2478/jaiscr-2020-0007

20. Dziwiński, P., Trippner, P., Paszkowski, J., Hayashi, Y.: Hardware implementation of a Takagi-Sugeno neuro-fuzzy system optimized by a population algorithm. J. Artif. Intell. Soft Comput. Res. **11**(3), 243–266 (2021)

21. Gabryel, M., Scherer, M.M., Sułkowski, Ł, Damaševičius, R.: Decision making support system for managing advertisers by ad fraud detection. J. Artif. Intell. Soft Comput. Res. **11**(4), 331–339 (2021)

22. He, L., Tan, H., Huang, Z.-C.: Online handwritten signature verification based on association of curvature and torsion feature with Hausdorff distance. Multimedia Tools Appl. **78**(14), 19253–19278 (2019). https://doi.org/10.1007/s11042-019-7264-6

23. Homepage of Association BioSecure (2022). http://biosecure.wp.imtbs-tsp.eu. Accessed 1 Mar 2022

24. Hu, H., Zheng, J., Zhan, E., Tang, J.: Online signature verification based on a single template via elastic curve matching. Sensors **19**, 4858 (2019). https://doi.org/10.3390/s19224858

25. Jeong, Y.S., Jeong, M.K., Omitaomu, O.A.: Weighted dynamic time warping for time series classification. Pattern Recogn. **44**, 2231–2240 (2011)

26. Korytkowski, M., Senkerik, R., Scherer, M.M., Angryk, R.A., Kordos, M., Siwocha, A.: Efficient image retrieval by fuzzy rules from boosting and metaheuristic. J. Artif. Intell. Soft Comput. Res. **10**(1), 57–69 (2020). https://doi.org/10.2478/jaiscr-2020-0005

27. Laskowski, Ł: Hybrid-maximum neural network for depth analysis from stereo-image. In: Rutkowski, L., Scherer, R., Tadeusiewicz, R., Zadeh, L.A., Zurada, J.M. (eds.) ICAISC 2010. LNCS (LNAI), vol. 6114, pp. 47–55. Springer, Heidelberg (2010). https://doi.org/10.1007/978-3-642-13232-2_7

28. Łapa, K., Cpałka, K., Galushkin, A.I.: A new interpretability criteria for neuro-fuzzy systems for nonlinear classification. In: Rutkowski, L., Korytkowski, M., Scherer, R., Tadeusiewicz, R., Zadeh, L.A., Zurada, J.M. (eds.) ICAISC 2015. LNCS (LNAI), vol. 9119, pp. 448–468. Springer, Cham (2015). https://doi.org/10.1007/978-3-319-19324-3_41

29. Łapa, K., Cpałka, K., Laskowski, Ł, Cader, A., Zeng, Z.: Evolutionary algorithm with a configurable search mechanism. J. Artif. Intell. Soft Comput. Res. **10**(3), 151–171 (2020). https://doi.org/10.2478/jaiscr-2020-0011

30. Mańdziuk, J., Żychowski, A.: Dimensionality reduction in multilabel classification with neural networks. In: International Joint Conference on Neural Networks (IJCNN 2019), pp. 1–8 (2019). https://doi.org/10.1109/IJCNN.2019.8852156

31. Niksa-Rynkiewicz, T., Szewczuk-Krypa, N., Witkowska, A., Cpałka, K., Zalasiński, M., Cader, A.: Monitoring regenerative heat exchanger in steam power plant by making use of the recurrent neural network. J. Artif. Intell. Soft Comput. Res. **11**(2), 143–155 (2021). https://doi.org/10.2478/jaiscr-2021-0009

32. Nowicki, R.K., Seliga, R., Żelasko, D., Hayashi, Y.: Performance analysis of rough set-based hybrid classification systems in the case of missing values. J. Artif. Intell. Soft Comput. Res. **11**(4), 307–318 (2021)

33. Okulewicz, M., Mańdziuk, J.: The impact of particular components of the PSO-based algorithm solving the dynamic vehicle routing problem. Appl. Soft Comput. **58**, 586–604 (2017). https://doi.org/10.1016/j.asoc.2017.04.070

34. Ren, Y., Wang, C., Chen, Y., Chuah, M.C., Yang, J.: Signature verification using critical segments for securing mobile transactions. IEEE Trans. Mob. Comput. **19**(3), 724–739 (2020). https://doi.org/10.1109/TMC.2019.2897657

35. Rutkowski, T., Łapa, K., Jaworski, M., Nielek, R., Rutkowska, D.: On explainable flexible fuzzy recommender and its performance evaluation using the Akaike information criterion. In: Gedeon, T., Wong, K.W., Lee, M. (eds.) ICONIP 2019. CCIS, vol. 1142, pp. 717–724. Springer, Cham (2019). https://doi.org/10.1007/978-3-030-36808-1_78

36. Scherer, R., Rutkowski, L.: Neuro-fuzzy relational classifiers. In: Rutkowski, L., Siekmann, J.H., Tadeusiewicz, R., Zadeh, L.A. (eds.) ICAISC 2004. LNCS (LNAI), vol. 3070, pp. 376–380. Springer, Heidelberg (2004). https://doi.org/10.1007/978-3-540-24844-6_54

37. Scherer, R., Rutkowski, L.: Neuro-fuzzy relational systems. In: Proceedings of FSKD 2002, pp. 44–48 (2002)

38. Scherer, R., Rutkowski, L.: Relational equations initializing neuro-fuzzy system. In: Proceedings of 10th Zittau Fuzzy Colloquium, Zittau, Germany, pp. 18–22 (2002)

39. Scherer, R.: Neuro-fuzzy systems with relation matrix. In: Rutkowski, L., Scherer, R., Tadeusiewicz, R., Zadeh, L.A., Zurada, J.M. (eds.) ICAISC 2010. LNCS (LNAI), vol. 6113, pp. 210–215. Springer, Heidelberg (2010). https://doi.org/10.1007/978-3-642-13208-7_27

40. Słowik, A.: Application of evolutionary algorithm to design minimal phase digital filters with non-standard amplitude characteristics and finite bit word length. Bull. Pol. Acad. Sci.-Tech. Sci. 59(2), 125–135 (2011). https://doi.org/10.2478/v10175-011-0016-z

41. Słowik, A.: Steering of balance between exploration and exploitation properties of evolutionary algorithms - mix selection. In: Rutkowski, L., Scherer, R., Tadeusiewicz, R., Zadeh, L.A., Zurada, J.M. (eds.) ICAISC 2010. LNCS (LNAI), vol. 6114, pp. 213–220. Springer, Heidelberg (2010). https://doi.org/10.1007/978-3-642-13232-2_26

42. Słowik, A., Białko, M.: Design and optimization of combinational digital circuits using modified evolutionary algorithm. In: Rutkowski, L., Siekmann, J.H., Tadeusiewicz, R., Zadeh, L.A. (eds.) ICAISC 2004. LNCS (LNAI), vol. 3070, pp. 468–473. Springer, Heidelberg (2004). https://doi.org/10.1007/978-3-540-24844-6_69

43. Słowik, A., Białko, M.: Modified version of roulette selection for evolution algorithms – the fan selection. In: Rutkowski, L., Siekmann, J.H., Tadeusiewicz, R., Zadeh, L.A. (eds.) ICAISC 2004. LNCS (LNAI), vol. 3070, pp. 474–479. Springer, Heidelberg (2004). https://doi.org/10.1007/978-3-540-24844-6_70

44. Słowik, A., Białko, M.: Partitioning of VLSI circuits on subcircuits with minimal number of connections using evolutionary algorithm. In: Rutkowski, L., Tadeusiewicz, R., Zadeh, L.A., Żurada, J.M. (eds.) ICAISC 2006. LNCS (LNAI), vol. 4029, pp. 470–478. Springer, Heidelberg (2006). https://doi.org/10.1007/11785231_50

45. Slowik, A., Bialko, M.: Design and optimization of IIR digital filters with non-standard characteristics using continuous ant colony optimization algorithm. In: Darzentas, J., Vouros, G.A., Vosinakis, S., Arnellos, A. (eds.) SETN 2008. LNCS (LNAI), vol. 5138, pp. 395–400. Springer, Heidelberg (2008). https://doi.org/10.1007/978-3-540-87881-0_39

46. Słowik, A., Białko, M.: Design of IIR digital filters with non-standard characteristics using differential evolution algorithm. Bull. Pol. Acad. Sci.-Tech. Sci. 55(4), 359–363 (2007)

47. Starczewski, J.T., Fijałkowska, J., Siwocha, A., Napoli, Ch.: Handwritten word recognition using fuzzy matching degrees. J. Artif. Intell. Soft Comput. Res. 11(3), 229–242 (2021)

48. Starczewski, J., Scherer, R., Korytkowski, M., Nowicki, R.: Modular type-2 neuro-fuzzy systems. In: Wyrzykowski, R., Dongarra, J., Karczewski, K., Wasniewski, J. (eds.) PPAM 2007. LNCS, vol. 4967, pp. 570–578. Springer, Heidelberg (2008). https://doi.org/10.1007/978-3-540-68111-3_59

49. Tolosana, R., et al.: SVC-onGoing: signature verification competition. Pattern Recognit. **127**, 108609 (2022). https://doi.org/10.1016/j.patcog.2022.108609

50. Zalasiński, M., Cpałka, K.: A new method of on-line signature verification using a flexible fuzzy one-class classifier. Academic Publishing House EXIT, pp. 38–53 (2011)

51. Zalasiński, M., Cpałka, K.: Novel algorithm for the on-line signature verification using selected discretization points groups. In: Rutkowski, L., Korytkowski, M., Scherer, R., Tadeusiewicz, R., Zadeh, L.A., Zurada, J.M. (eds.) ICAISC 2013. LNCS (LNAI), vol. 7894, pp. 493–502. Springer, Heidelberg (2013). https://doi.org/10.1007/978-3-642-38658-9_44

52. Zalasiński, M., Cpałka, K., Hayashi, Y.: New method for dynamic signature verification based on global features. In: Rutkowski, L., Korytkowski, M., Scherer, R., Tadeusiewicz, R., Zadeh, L.A., Zurada, J.M. (eds.) ICAISC 2014. LNCS (LNAI), vol. 8468, pp. 231–245. Springer, Cham (2014). https://doi.org/10.1007/978-3-319-07176-3_21

53. Zalasiński, M., Cpałka, K., Hayashi, Y.: New fast algorithm for the dynamic signature verification using global features values. In: Rutkowski, L., Korytkowski, M., Scherer, R., Tadeusiewicz, R., Zadeh, L.A., Zurada, J.M. (eds.) ICAISC 2015. LNCS (LNAI), vol. 9120, pp. 175–188. Springer, Cham (2015). https://doi.org/10.1007/978-3-319-19369-4_17

54. Zalasiński, M., Cpałka, K., Laskowski, L, Wunsch, D.C., Przybyszewski, K.: An algorithm for the evolutionary-fuzzy generation of on-line signature hybrid descriptors. J. Artif. Intell. Soft Comput. Res. **10**(3), 173–187 (2020). https://doi.org/10.2478/jaiscr-2020-0012

55. Zalasiński, M., Łapa, K., Cpałka, K.: New algorithm for evolutionary selection of the dynamic signature global features. In: Rutkowski, L., Korytkowski, M., Scherer, R., Tadeusiewicz, R., Zadeh, L.A., Zurada, J.M. (eds.) ICAISC 2013. LNCS (LNAI), vol. 7895, pp. 113–121. Springer, Heidelberg (2013). https://doi.org/10.1007/978-3-642-38610-7_11

56. Zalasiński, M., Łapa, K., Cpałka, K., Przybyszewski, K., Yen, G.G.: On-line signature partitioning using a population based algorithm. J. Artif. Intell. Soft Comput. Res. **10**(1), 5–13 (2020). https://doi.org/10.2478/jaiscr-2020-0001

Author Index

Printed in the United States
by Baker & Taylor Publisher Services